The Maillard Reaction
Interface between Aging, Nutrition and Metabolism

The Maillard Reaction
Interface between Aging, Nutrition and Metabolism

Edited by

Merlin C Thomas and Josephine Forbes
Baker IDI Heart and Diabetes Institute, Melbourne, Australia

RSC Publishing

The proceedings of the International Symposium on the Maillard Reaction held at Palm Cove, Queensland, Australia on 29-31 August 2009.

Special Publication No. 322

ISBN: 978-1-84973-079-2

A catalogue record for this book is available from the British Library

Published by The Royal Society of Chemistry,
Thomas Graham House, Science Park, Milton Road,
Cambridge CB4 0WF, UK

Registered Charity Number 207890

For further information see our web site at www.rsc.org

PREFACE: WE ARE WHAT WE EAT

Merlin C Thomas[1]

[1]Baker IDI Heart and Diabetes Institute, Melbourne, AUSTRALIA

We are what we eat. Like our food, we are all made of sugar, protein, fat and a host of other chemicals. And like what we eat, the resulting product depends not on any individual element, but rather the combination of ingredients that interact in a complex chemistry that contributes colour, texture and aroma to our lives. Get the recipe right and the result can be propitious. But get the mix wrong and these interactions can alter the quality, function, and structure of our futures.

One of the most important of these interactions is **browning,** also known as the **Maillard Reaction**, named in honour of Louis-Camille Maillard who first described its chemistry nearly a centruy ago. Browning occurs as a non-enzymatic chemical reaction between sugars and protein that occurs during prolonged or high-temperature cooking and fermentation. For example, the brown and hard crust of bread occurs as sugars combine with proteins during processing. Similarly, the curing and burning of tobacco also involves this chemistry. Cooking slowly away at 37^{0}C, the same chemistry is ultimately possible inside the human body, making us all a little crusty.

Inside the body, the longer a protein lives, the greater the chance it will meet a reducing sugar and become modified by it. This is why the chemical products of these reactions are also known as AGEs (**Advanced Glycation End-products**) as they can be used as a marker of ageing. However, AGEs are more than just markers of the passage of time. AGEs are in fact key elements of the ageing process. Over a lifetime of cooking, the amount and variety of AGE-modified tissue progressively increases. High sugar levels, elevated fats and increased oxidative stress all act to hasten the formation of AGEs, and with it the progression of ageing and disease.

AGE-modification of any protein has the capacity to alter its structure and or function in a number of ways. Classically, AGEs represent inter or intramolecular **cross-links** that reduce the flexibility of structures, tying and fixing proteins in certain shapes. Other modifications of nucleophilic targets are able to interfere with charge dependent functions or other ligand interactions. In addition, some modifications are able to be detetcted by multiligand AGE receptors. Over recent years a number of AGE-receptors have been identified. Some of these are clearance receptors, whose job it is to remove AGEs from the body. Other receptors are activated by AGEs, triggering a cascade of adaptive responses, that when persistent contribute to ageing and disease. The best known of these is appropriately known as **RAGE (Receptor for Advanced Glycation End-products)**. Atherosclerosis, cancer, dementia, mental illness and diabetes are all influenced by activation of RAGE.

Many of the processed foods we eat will contain AGEs in varying amounts and infinite complexity. This has a number of consequences, both positive and negative. For example, browning produces flavours, textures and aromas that give many foods their appeal. Incorporating AGEs into foods and their processing helps boost flavour. For example, the crunch of crisps is conveyed by cooking at very high temperature and low humidity. Many of the most successful fast food producers have known this for many years, and with increasing demand as well as competition, AGE in levels in the Western diet have substantially increased over the last fifty years. Browning and glycoxidation may also occur resulting during long-term storage, causing foodstuffs to become stiffer and less digestible. Indeed, AGE-modifications of stored food can serve as useful parameters of food quality.

There appears to be a strong relationship between the levels of AGEs in our blood and the amount of AGEs consumed in our diets. Indeed, it could be said that you are only as old as what you eat. However, the exact nature of this association is controversial. Since intact AGE-modified proteins do not survive gastrointestinal proteolysis, it is not protein that our system receives but rather AGE-peptides or amino acids, which are transported across the intestinal wall, enter the circulation and are subsequently cleared largely by the kidney. Consequently, levels of AGEs may increase as a both result of increased intake or decreased clearance, such as in patients with renal impairment or cirrhosis. An alternative hypothesis is that AGE-precursors, such as methylglyoxal and other reactive carbonyl compounds may be taken up from the diet, subsequently forming AGEs in plasma or tissues. The levels of these reactive precursors may also correspond to the level of AGEs in the diet, so the association between plasma AGEs increase and AGE content of the diet may simply be an indirect one.

Regardless of the mechanism, it is likely that modulation of AGEs in our diet will in the future provide an important adjunct to interventions directed towards preventing the accumulation of AGEs in diabetes and other degenerative conditions, analogous to the role of a low cholesterol diet. It is very possible that within our lifetime dietary AGEs will capture the attention of the public much as cholesterol and more recently trans-fats have become a target of universal rancor. The good news is that the dietary changes required tom reduce AGEs in the diet may be little different from recommendations currently advocated for a healthy diet such as increased consumption of fresh foods or limited cooking or only brief applications of heat, in the presence of ample water or humidity.

This book details research material presented at the International Maillard Reaction Society meeting that was held in Palm Cove, Australia in 2009. It covers new discoveries and investigations into the complex chemistry of the Maillard reaction and its implications for both food production, health and disease. It is a documentation of the state of the art in a rapidly expanding field. For more information on the Maillard reaction and plans for subsequent meetings go to www.imars.org.

Contents

THE MAILLARD REACTION – A JOURNEY FROM THE DISCOVERY OF ADVANCED GLYCATION ENDPRODUCTS TO THEIR CHIEF CELLULAR RECEPTOR, RAGE: A MECHANISM UNDERLYING DIABETIC COMPLICATIONS AND THE INFLAMMATORY RESPONSE

R, Ramasamy[1], S.F. Yan[1] and A.M. Schmidt[1]

[1]College of Physicians & Surgeons, Columbia University, New York, New York, 10032

1 INTRODUCTION

An emerging theme in the biology of chronic disease is that modifications of proteins may play major roles in re-routing biological reactions from preservation of homeostasis to induction of tissue injury. Atop the list of potential modifications in diabetes that may significantly alter cellular properties is the process of glycation. From Maillard's discovery that heating proteins particularly in the presence of sugars resulted in their "browning," was born the field of AGEs.[1] In recent years, the identification of the diverse settings in which AGEs may form has greatly expanded our understanding of how proteins subjected to fundamental post-translational modifications of proteins may experience both loss- and gain-of function properties.[2]

The products of non-enzymatic glycation and oxidation of proteins, the advanced glycation end-products (AGEs), may be formed by multiple means, such as hyperglycaemia, aging, inflammation, oxidation (such as of oxidized low density lipoprotein, oxLDL), hypoxia or ischaemia/reperfusion (I/R) injury, or by specific forms of food preparation.[2] AGEs may exert diverse effects in biological systems; the discovery that the Receptor for AGE (RAGE) was a signal transduction receptor for this class of modified species suggested specific means by which modified proteins gain functions such as the ability to modulate transcription, translation and activity of proteins in biological systems.[3] AGEs are a heterogeneous class of substances and therefore a particular effort is to determine which specific form(s) of AGEs may interact with RAGE.

In this review, we discuss the growing list of settings in which AGEs may form and interact with RAGE. Insights into roles for AGE-RAGE action in prototypical complications of diabetes and inflammation are also considered. Taken together, a growing body of evidence suggests that antagonism of RAGE may be a logical target for therapeutic intervention in the chronic diseases in which AGEs accumulate, such as diabetes, aging and tissue-damaging inflammatory responses.

2 THE DIVERSE ROLES OF RAGE

2.1 Different Settings for AGE Production and Accumulation

The formation of AGEs, although accelerated in hyperglycaemia, also occurs over extended periods of time even in essentially euglycaemic states, such as in natural aging. Proteins modified in the presence of reducing sugars such as glucose, glucose-6-phosphate or ribose resulted in species that interact with RAGE.[4] More recently, specific AGEs such as carboxy-methyl lysine (CML) AGEs have been shown by numerous groups to interact with RAGE.[5]

Oxidized LDL (ox LDL) species contain AGE epitopes and evidence that they stimulate signal transduction via RAGE emerged from experiments in which incubation of primary wild-type murine aortic endothelial cells with oxLDL activated signal transduction pathways, but in contrast, exposure of RAGE null endothelial cells to oxLDL failed to stimulate signalling or evoke such changes in gene expression.[6] In addition, further evidence that AGE epitopes were responsible for changes in endothelial gene expression was suggested by studies in which pre-treatment of the endothelial monolayers with anti-AGE IgG prior to exposure to oxLDL suppressed signal transduction.[6] Of note, a recently identified ligand family of RAGE, the advanced oxidation protein products (AOPPs), also activate RAGE, but not via AGE epitopes, as anti-AGE antibodies had no impact on AOPP-stimulated RAGE signalling.[7]

Recent experiments highlighted that exposure of wild-type primary murine aortic endothelial cells to hypoxia generated AGE-immunoreactive epitopes, as assessed by ELISA experiments using anti-AGE IgG.[8] *In vitro* or *in vivo,* pre-treatment of endothelial cells or wild-type mice prior to hypoxia with either anti-AGE IgG or aminoguanidine, an inhibitor of AGE formation, suppressed hypoxia-mediated activation of Protein kinase C, phosphorylation of mitogen activated protein kinases (MAP kinases), and up-regulation of egr-1 transcripts and protein.[8] These actions occurred via RAGE, as mice and endothelial cells devoid of RAGE failed to upregulate Egr-1 in hypoxia versus wild-type mice and primary endothelial cells.[8]

One consequence of inflammatory conditions is the generation of AGEs. At least in part via the action of the myeloperoxidase pathway, or via generation of reactive oxygen species (ROS), AGE formation may occur in non-diabetic wounds or atherosclerosis, for example.[9] Of note, increased ROS may support AGE formation via depletion of anti-oxidant defenses. In the presence of decreased levels of glutathione, the enzyme glyoxalase-1 is less efficient, thereby reducing detoxification of the pre-AGE methylglyoxal (MGO).[10]

A burgeoning literature suggests that specific forms of food preparation, such as in heating procedures, may lead to excessive amounts of AGEs in the diet.[11] In animal models, low AGE diets were associated with less cellular perturbation and decreased manifestations of vascular disease compared to high AGE diets.[12,13] A recent study suggested that a food-derived AGE, pronyl glycine, was a novel ligand for RAGE.[14] Treatment of Caco-2 intestinal epithelial cells with pronyl glycine AGE activated MAP kinase signalling via RAGE.[14] Taken together, these considerations indicate that stimuli to AGE generation are diverse and not limited to hyperglycaemic states. Even in euglycaemia, such factors as aging, hyperlipidaemia, inflammation, oxidative stress and modifications of food ingredients by preparation techniques may produce AGE species. Although the full range of these species capable of binding RAGE is likely yet to be fully elucidated, heterogeneous AGE mixtures, CML-AGEs and prolyl glycine, for example, may stimulate signalling mechanisms via RAGE.[5, 14] Apart from distinct AGEs, however, the repertoire of RAGE ligands is more broad and includes ligands specifically linked to inflammation and cellular migration.

2.2 Multi-ligand Nature of RAGE

At least four other classes of RAGE ligands have been identified. S100/calgranulins are members of a family of molecules which have distinct intracellular functions, such as calcium binding. In the extracellular space, multiple S100/calgranulins may bind RAGE, such as S100A12, S100B, S100P, S100A6, and S100A4.[15-17] Release of S100/calgranulins to the extracellular space, we predict, is a key control mechanism by which activated, dead or injured cells may release these molecules.[17] Although initially most probably intended as a cellular defense strategy, if sustained, the release and paracrine/autocrine action of these molecules may intensify inflammation and tissue damage, thereby facilitating chronic disease.[15]

High mobility group box-1 (HMGB-1) ligands also bind to RAGE.[18,19] Although these molecules may also bind certain toll receptors (TLRs), such as TLRs 2 and 4, HMGB1 acts on RAGE-expressing cells to stimulate signalling[20]. HMGB1 is a member of the non-histone DNA binding family of molecules normally present in the nucleus.[20] Akin to S100/calgranulins, HMGB1 molecules may be released by necrosis, thereby freeing them to act in the extracellular space.[21] Once available in the extracellular space, these molecules may act on a diverse array of inflammatory cells, vascular cells, cardiomyocytes, and neurons as examples, thereby propagating inflammatory mechanisms, at least in part via cell surface receptors such as RAGE. In the context of inflammation, Mac-1 has been shown to be a ligand for RAGE.[22] Further, amyloid-β peptide and β-sheet fibrils in general interact with RAGE.[23] As amyloid often forms and accumulates in *response* to chronic inflammation, it is plausible that this class of ligands represents yet an additional means by which RAGE signalling may be amplified and sustained in chronic inflammation. In the section to follow, we review the evidence that RAGE signalling importantly contributes to the pathogenesis of the inflammatory response.

2.3 RAGE and the Inflammatory Response – its Natural Function?

We and others had long speculated that AGEs were not the only ligands of RAGE and indeed, the discovery that S100/calgranulins in particular were RAGE ligands evoked the initial suggestion that RAGE was involved in inflammatory responses[15]. Work of Tracey and colleagues implicated HMGB1 in late responses to sepsis[24]; hence, the finding that RAGE transduced the signals of these classes of ligands firmly linked RAGE to inflammation. RAGE is expressed on monocytes/macrophages and ligand-RAGE interaction contributes to their biology, including migration, activation and delayed apoptosis in the face of ligand challenge.[25,26] RAGE is also expressed on T and B lymphocytes, and dendritic cells, thereby suggesting important roles for ligand-RAGE interaction in adaptive immune responses.[27]

This concept was tested in a murine model of orthotopic allogeneic heart transplantation. Using allo-mismatched donors and recipients, we found that administration of sRAGE prolonged allograft survival in a dose-dependent manner compared to vehicle treatment. Histology of the grafts revealed a marked reduction in T lymphocyte and macrophage influx into the tissue in the presence of sRAGE.[28] To precisely address the specific RAGE-dependent mechanisms in T lymphocytes, OTII T lymphocytes (CD-4 like T lymphocytes expressing T cell receptors recognizing ovalbumin) were employed. OTII mice were bred into the RAGE null background. In both in vitro and in vivo studies,

RAGE deficient OTII cells displayed significantly less proliferation and cytokine production in response to ovalbumin vs. RAGE-expressing OTII controls.[29] We concluded from those experiments that RAGE was required for effective T lymphocyte priming responses. Further studies using an islet allograft model suggested that RAGE activation in T cells may be a key factor in the early events linked to Th1 differentiation.[30] Thus, RAGE may have multiple roles in diabetic complications due to its actions in distinct steps in these processes. From transducing the effects of glycated ligands, to amplifying inflammatory responses, RAGE may exert significant impact in diabetes complications. In the sections to follow, we consider roles for RAGE in prototypic complications of diabetes and how AGEs and inflammatory mechanisms may contribute.

2.4 RAGE and Diabetic Nephropathy

Diabetes is a major cause of end-stage renal failure. Multiple publications in both animal models and human diabetes have shown accumulation of broad classes of AGEs in the diabetic kidney.[31] Hence, to test the role of RAGE in the pathogenesis of diabetes-associated nephropathy, we first established that in both human and rodent diabetes, RAGE was principally expressed in glomerular epithelial cells (podocytes) and in glomerular endothelial cells.[32] By immunohistochemistry of *in situ* tissue preparations of renal cortex, there was no evidence that RAGE was expressed to any appreciable degrees in tubular cells in either the diabetic or non-diabetic state.[32] Given the emerging central role for the podocyte in the earliest initiating mechanisms in diabetic nephropathy[33], it was logical to test the premise that RAGE contributed to the pathogenesis of diabetes-associated nephropathy.

The first studies to address this concept employed pharmacological antagonists of RAGE, such as soluble RAGE. Soluble RAGE, composed of the extracellular ligand-binding domain of RAGE, binds ligands and blocks their interaction with cell surface receptors. Soluble RAGE is generated in a baculovirus expression system and purified to homogeneity and rendered free of any contaminating lipopolysaccharide. Soluble RAGE was administered to db/db mice from the onset of hyperglycaemia through at least six months.[34] Administration of sRAGE suppressed indices of mesangial expansion, thickening of the glomerular basement membrane and microalbuminuria.[34] In other studies, distinct pharmacological antagonists of ligand-RAGE interaction were tested, such as neutralizing anti-RAGE antibodies. In mouse models of type 1 and 2 diabetes, anti-RAGE antibodies beneficially impacted the course of renal disease.[35,36]

More recently, genetic approaches have been employed to test the role of RAGE in diabetes-associated nephropathy. Myint and colleagues studied homozygous RAGE null mice and found that compared to wild-type animals, RAGE deficient animals displayed suppression of the following major endpoints such as kidney enlargement, increased glomerular cell number, mesangial expansion, advanced glomerulosclerosis, increased albuminuria and increased serum creatinine.[37]

As a definite pitfall of most available mouse models of diabetes is the failure to progress to highly advanced and more human-like stages of disease, Inagi and colleagues created transgenic mice that over-expressed megsin (a glomerular-specific serpin), RAGE and iNOS to develop a model system predicted to develop severe nephropathic changes.[38] These authors found that compared with the single- or two- gene transgenic mice, the triple transgenic mice (over-expression of RAGE) developed severe albuminuria, glomerular hypertrophy, diffuse mesangial expansion, inflammatory cell infiltration and interstitial fibrosis, with 30-40% of the glomeruli displaying nodule-like lesions at age 16 weeks of age, a point that was relatively early in terms of other available mouse models.[38] This study reinforced that RAGE appeared to be one of the "multiple hits" responsible for advanced diabetes-associated nephropathy, at least in mouse models.

Of note, a recent model system, the OVE26 mouse, has been reported by Epstein and colleagues in which diabetes was induced by transgene-mediated introduction of calmodulin in pancreatic beta cells.[39] Advanced stages of nephropathy develop, including advanced albuminuria and glomerulosclerosis in parallel with a reduced glomerular filtration rate. In parallel, extensive podocyte loss and tubulointerstitital fibrosis in aging OVE26 mice has been reported.[39,40] The observed podocyte loss may be central to the pathogenesis of advanced glomerular disease. Studies are ongoing to test the role of RAGE in OVE26 mice. The RAGE null mouse has been bred into this background and studies are in progress to assess the impact of RAGE deletion on the renal function and pathology. Preliminary data suggest that deletion of RAGE in OVE26 mice is associated with reduced glomerulosclerosis and albuminuria, in parallel with reduced loss of renal function as assessed by inulin clearance at advanced stages of disease.[41]

Evidence suggesting roles for RAGE in inflammatory mechanisms in diabetes-associated nephropathy emerged from experiments in which numbers of macrophages, found to be increased in db/db glomerulus compared to m/db non-diabetic controls, were reduced in db/db mice treated with sRAGE.[34] Factors such as transforming growth factor-β and connective tissue growth factor, both increased in the renal cortex of type 1 and type 2 diabetic mice, were reduced in mice treated with sRAGE or in RAGE deficient diabetic mice.[34]

To address whether RAGE-dependent oxidative stress might contribute to podocyte dysfunction in the kidney, we tested a distinct model, that in which administration of doxorubicin (adriamycin) induces extensive oxidative damage particularly to the podocyte. Deficiency of RAGE was protective against activation of NADPH oxidase, podocyte effacement, albuminuria and glomerulosclerosis.[42] Although not specifically studied in our report, others have shown that inflammation is a key driver of glomerular injury stimulated by adriamycin, particularly in lymphocytes.[43,44] RAGE is expressed on both T and B lymphocytes, and therefore it will be interesting to elucidate if at least some of the protection imparted by RAGE deletion in this model occurred in part due to mitigation of inflammatory signalling.

Taken together, extensive data in animal models of early and progressive glomerular disease in diabetes suggest that RAGE contributes to the pathogenesis of nephropathy, at least in part via the impact of AGEs, oxidative stress and likely inflammatory mechanisms. Importantly, these same factors are critical mediators of a distinct complication of diabetes, accelerated macrovascular disease.

2.5 The problem of diabetes-associated accelerated atherosclerosis.

The first charge in assessing whether RAGE might play relevant roles in the pathogenesis of the macrovascular disease of diabetes was to determine if RAGE was expressed in human atherosclerotic lesions. Multiple experiments have shown that AGEs, S100/calgranulins and HMGB1 accumulate in human atherosclerotic plaques, to an accelerated degree in diabetes.[45,46] RAGE expression was identified on inflammatory cells as well as vascular cells, suggesting that the interplay of RAGE action in these cell types might contribute to the mechanisms initiating and/or sustaining atherosclerosis. To provide definitive support for the RAGE hypothesis in vascular pathology, studies were performed in animal models using a number of distinct strategies to address this question.

The first test of the RAGE hypothesis in macrovascular complications of diabetes employed soluble RAGE. Initially, a model system of accelerated atherosclerosis was required. Hence, mice deficient in apolipoprotein E, which develop spontaneous atherosclerosis by virtue of hypercholesterolaemia on chow diet, were rendered diabetic with streptozotocin at approximately 6 weeks of age. Acceleration of atherosclerosis at the aortic root was observed versus euglycaemic control animals at 14 and 20 weeks of age.[47,48] Examination of the atherosclerotic lesions revealed that AGEs, S100/calgranulins and RAGE epitopes were increased in the diabetic atherosclerotic plaques, particularly in the intimal macrophages and smooth muscle cells. Markers of inflammation, including vascular cell adhesion molecule-1 (VCAM-1), monocyte chemoattractant peptide (MCP-1), cyclooxygenase 2 (cox-2), matrix metalloproteinases (MMPs), tissue factor and oxidative stress were increased in the diabetic tissue, even at sites within the aorta where frank lesions were not yet evident.[47,48]

Experiments using sRAGE tested both early initiation and progression of established atherosclerosis. In the first set of experiments, as a means to test if RAGE contributed to the early initiation of atherosclerosis, sRAGE was administered to diabetic apo E null mice immediately after the development of hyperglycaemia. Compared to vehicle-treated (albumin) mice, sRAGE-treated mice displayed a dose-dependent suppression of accelerated atherosclerotic lesion area and complexity after approximately 6 weeks of treatment.[47] Treatment with sRAGE resulted in lower levels of RAGE ligands (AGEs), in parallel with markedly decreased vascular inflammation and oxidative stress.[47] Soluble RAGE had no apparent effect on the levels of glucose or cholesterol/triglyceride compared to vehicle treatment. These data implied that RAGE was not the proximate cause driving atherosclerosis in these mice, but that RAGE action amplified the adverse vascular and inflammatory consequences of hyperglycaemia and hyperlipidaemia. In addition to the above murine model of type 1 diabetes, soluble RAGE was also tested in a murine model of type 2 diabetes.[49] When sRAGE was administered to db/db mice bred into the apo E null background, diabetes accelerated early atherosclerosis in a manner suppressed by administration of sRAGE, without effect on levels of glucose, lipids or insulin.[49]

In the second set of studies, as a means to test if interruption of ligand-RAGE action in established atherosclerosis might modify the course of established vascular pathology, sRAGE was administered to diabetic apo E null mice beginning six weeks *after* the establishment of sustained hyperglycaemia. When diabetic mice were sacrificed six weeks after the initiation of sRAGE, those animals receiving vehicle murine serum albumin) displayed marked progression of atherosclerosis compared to all non-diabetic apo E null mice at the aortic root.[48] However, the diabetic animals treated with sRAGE demonstrated stabilization of the lesions, with no apparent progression compared to the time point at which sRAGE was begun.[48] Consistent with studies in which sRAGE was begun immediately upon the development of hyperglycaemia, multiple markers of vascular inflammation were significantly lower in the sRAGE-treated diabetic mice vessels compared to vehicle. These experiments indicated that RAGE-dependent inflammation and oxidative stress were operative in both initiation as well as the progression phases of accelerated atherosclerosis in diabetic rodents. In addition to pharmacological approaches to block ligand-RAGE interactions, additional strategies, including genetic manipulation of RAGE expression and action, have been employed to test the effects of RAGE in vascular pathology. Soro-Paavonen and colleagues showed that diabetic homozygous RAGE null mice in the apo E null background displayed reduced atherosclerosis.[50]

As the ligands of RAGE, such as the AGEs and S100/calgranulins, accumulate in non-diabetic aorta in a hyperlipidaemic environment, the impact of genetic modulation of RAGE in non-diabetic animals was studied as well. RAGE null mice were bred into the apo E null background, and transgenic mice expressing dominant negative (DN) RAGE particularly in endothelial cells (EC) (as driven by the preproendothelin-1 promoter) were used [6]. The DN RAGE construct is composed of the extracellular and membrane-spanning domains of RAGE, but is devoid of the short cytoplasmic domain.[6] At age 14 weeks, compared to apo E null mice, RAGE null mice and hemizygous transgenic mice expressing DN RAGE in endothelial cells in the apolipoprotein E null background displayed significantly less atherosclerosis.[6] Additionally, when homozygous RAGE null mice were bred into the apo E null background, reduced atherosclerosis was observed versus RAGE-expressing apo E null animals.[6] In both sets of transgenic animals, RAGE modulation was not associated with differences in plasma cholesterol or triglyceride in apo E null mice.[6] Recently, Sun and colleagues have corroborated these key findings by breeding RAGE null mice into the low density lipoprotein receptor null (LDLR null) background. In the non-diabetic state, deficiency of RAGE reduced atherosclerosis.[51]

Taken together, these reports implicate RAGE and its ligands in the pathogenesis of vascular dysfunction and accelerated atherosclerosis typical of diabetes. By virtue of the lack of changes in glucose or lipid levels, these data have profound translational impact. Multi-modality strategies to attack the macro- and microvascular complications of diabetes, including glucose and lipid control, together with RAGE antagonism, should impact both the immediate and downstream consequences of hyperglycaemia.

3 CONCLUSIONS

AGEs may form in diverse settings, such as hyperglycaemia, natural aging, inflammation, food-preparation methods, oxidative stress and hypoxia or ischaemia/reperfusion, and at least certain AGEs may signal via RAGE. Although initial studies pinpointing RAGE as an AGE receptor used heterogeneous mixtures of AGEs, CML-AGE and pronyl glycine AGEs are specific AGE ligands of RAGE. Chief consequences of RAGE activation are the attraction and activation of inflammatory cells to sites of acute or chronic AGE accumulation, and target cell injury, such as damage to neurons, cardiomyocytes or podocytes, as examples. Consequently, release of S100/calgranulins and HMGB1 by activated inflammatory or injured target cells may act as "second messengers," thereby sustaining RAGE activation and, perhaps leading to engagement of other receptors such as toll receptors. If left unchecked, such means of cellular activation may then cause functional and pathological derangements in organs such as the kidney and great vessels (**Figure 1**).

Importantly, beyond diabetic nephropathy and accelerated atherosclerosis, multiple publications have reported roles for RAGE in the pathogenesis of distinct complications of diabetes, such as retinopathy, neuropathy, impaired wound healing, ischaemia/reperfusion injury in the heart, and accelerated periodontal disease, as examples.[52-58] Intriguingly, the effects of RAGE antagonism or deletion are independent of glucose, lipids or insulin. Thus, although other hyperglycaemia-mediated pathways such as action of the polyol pathway (aldose reductase), protein kinase C isoforms, mitochondrial oxidative stress and the hexosamine pathway are presumably intact,[59] blockade of RAGE has been shown to exert significant effects on suppressing vascular and end organ perturbation and damage. These considerations suggest that RAGE antagonism in concert with complementary pathway targets may affect optimal suppression of diabetes-associated complications.

As clinical trials proceed, testing RAGE antagonism, it will be particularly important to establish if chronic antagonism of the ligand-RAGE axis suppresses natural functions of the receptor. Although homozygous RAGE null mice display no adverse phenotype, and long-term administration of blocking antibodies to RAGE or sRAGE in animal models imparted no adverse effects, it remains to be seen if such blockade suppresses protective roles for RAGE in acute or chronic inflammatory challenges. To date, most studies testing

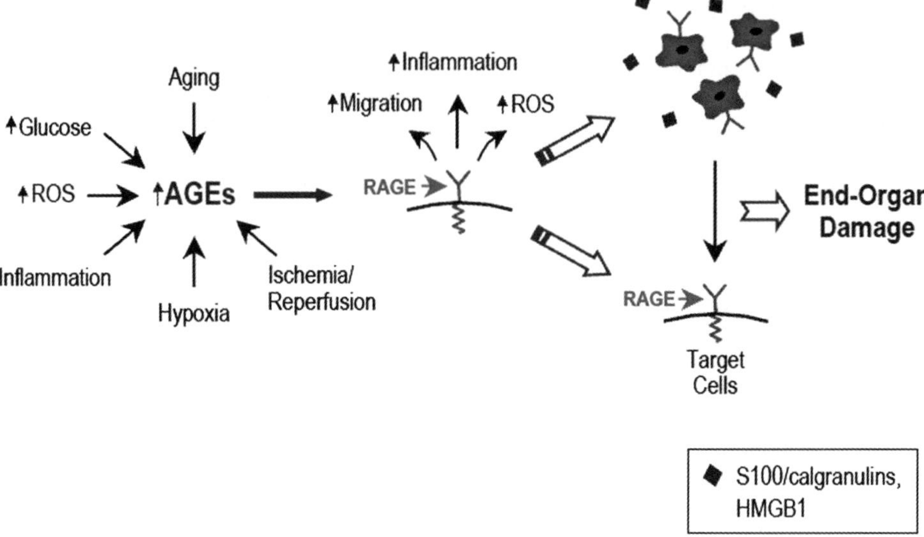

Figure 1. *AGEs form in diverse settings, activate RAGE, and initiate and sustain end-organ damage. AGEs may be formed by a range of mechanisms, such as hyperglycaemia, aging, inflammation, increased reactive oxygen species, hypoxia, ischaemia/reperfusion and in food preparation (such as heating. Once formed, at least certain AGEs may engage RAGE, thereby stimulating signal transduction and activating mechanisms leading to increased generation of inflammatory mediators, increased reactive oxygen species and cellular migration. Direct damage to target cells and/or influx of inflammatory cells to AGE foci lead to release of "second messengers" such as S100/calgranulins and HMGB1. These molecules may bind RAGE or perhaps other receptors such as toll receptors, thereby leading to end-organ damage if RAGE action is not interrupted. AGE-RAGE action may be relevant in the pathogenesis of the complications of diabetes, inflammatory responses, ischaemia/reperfusion (even in the absence of diabetes), aging, and unhealthy diets. Stopping the cycle of AGE-RAGE and second messenger-RAGE signalling may be essential in curtailing the consequences of chronic disease and aging.*

sepsis, extensive liver injury or massive infection have failed to reveal innate roles for RAGE in survival mechanisms.[60-62] Quite the contrary, RAGE deficient mice display significantly improved survival and less organ damage, but without failure to clear pathogens when exposed to bacterial and viral challenges.[63,64] In one study to date addressing bacterial challenge and the RAGE axis, inoculation of mice with massive doses of *Escherichia coli* uncovered possible beneficial roles for RAGE.[65] Of note, however, mono-infection with injection of *E. coli* in the peritoneum is not likely to occur naturally *in vivo*. Hence, the applicability of this model to human infection may not be certain. Taken together, propelled by promising findings in pre-clinical biology, RAGE antagonism marches on in clinical trials.

Acknowledgements

The authors gratefully acknowledge support for these studies from the US Public Health Service and the Juvenile Diabetes Research Foundation.

References

1   Maillard L.C. *Cornput. Rend* 1912, **154**, 66-68.
2   Negre-Salvayre A, Salvaryre R, Auge N, Pamplona R, Portero-Otin M. *Antioxid Redox Signal* 2009, **11**, 3071-3109.
3   Yan SF, Du Yan S, Ramasamy R, Schmidt AM. Ann Med 2009, in press.
4   Schmidt AM, Yan SD, Brett J, Mora R, Nowygrod R, Stern D. *J Clin Invest* 1993, **91**, 2155-2168.
5   Kislinger T, Fu C, Huber B, Qu W, Taguchi A, Du Yan S, et al. *J Biol Chem* 1999, **274**, 31740-9.
6   Harja E, Bu D.X, Hudson BI, Chang JS, Shen X, Hallam K, et al. *J Clin Invest* 2008, **118**, 183-94.
7   Guo Z.J, Niu H.X, Hou F.F, Zhang L, Fu N, Nagai R, et al. *Antioxid Redox Signal* 2008, **10**, 1699-712.
8   Chang JS, Wendt T, Qu W, Kong L, Zou YS, Schmidt AM, et al. *Circ Res* 2008, **102**, 905-913.
9   Anderson M.M, Requena J.R, Crowley J.R, Thorpe S.R, Heinecke JW. *J Clin Invest* 1999, **104**, 103-113.
10   Barati M.T, Merchant M.L, Kain AB, Jevans A.W, McLeish K.R, Klein J.B. *Am J Physiol Renal Physiol* 2007, **293**, F1157-F1165.
11   Foerster A, Henle T. *Biochem Soc Trans* 2003, **31**, 1383-1385
12   Lin R.Y, Choudhury R.P, Cai W, Lu M, Fallon J.T, Fisher E.A, et al. *Atherosclerosis* 2003, **168**, 213-220.
13   Cai W, He J.C, Zhu L, Chen X, Wallenstein S, Striker G.E, et al. *Am J Pathol* 2007, **170**, 1893-1902.
14   Somoza V, Lindenmeier M, Hofmann T, Frank O, Erbersdobler H.F, Baynes J.W, et al. *Ann N Y Acad Sci* 2005, **1043**, 492-500.
15   Hofmann MA, Drury S, Fu C, Qu W, Taguchi A, Lu Y, et al. *Cell* 1999, **97**, 889-901.
16   Leclerc E, Fritz G, Vetter SW, Heizmann CW. 2009, **1793**, 993-1007.
17   Donato R, Sorci G, Riuzzi F, Arcuri C, Bianchi R, Brozzi F, et al. *Biochim Biophys Acta* 2009, **1793**, 1008-1022.
18   Hori O, Brett J, Slattery T, Cao R, Zhang J, Chen J.X, et al. *J Biol Chem* 1995, **270**, 25752-61.
19   Taguchi A, Blood D.C, del Toro G, Canet A, Lee D.C, Qu W, et al. *Nature* 2000, **s405**, 354-60.
20   van Beijnum J.R, Buurman W.A, Griffioen A.W. *Angiogenesis* 2008, **11**, 91-99.
21   Scaffidi P, Misteli T, Bianchi M.E. *Nature* 2002, **418**, 191-195.
22   Chavakis T, Bierhaus A, Al-Fakhri N, Schneider D, Witte S, Linn T, et al. *J Exp Med* 2003, **198**, 1507-1515.
23   Yan SD, Zhu H, Zhu A, Golabek A, Du H, Roher A, et al. *Nat Med* 2000, **6**, 643-651.
24   Wang H, Bloom O, Zhang M, Vishnubhakat JM, Ombrellino M, Che J, et al. *Science* 1999, **285**, 248-251.
25   Miyata T, Hori O, Zhang J, Yan SD, Ferran L, Iida Y, et al. *J Clin Invest* 1996, 98, 1088-1094.

26   Hou F.F, Miyata T, Boyce J, Yuan Q, Chertow G.M, Kay J, et al. *Kidney Int* 2001, **59**, 990-1002.

27   Chen Y, Yan S.S, Colgan J, Zhang H.P, Luban J, Schmidt A.M, et al. *J Immunol* 2004, **173**, 1399-1405.

28   Moser B, Szabolcs M.J, Ankersmit H.J, Lu Y, Qu W, Weinberg A, et al. *Am J Transplant* 2007, **7**, 293-302.

29   Moser B, Desai D.D, Downie M.P, Chen Y, Yan SF, Herold K, et al. *J Immunol* 2007, **179**, 8051-8058.

30   Chen Y, Akirav EM, Chen W, Henegariu O, Moser B, Desai D, et al. *J Immunol* 2008, **181**, 4272-4278.

31   Horie K, Miyata T, Maeda K, Miyata S, Sugiyama S, Sakai H, et al. *J Clin Invest* 1997, **100**, 2995-3004.

32   Tanji N, Markowitz GS, Fu C, Kislinger T, Taguchi A, Pischetsrieder M, et al. *J Am Soc Nephrol* 2000, 11, 1656-1666.

33   Lewko B, Stepinski J. *J Cell Physiol* 2009, **221**, 288-295.

34   Wendt TM, Tanji N, Guo J, Kislinger TR, Qu W, Lu Y, et al. *Am Pathol* 2003, 162, 1123-1137.

35   Flyvbjerg A, Denner L, Schrijvers BF, Tilton RG, Mogensen TH, Paludan SR, et al. *Diabetes* 2004, **53**, 166-172

36   Jensen LJ, Denner L, Schrijvers BF, Tilton RG, Rasch R, Flyvbjerg A.
1.   *J Endocrinol* 2006, **188**, 493-501

37   Myint K.M, Yamamoto Y, Doi T, Kato I, Harashima A, Yonekura H, et al. *Diabetes* 2006, **55**, 2510-2522.

38   Inagi R, Yamamoto Y, Nangaku M, Usuda N, Okamato H, Kurokawa K, et al. *Diabetes* 2006, **55**, 356-366

39   Zheng S, Noonan W.T, Metreveli N.S, Coventry S, Kralik P.M, Carlson EC, et al. *Diabetes* 2004, **53**, 3248-3257

40   Teiken J.M, Audettey J.L, Laturnus D.I, Zheng S, Epstein P.N, Carlson E.C. *Anat Rec* 2008, 291, 114-121

41   Reiniger N, Lau K, McCalla D, Eby B, Cheng B, Rosario R, et al. *J Am Soc Nephrol* 2009, SA-PO2913.

42   Guo J, Ananthakrishnan R, Qu W, Lu Y, Reiniger N, Zeng S, et al. *J Am Soc Nephrol* 2008, 19, 961-972

43   Wang Y, Wang YP, Tay YC, Harris DC. *Kidney Int* 2000, **58**, 1797-1804.

44   Wang Y, Wang YP, Tay YC, Harris DC *Kidney Int* 2001, **59**, 941-949.

45   Burke A.P, Kolodgie F.D, Zieske A, Fowler D.R, Weber D.K, Varghese P.J, et al. *Arterioscler Thromb Vasc Biol* 2004, **24**, 1266-1271.

46   Cipollone F, Iezzi A, Fazia M, Zucchelli M, Pini B, Cuccurullo C, et al. *Circulation* 2003, 108, 1070-1077.

47   Park L, Raman K.G, Lee K.J, Lu Y, Ferran L.J, Jr., Chow W.S, et al. *Nat Med* 1998,4,1025-31.

48   Bucciarelli L.G, Wendt T, Qu W, Lu Y, Lalla E, Rong L.L, et al. *Circ* 2002,**106**,2827-35.

49   Wendt T, Harja E, Bucciarelli L, Qu W, Lu Y, Rong L.L, et al. *Atherosclerosis* 2006, **185**, 70-7.

50   Soro-Paavonen A, Watson AM, Li J, Paavonen K, Koitka A, Calkin AC, et al. *Diabetes* 2008, **57**, 2461-9.

51   Sun L, Ishida T, Yasuda T, Kojima Y, Honjo T, Yamamoto Y, et al. *Cardiovasc Res* 2009, **82**, 371-81.

52 Bucciarelli L.G, Ananthakrishnan R, Hwang Y.C, Kaneko M, Song F, Sell D.R, et al. *Diabetes* 2008, **57**, 1941-1951.

53 Bucciarelli L.G, Kaneko M, Ananthakrishnan R, Harja E, Lee L.K, Hwang Y.C, et al. *Circ* 2006, **113**, 1226-34.

54 Barile G.R, Pachydaki S.I, Tari S.R, Lee SE, Donmoyer C.M, Ma W, et al. *Invest Ophthalmol Vis Sci.* 2005, **46**, 2916-24.

55 Bierhaus A, Haslbeck K.M, Humpert PM, Liliensiek B, Dehmer T, Morcos M, et al. *J Clin Invest* 2004, **114**, 1741-1751.

56 Toth C, Rong L.L, Yang C, Martinez J, Song F, Ramji N, et al. *Diabetes* 2008, **57**, 1002-1017.

57 Goova M.T, Li J, Kislinger T, Qu W, Lu Y, Bucciarelli L.G, et al. *Am J Pathol* 2001, **159**, 399-403.

58 Lalla E, Lamster I.B, Feit M, Huang L, Spessot A, Qu W, et al. *J Clin Invest* 2000, **105**, 1117-1124.

59 Calcutt N.A, Cooper M.E, Kern T.S, Schmidt A.M. *Nat Rev Drug Discov* 2009, **8**, 417-429.

60 Liliensiek B, Weigand M.A, Bierhaus A, Nicklas W, Kasper M, Hofer S, et al. *J Clin Invest* 2004, **113**, 1641-50.

61 Lutterloh EC, Opal SM, Pittman DD, Keith JC, Jr., Tan XY, Clancy BM, et al. *Crit Care* 2007, **11**, R122.

62 Cataldegirmen G, Zeng S, Feirt N, Ippagunta N, Dun H, Qu W, et al. *J Exp Med* 2005, **201**, 473-484.

63 van Zoelen M.A, Schouten M, de Vos A.F, Florquin S, Meijers J.C, Nawroth P.P, et al. *J Immunol* 2009, **182**, 4349-56.

64 van Zoelen M.A, van der Sluijs K.F, Achouiti A, Florquin S, Braun-Pater J.M, Yang H, et al. Receptor for advanced glycation end products is detrimental during influenza A virus pneumonia. *Virology* 2009, **391**, 265-73.

65 van Zoelen M.A, Schmidt A.M, Florquin S, Meijers J.C, de Beer R, de Vos A.F, et al. *J Infect Dis* 2009, **200**, 765-73.

ACTIVATION OF CELLULAR ANTIOXIDATIVE DEFENSE MECHANISMS BY CARBOXYMETHYL-LYSINE-RECEPTOR FOR ADVANCED GLYCATION END-PRODUCTS (RAGE) INTERACTION

S, Foth[1] and V, Somoza[2]

[1]German Research Center for Food Chemistry, Garching, Germany
[2] Research Platform of Molecular Food Science, University of Vienna, Austria

1 INTRODUCTION

Thermal treatment during the production and processing of foods leads to a variety of desired changes in color, taste and aroma but also enhances the formation of Advanced Glycation End Products (AGE), formed by the reaction of amino acids and reducing carbohydrates. Carboxymethyl-lysine (CML), a lysine derivative and major AGE, has been detected in dairy products and in a multitude of other foods [1]. Various studies indicate that food derived AGEs are absorbed in the intestines [2-4], with the transephitelial uptake based on passive diffusion [5].

Formation of AGEs does also occur *in vivo* [6]. High tissue concentrations of CML are associated with the progression of numerous diseases, such as diabetes [7] and its consequent disorders like retinopathy [8] or nephropathy [9], but also Alzheimer's disease [10] and aging [11]. A possible mechanism by which AGEs may provoke their adverse biological activity is *via* their interaction with plasma membrane receptors. Interaction with AGEs has been described for various receptors so far, including oligosaccharyl transferase complex protein 48 (AGE-R1), 80K-H protein (AGE-R2), galectin-3 (AGE-R3) and the receptor for AGEs (RAGE) [12], the last being most intensively investigated so far. Raised protein expressions of RAGE have been described in tissues of subjects on diabetes mellitus or Alzheimer's disease.

The RAGE receptor is a member of the immunoglobulin superfamily, consisting of three extracellular domains and one cytosolic domain, that activates several intracellular response pathways. CML is a known ligand of RAGE [13] which results in an activation of NF-kB and the formation of pro inflammatory cytokines [14]. It has also been shown that food-derived AGEs, like CML-enriched casein (CasCML), activate p42/44 MAP kinase in Caco-2 cells [15] *via* interaction with RAGE [16]. In a model of transfected HEK-293 expressing either the full length of RAGE or the receptor lacking its cytosolic domain, further confirmation for the RAGE-mediated activation of MAP kinases by CML and CasCML was obtained [16]. Thus, is can be hypothesized that CML activates cellular signaling pathways which are involved in an inflammatory response.

AGE-RAGE interactions have also been demonstrated to induce NADPH oxidase and, in consequence, to increase cellular concentrations of reactive oxygen species (ROS) [17]. ROS themselves may initiate the oxidative generation of AGEs (Thornalley, 1999), and, thus, induce a virtuous circle. As a consequence, intracellular ROS release after AGE-RAGE interaction has been associated with subsequent accumulation of AGEs and related pathologies [18].

Mitochondria are one of the primary sources of ROS in living organisms. In single electron transfer reactions, superoxide anion radicals are generated. These can be transformed into less reactive hydrogen peroxide and molecular oxygen by the catalytic activity of the enzyme superoxide dismutase. Coughlan et al. (2009) showed that activation of RAGE induced mitochondrial superoxide formation under hyperglycemic conditions [19]. One cellular response to increased superoxide formation is an induction and/or increased activity of the enzyme superoxide dismutase (SOD). However, at high concentrations of AGEs, the SOD protein can be glycated as well, resulting in enzyme inactivation [20]. Here, we investigated the influence of free and protein-bound CML on the mitochondrial formation of ROS in human kidney (HEK293) and intestinal (Caco-2) cells using the fluoroprobe MitoSOX^{red} [21]. The cell's response to increased ROS release was studied by analyzing the catalytic activities and the protein expression of SOD. Furthermore, we demonstrated for the first time, that the receptor for advanced glycation end products, RAGE, plays a pivotal role not only in CML-mediated ROS-release but also in SOD induction.

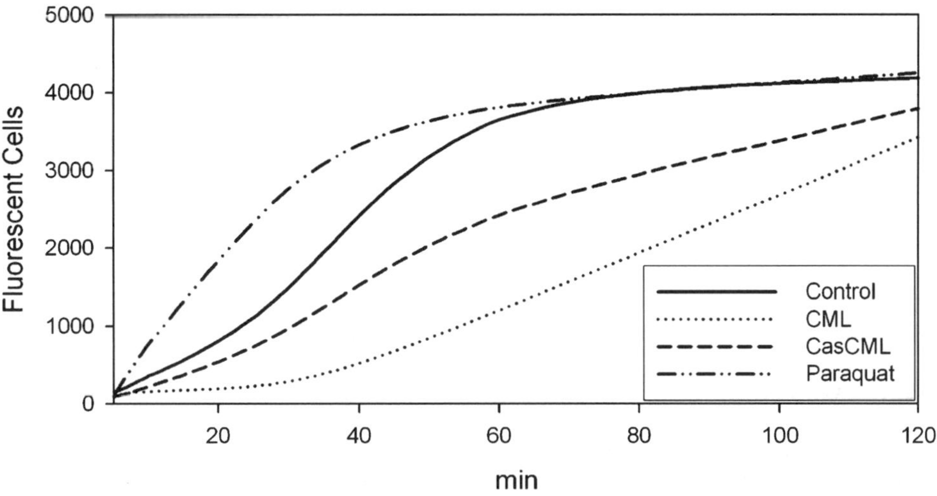

Figure 1 *Number of MitoSOX-fluorescent cells treated with either pure HBBS (Control), 1.1 mM CML or casein-bound CML, or 1 mM paraquat*

2 MATERIAL AND METHODS

2.1 Materials

All reagents were applied in the highest purity available. CML was purchased from the PolyPetide Group (Strasbourg, France). Casein was CML-enriched as described before [22], resulting in a 72 % CML-modification of lysine. For all experiments, pure and casein-bound CML were used in a concentration of 1.1 mM.

2.2 Cell culture

Caco-2 cells and HEK-293 cells expressing either the full length of RAGE (HEK-FL) or its non-cytoplasmatic domain solely ($\Box$-cyto, HEK-DC) were cultured to confluence under standard conditions (37° C, 5 % CO_2).

2.3 Measurement of the mitochondrial superoxide formation in living cells

Experiments were performed on the basis of the experimental protocol published by Mukhopadhyay et al. (2007), using MitoSOX red (invitrogen), a selective fluorescent dye that allows the determination of mitochondrial superoxide in living cells.
Cells were treated with 5 mM MitoSOX prior to incubation with pure Hank's Buffered Saline Solution (HBBS, control) or HBBS containing either CML, CasCML or 1 mM paraquat (positive control). The number of cells showing MitoSOX-fluorescence out of a total of 10,000 cells over a time period of two hours was determined using a flow cytometer (BD Biosciences) at a wavelength of λ_{Ex} = 510 nm for excitation and an emission wavelength of λ_{Em} = 580 nm.

2.4 Protein expression of CuZn- and MnSOD

Cells were lyzed after incubation with CML, CasCML or paraquat for various time periods, ranging from 2 min up to 6 h. After Western Blotting, CuZn- and MnSOD were detected by a combination of an appropriate selective primary antibody (polyclonal, rabbit, StressGen) and secondary antibody (goat, Abcam) using chemoluminescence. Quantification was performed using the Kodak 1D software.

2.5 Activity of SOD

A cell's lysate was obtained as described above and was applied to a SOD Assay Kit-WST (Dojindo) following the manufacturer's instruction.

2.6 Statistics

In bar diagrams, mean values and SEM of six to eight replicates are shown and all data refer to non-treated control cells, the results of which were set as 100 %. Statistical significances were calculated by Student's t-test, with a $p \leq 0.05$ considered as statistically significant (*: $p \leq 0.05$; **: $p \leq 0.01$; ***: $p \leq 0.001$).

3 RESULTS

3.1 Mitochondrial superoxide in living cells

ROS formation in Caco-2 and in HEK-FL cells was analyzed after treatment with either free or casein-bound CML, or paraquat as positive control by flow cytometry using MitoSOX labeling. The number of cells showing superoxide-related fluorescence increased in each of the experiments, even when cells were treated with HBBS (control) solely for about 60 min (figure 1). However, this reaction was accelerated by paraquat as a compound that generates superoxide radical anions at high reaction rates (Smith et al., 1978). For further evaluation, data were displayed as percentage of HBBS-treated controls (figure 2).

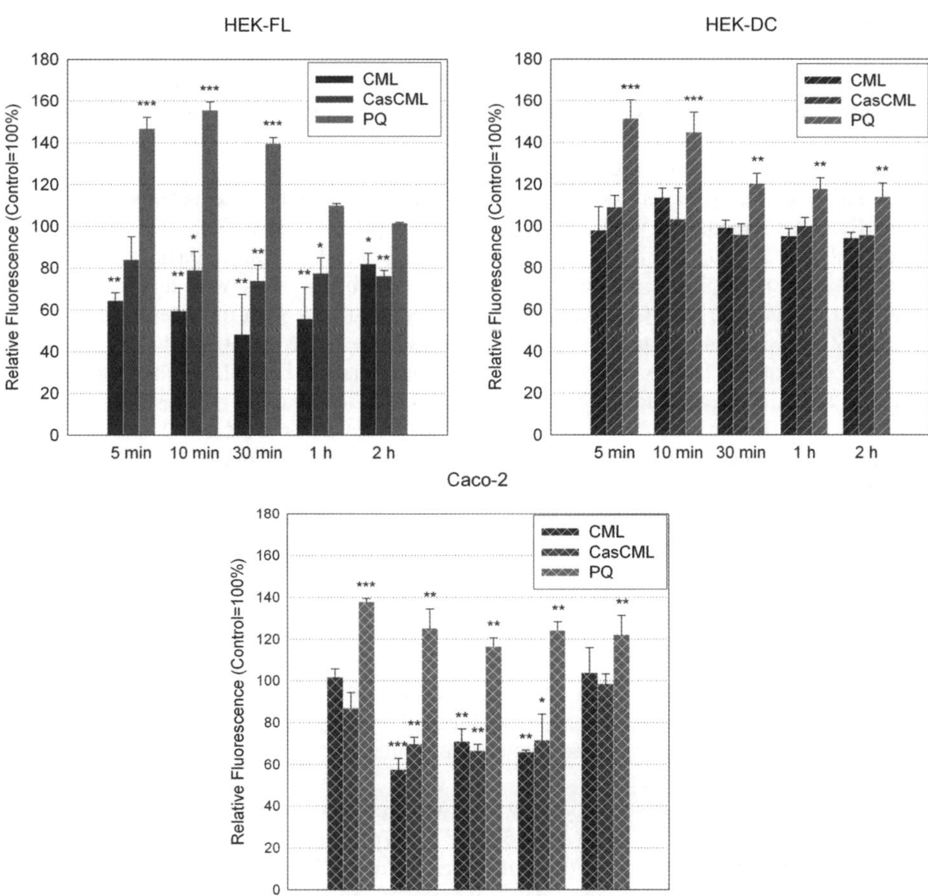

Figure 2   *Relative MitoSOX Fluorescence in HEK-FL, HEH-DC and Caco-2 after incubation with either 1.1 mM CML, 1.1 mM CasCML or 1 mM paraquat (untreated control = 100 %)*

Treatment with paraquat increased the superoxide formation significantly in all cell lines after exposure times up to 2 h, with a maximum increase of up to 60 % after 5 to 10 min of exposure. In contrast, MitoSOX-fluorescence in HEK-FL and Caco-2 cells was significantly decreased after treatment with free or casein-bound CML. In Caco-2 cells, fluorescent cell counts decreased up to 42.5 ± 5.5 % (p<0.001 and up to 33.4 ± 3.0 % (p<0.01) after treatment with CML or CasCML, respectively. In HEK-FL cells, a similar result was obtained, with a decrease of up to 51.7 ± 19.0 % (p<0.01) and up to 26.2 ± 7.6%. (p<0.01) by CML and CasCML, respectively. Here, differences between free and casein-bound CML reached statistical significance (p<0.01). In HEK-DC cells, in contrast, neither treatment with CML nor CasCML changed the number of fluorescent cell counts at a level of statistical significance, with a maximum decrease of less than 10

3.2 Protein expression of CuZn- and MnSOD

Protein expression of the cytosolic CuZnSOD and the mitochondrial MnSOD was determined after incubating cells with CML, CasCML or paraquat. Paraquat significantly raised the expression of CuZn and MnSOD in all cells by up to 20 % within 10 min (data not shown). All other results are given in figure 3.

In HEK-FL cells, SOD protein expression was significantly affected by CML and CasCML. After incubation with CML for 5 min, expression of CuZnSOD and MnSOD rapidly increased by up to 17.4 ± 2.8 % (p<0.01) and 18.3 ± 3.9 % (p<0.01), respectively. After 30 min of exposure, protein expression of both enzymes was down regulated. Treatment of the cells with CasCML resulted in a similar, although time delayed, increase of CuZnSOD (13.0 ± 4.0%, p<0.01) and MnSOD (17.5 ± 3.1 %, p<0.001).

In Caco-2 cells, SOD protein expression was less pronounced, but still reached the level of statistical significance. Treatment of the cells with CML for up to 30 min resulted in an 8.9 ± 1.7 % (p<0.01) and 11.8 ± 0.7 % (p<0.001) increase for CuZnSOD and MnSOD, respectively. Similar results were obtained for CasCML, with an increase of 12.2 ± 5.1 % and 6.0 ± 1.9 %, p<0.05 for the CuZnSOD and MnSOD, respectively. In contrast, neither CML nor CasCML had a significant effect on the expression of SOD proteins in HEK-DC cells. Significant differences between the responses to CML and CasCML could not be found in any cell line.

3.3 Activity of SOD

Besides the protein expression, the catalytic activity of SOD was analyzed. The assay applied to the cell lysates is based on an indirect method to determine the total SOD activity using a tetrazolium salt. Results are presented in figure 4 as percentage of non-treated control cells (set as 100 %).

In HEK-FL and Caco-2 cells incubated with CML, the activity of SOD was significantly increased by 25.4 ± 7.1 % (p<0.01) and 56.4 ± 2.8 % (p<0.001), respectively. Although the maximum response to the incubation with casein-bound CML was delayed compared to that demonstrated for CML, a significant increase of the activity SOD by 79.9 ± 2.8 % (p<0.001) and 31.1 ± 0.1 % (p<0.001) was shown for HEK-FL and Caco-2 cells, respectively. Thus, in Caco-2 cells, the activation of the SOD was clearly more pronounced after treatment with CML compared to that with CasCML. In contrast, the opposite effect was demonstrated for HEK-FL cells (p<0.001). In contrast, SOD activity did not change in HEK-DC cells under the experimental conditions given.

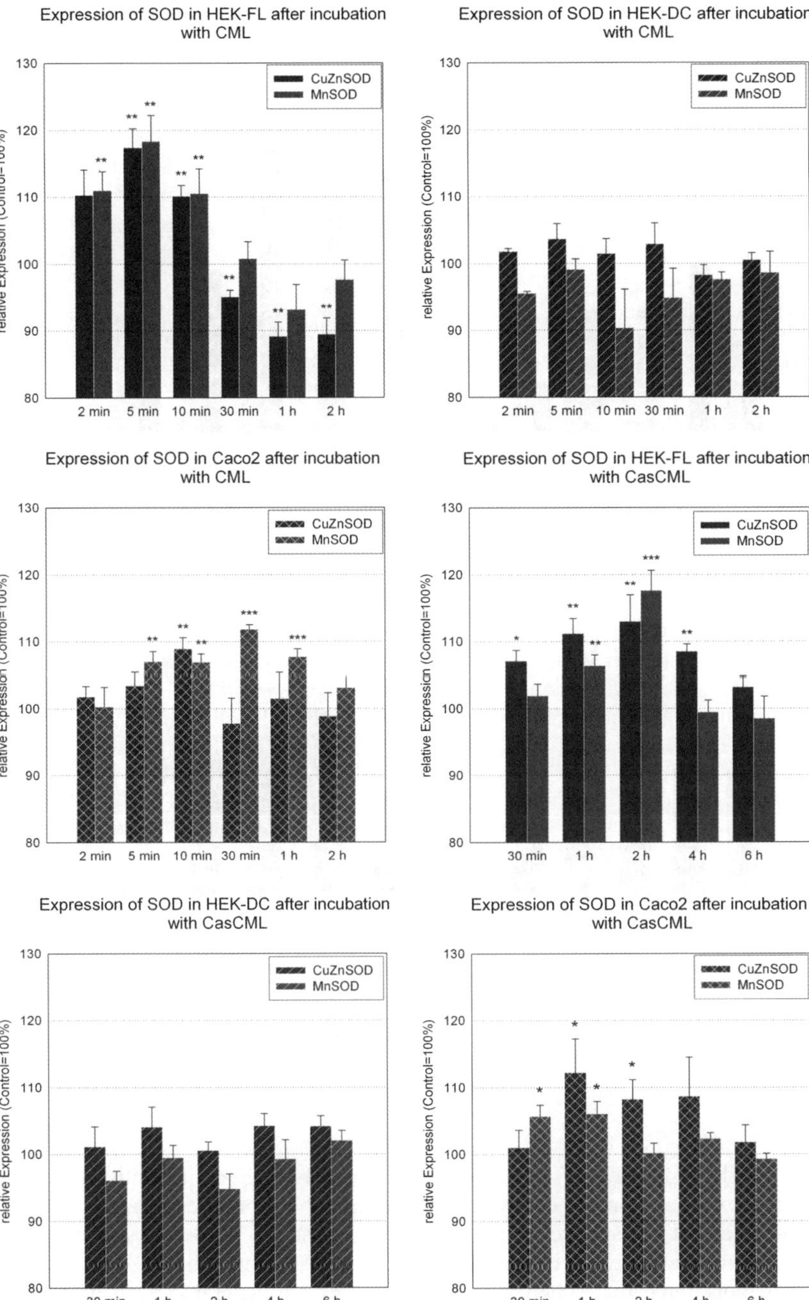

Figure 3 *Relative expression of mitochondrial MnSOD and cytosolic CuZnSOD in HEK-FL, HEH-DC and Caco-2 after incubation with 1.1 mM CML or CasCML (untreated control = 100 %)*

                                        The Maillard Reaction

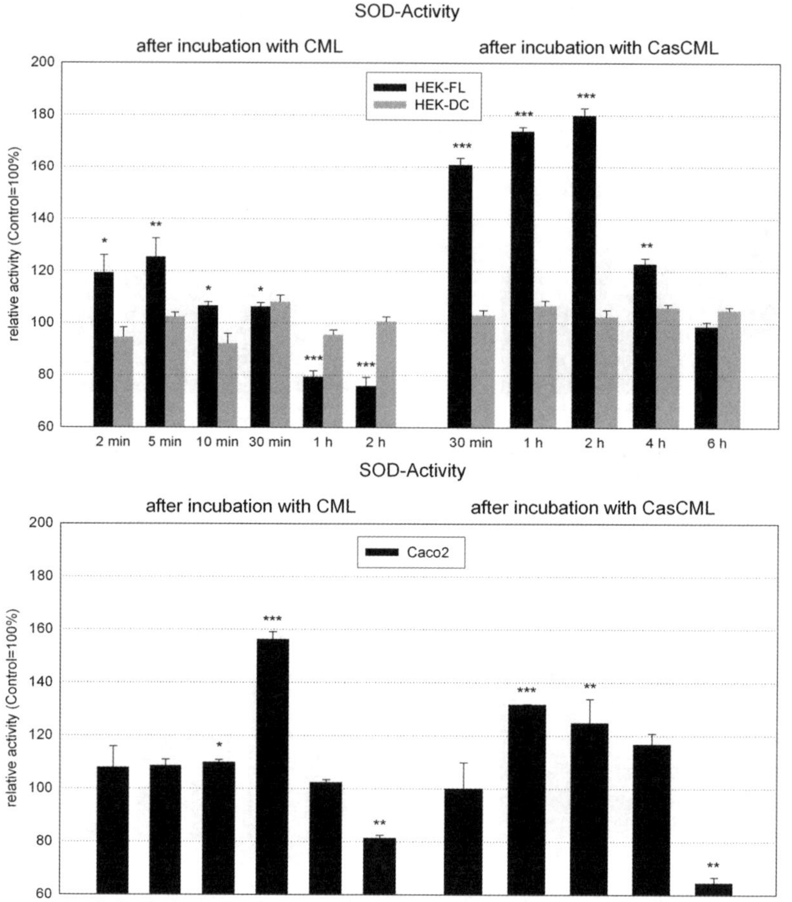

Figure 4   *Catalytic activity of total SOD in HEK-FL, HEH-DC and Caco-2 after incubation with 1.1 mM CML or 1.1 mM CasCML (untreated control = 100 %)*

4 DISCUSSION

We analyzed the influence of free and protein-bound CML on the mitochondrial formation of superoxide and on the catalytic activities and the protein expression of SOD as indicators of the cell's response to an increased release of ROS. Measurement of mitochondrial superoxide in living cells using the fluorescent label MitoSOX in Caco-2 and HEK-293 demonstrated that superoxide radical anions are generated even in control cells which have been treated with phosphate buffer solely. This result indicates that mitochondria release superoxide radical anions from oxidative phosphorylation in the absence of any experimental stimulus.

Treatment with the cells with the redox cycler paraquat, as expected, increased superoxide radical anion formation significantly. Incubation with CML or CasCML resulted in a major decrease of MitoSOX-fluorescent cells in HEK-FL and Caco-2, with CML causing the more explicit effects. Since mitochondrial superoxide anion radical formation in HEK-DC cells was not affected, the data clearly point to a RAGE-dependent pathway of ROS formation. To our knowledge, this reduction of mitochondrial superoxide by AGEs has not been described before.

One reason for this result might be the activation of SOD by free as well as casein-bound CML. We found that the expression of cytosolic SOD and mitochondrial SOD was significantly increased in Caco-2 and HEK-FL cells. The maximum cellular response to CML was faster than that to CasCML, and of a larger magnitude. We further found an enhanced catalytic activity of total SOD in Caco-2 and HEK-FL cells. Again, the response to CML treatments was faster, although in HEK-FL cells, most pronounced effects were induced by CasCML. Still, neither free nor casein-bound CML induced a significant change of SOD protein expression or catalytic activity in HEK-DC cells. This result, again, points to an involvement of RAGE and may explain why mitochondrial superoxide was not reduced.

In summary, our data clearly indicate that particularly CML, but also casein-bound CML, can activate the SOD as one of the major antioxidant enzymes *via* interaction with RAGE. Increased SOD activity may (1) counteract the formation of superoxide radical anions in response to RAGE-mediated NADPH-oxidase activity and may also (2) reduce the amount of AGEs that is formed from oxidative pathways. Therefore, interaction of an AGE such as CML with RAGE may activate cellular defense mechanisms against ROS. Rising levels of AGEs may, therefore, be based on a combination of the dietary intake, *in vivo* formation and cellular defense mechanisms.

Acknowledgements

The authors like to thank International Maillard Society for the generous travel award, which allowed us to present our work at the 9th International Symposium on the Maillard Reaction, held in Cairns / Australia, August 29 – 31, 2009.

References

1. S. Drusch, V. Faist and H. F. Ebersdobler, *Food Chemistry*, 1999, **65**, 547-553.
2. V. Faist and H. F. Erbersdobler, *Annals of nutrition & metabolism*, 2001, **45**, 1-12.
3. V. Somoza, E. Wenzel, C. Weiss, I. Clawin-Radecker, N. Grubel and H. F. Erbersdobler, *Molecular nutrition & food research*, 2006, **50**, 833-841.
4. K. Sebekova and V. Somoza, *Molecular nutrition & food research*, 2007, **51**, 1079-1084.
5. S. Grunwald, R. Krause, M. Bruch, T. Henle and M. Brandsch, *The British journal of nutrition*, 2006, **95**, 1221-1228.
6. M. Brownlee, *Annual review of medicine*, 1995, **46**, 223-234.
7. D. G. Dyer, J. A. Dunn, S. R. Thorpe, K. E. Bailie, T. J. Lyons, D. R. McCance and J. W. Baynes, *The Journal of clinical investigation*, 1993, **91**, 2463-2469.
8. T. Murata, R. Nagai, T. Ishibashi, H. Inomuta, K. Ikeda and S. Horiuchi, *Diabetologia*, 1997, **40**, 764-769.
9. T. Miyata, S. Sugiyama, D. Suzuki, R. Inagi and K. Kurokawa, *Kidney Int Suppl*, 1999, **71**, S54-56.
10. A. Takeda, M. Wakai, H. Niwa, R. Dei, M. Yamamoto, M. Li, Y. Goto, T. Yasuda, Y. Nakagomi, M. Watanabe, T. Inagaki, Y. Yasuda, T. Miyata and G. Sobue, *Acta neuropathologica*, 2001, **101**, 27-35.

11. E. D. Schleicher, E. Wagner and A. G. Nerlich, *The Journal of clinical investigation*, 1997, **99**, 457-468.
12. A. M. Schmidt, M. Vianna, M. Gerlach, J. Brett, J. Ryan, J. Kao, C. Esposito, H. Hegarty, W. Hurley, M. Clauss and et al., *The Journal of biological chemistry*, 1992, **267**, 14987-14997.
13. T. Kislinger, C. Fu, B. Huber, W. Qu, A. Taguchi, S. Du Yan, M. Hofmann, S. F. Yan, M. Pischetsrieder, D. Stern and A. M. Schmidt, *The Journal of biological chemistry*, 1999, **274**, 31740-31749.
14. S. D. Yan, A. M. Schmidt, G. M. Anderson, J. Zhang, J. Brett, Y. S. Zou, D. Pinsky and D. Stern, *The Journal of biological chemistry*, 1994, **269**, 9889-9897.
15. H. Zill, R. Gunther, H. F. Erbersdobler, U. R. Folsch and V. Faist, *Biochemical and biophysical research communications*, 2001, **288**, 1108-1111.
16. H. Zill, S. Bek, T. Hofmann, J. Huber, O. Frank, M. Lindenmeier, B. Weigle, H. F. Erbersdobler, S. Scheidler, A. E. Busch and V. Faist, *Biochemical and biophysical research communications*, 2003, **300**, 311-315.
17. M. P. Wautier, O. Chappey, S. Corda, D. M. Stern, A. M. Schmidt and J. L. Wautier, *American journal of physiology*, 2001, **280**, E685-694.
18. A. L. Tan, K. C. Sourris, B. E. Harcourt, V. Thallas-Bonke, S. Penfold, S. P. Andrikopoulos, M. C. Thomas, R. C. O'Brien, A. Bierhaus, M. E. Cooper, J. M. Forbes and M. T. Coughlan, *Am J Physiol Renal Physiol*, 2009.
19. M. T. Coughlan, D. R. Thorburn, S. A. Penfold, A. Laskowski, B. E. Harcourt, K. C. Sourris, A. L. Tan, K. Fukami, V. Thallas-Bonke, P. P. Nawroth, M. Brownlee, A. Bierhaus, M. E. Cooper and J. M. Forbes, *J Am Soc Nephrol*, 2009, **20**, 742-752.
20. N. Taniguchi, M. Takahashi, H. Sakiyama, Y. S. Park, M. Asahi, Y. Misonou and Y. Miyamoto, *Annals of the New York Academy of Sciences*, 2005, **1043**, 521-528.
21. P. Mukhopadhyay, M. Rajesh, K. Yoshihiro, G. Hasko and P. Pacher, *Biochemical and biophysical research communications*, 2007, **358**, 203-208.
22. V. Faist, C. Muller, S. Drusch and H. F. Erbersdobler, *Die Nahrung*, 2001, **45**, 218-221.

THE PATHOGENIC POTENTIAL OF DIFFERENT SIZED AGE MODIFIED MOLECULES VIA RAGE SIGNALLING PATHWAYS

SA Penfold[1], KC Sourris[1], MT Coughlan[1], AL Tan[1], DM Kaye[1], ME Cooper[1], JM Forbes[1]

[1]Glycation and Diabetes Complications, Baker IDI Heart and Diabetes Institute, Melbourne, Australia

1   INTRODUCTION

Advanced glycation end-products (AGEs) are a heterogenous group of modifications on proteins that are formed as a result of non-enzymatic reactions, including the Maillard reaction [1]. The formation of AGEs is accelerated under diabetic conditions due to chronic hyperglycemia, oxidative stress and diminished clearance of AGE precursors [2, 3]. Therapeutic inhibition of AGE formation is known to reduce cellular dysfunction and the progression of diabetic complications, including kidney disease [4-6]. AGEs have a wide range of chemical, cellular, and tissue effects that contribute to the development and progression of diabetic complications. Some of these effects are mediated via binding to the receptor for AGE (RAGE) and activation of downstream signalling pathways including oxidative stress, nuclear factor κB (NFκB) [7], protein kinase C (PKC) [8], and vascular endothelial growth factor (VEGF) [9] and each may promote the formation of AGEs in the vicious cycle associated with progressive kidney damage. RAGE is also activated by a range of other ligands, including the pro-inflammatory calcium-binding S100/calgranulins [10], high-mobility group box 1 (HMGB1) [11] [12], and β-amyloid sheets [13]. The purpose of this study was to investigate the activation of RAGE-dependent signalling pathways in primary renal mesangial cells by different RAGE ligands including AGE peptides and AGE modified proteins, calcium-binding S100 proteins and HMGB1

2   MATERIALS AND METHODS

2.1   Preparation of AGE-BSA

AGE-BSA was prepared following the incubation of BSA in 500mM glucose, as previously described [14]. AGE-BSA (50mg/ml) and BSA preparations were enriched for specific molecular weights using Amicon Ultra- 15 Centrifugal Filter devices[15]. AGE fractions were separated using SDS-PAGE [15]. Carboxymethyllysine (CML) concentrations in AGE-BSA fractions were assessed by an in-house indirect ELISA as previously described in [16]. Tryptic digests were analysed by LC-MS/MS using a HCT ULTRA ion trap spectrometer (Bruker Daltonics, Bremen, German) coupled online with a 1200 capillary HPLC (Agilent technologies, Santa Clara, CA, USA) as previously described in [15]. AGE-BSA fractions were administered to cells at a dose equivalent to 100µg/ml of CML.

2.2   Cell Culture conditions

Mesangial cells (MC) were isolated from kidneys of Sprague Dawley rats, as previously described [14]. Cells were cultured in Dulbecco's modified Eagles medium (DMEM, with no phenol red), supplemented with 2mmol/l glutamine, 5000U/ml penicillin/streptomycin, and 20% fetal bovine serum. Cells were passaged weekly for up to 10 passages. For experiments, cells were cultured in high glucose conditions (25mmol/l D-glucose) of low glucose conditions (5 mmol/l) for one week in the presence or absence of HMGB1 (5ug), S100A (2ug), S100B (5ug), CML-BSA (1ug) AGE-BSA or BSA fractions. For priming experiments, mesangial cells were primed for two hours with HMGB1 (5ug) and then treated with either non fractioned or fractioned AGE-BSA or BSA for 1, 2,3,and 7 days under 5 or 25mmol/l D-glucose conditions.

2.3   Activation of pathogenic pathways

In each experiment, RAGE-dependent activation of important pathogenic pathways was estimated. Intracellular calcium influx was measured by Fura-2 [17]. PKC kinase activity was measured in cell lysates using the StressXpress PKC Kinase Activity Assay Kit (EKS-420A; Stressgen Bioreagents Corporation Victoria, BC, Canada) [15]. Secreted VEGF was measured in the concentrated medium by ELISA (R&D systems, Minneapoli, MN) [15]. NFκB DNA-binding activity was determined in cell lysates using an NFκB p65 transcription factor assay kit (TransAM, Active Motif; Carlsbad CA, USA) as per the manufacturer's instructions. Expression of AGER was measured by real-time RT-PCR. For all RT-PCR reactions, 6µg of total RNA extracted from cell lysates was used to synthesis cDNA with the Superscript first-strand synthesis system for RT-PCR (Life Technologies Inc., Grand Island, NY, USA) [14]. Cell surface RAGE protein expression was measured by immunostaining and flow cytometry [15].

2.4   RAGE binding experiments

Mesangial cells were incubated with radiolabelled ^{125}I AGE-BSA fractions 3 x 10^6 cpm as previously described in [15].Mesangial cells were infected with RAGE or soluble RAGE adenovirus or an empty vector (ADGO). Once expression was confirmed, cells were treated with AGE-BSA fractions and the induction of calcium signalling

2.5   Statistical Analysis

All statistical computations were performed using GraphPad Prism version 4.0a for Mac OS X (GraphPad Software, San Diego, California, USA). Values of experimental groups are shown as mean ± SEM unless otherwise stated. One-way ANOVA with Tukey's post-test analysis was used to determine statistical significance. Where appropriate, either two-tailed t-tests or Mann Whitney U tests were performed. A probability of $P < 0.05$ was considered to be statistically significant.

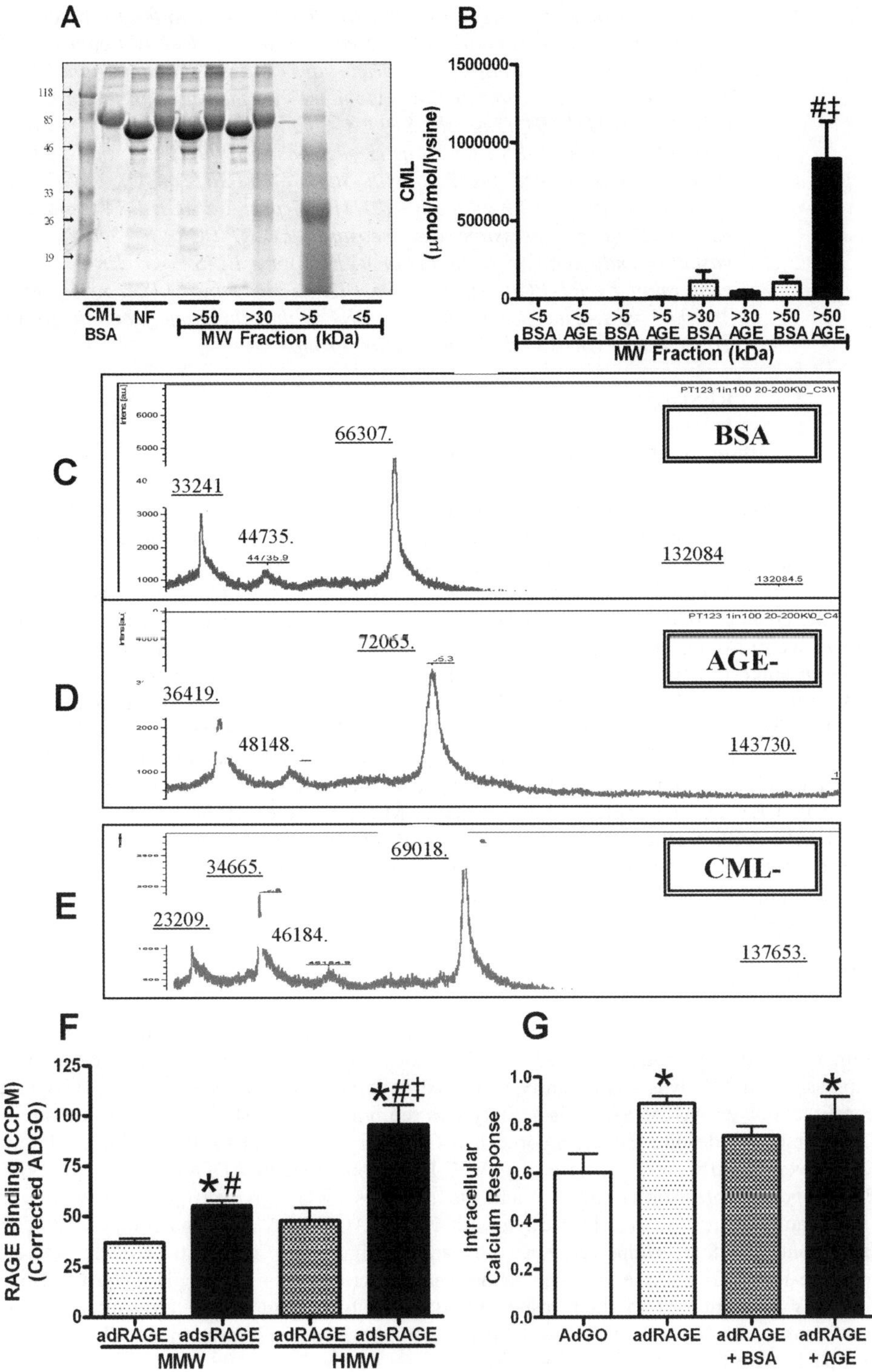
A
118
85
46
33
26
19
CML BSA
NF
>50
>30
>5
<5
MW Fraction (kDa)
B
CML (µmol/mol/lysine)
1500000
1000000
500000
0
#‡
<5 BSA
<5 AGE
>5 BSA
>5 AGE
>30 BSA
>30 AGE
>50 BSA
>50 AGE
MW Fraction (kDa)
C
Itens [au.]
6000
5000
40
3000
2000
1000
PT123 1in100 20-200K\0_C3\1
BSA
33241
44735.
44735.9
66307.
132084
132084.5
D
4000
3000
2000
1000
PT123 1in100 20-200K\0_C4
AGE-
36419.
48148.
72065.
143730.
E
4000
1000
600
CML-
23209.
34665.
46184.
46184.9
69018.
137653.
F
RAGE Binding (CCPM) (Corrected ADGO)
125
100
75
50
25
0
*#
*#‡
adRAGE adsRAGE
MMW
adRAGE adsRAGE
HMW
G
Intracellular Calcium Response
1.0
0.8
0.6
0.4
0.2
0.0
*
*
AdGO
adRAGE
adRAGE + BSA
adRAGE + AGE

Figure 1:   *Modulation of RAGE signalling by RAGE ligands of different molecular weight. Primary mesangial cells were exposed to RAGE ligands with different molecular weight for seven days under HG conditions. Cell surface RAGE expression by flow cytometry was measured in the presence or either A) HMGB1 (5ug/ml), S100 A (2ug/ml) + B (5ug/ml), CML-BSA (1ug/ml) or B) AGE-BSA fractions (Doses based 100ug CML) or C) AGE-BSA fractions primed for 2 h with 5ug of HMGB1 or D) AGE-BSA fractions treated for 2 h with 5ug/ml HMGB1 post 7 d AGE-BSA treatment. E) VEGF gene expression was measured by RT-PCR. F) Secretion of vascular endothelial growth factor (VEGF) by ELISA. G) PKC-α gene expression by RT-PCR. H) Relative PKC activity by ELISA. n=6/group. *p<0.05 vs L-Glucose, #p<0.05 vs BSA within the same fraction group, ‡p<0.05 vs AGE-BSA different fraction group.*

3   RESULTS

3.1   Fractions of AGE modified molecules

AGE-BSA and BSA were fractionated by molecular weight and subjected to SDS-PAGE under reducing conditions (Figure 1a). AGE-BSA had greater molecular weight than unmodified BSA fractions. CML-modified BSA (CML-BSA) was used as a positive control. Protein-bound CML was only detected in AGE-BSA fractions greater than 30kDa, with the greatest concentration in the HMW fraction, >50kDa (Figure 1b). The specific ion spray spectra from each group (Figure 1c-d) showed that there was a prominent peak at the albumin spike at approximately 72kDa in the AGE-BSA sample, compared to a smaller spike present in BSA sample at 66kDa (the molecular mass of BSA). Furthermore, a 69kDa spike was also identified in the spectrum of CML-modified BSA (Figure 1e), used as a positive control. The RAGE binding capacity of radiolabelled fractions from AGE-BSA and BSA was next examined in primary renal mesangial cells over-expressing human full length RAGE or secreted soluble RAGE. Medium molecular weight (>30<50KDa) and HMW AGE fractions (>50kDa) had significantly increased cellular binding to cells over-expressing soluble RAGE when compared to adenovirus-RAGE control (Figure 1f). In addition, the HMW AGE-BSA fraction demonstrated more cellular binding to adenovirus-RAGE than the MMW AGE-BSA fraction.

3.2   Modulation of RAGE signalling by RAGE ligands

To investigate which AGE-BSA fractions can induce pathogenic changes *in vitro*, primary renal MC were exposed for 7d under HG conditions. Cell surface RAGE expression on MC was significantly increased by treatment with S100B CML-BSA and non-Fractionated AGE-BSA. This activity was reproduced in HMW-AGE-BSA but not in LMW fractions despite the same content of CML (figure 2a and b). Intracellular calcium levels were significantly increased by AGE-BSA compared to ADGO (figure 1g). Both PKC-α gene expression (figure 2g) and membranous PKC activity (figure 1h) were also significantly increased in cells treated with HMW AGE-BSA fractions compared to the BSA control. Other groups remained unchanged (figure 2h). By contrast, there was an increase in VEGF gene expression in both non-fractionated, LMW-, HMW-AGE-BSA fractions as compared to their respective BSA controls (figure 2e). Exposure to non-fractionated, LMW and HMW AGE-BSA also elevated the secretion of VEGF when compared with fractioned BSA alone (figure 2f).

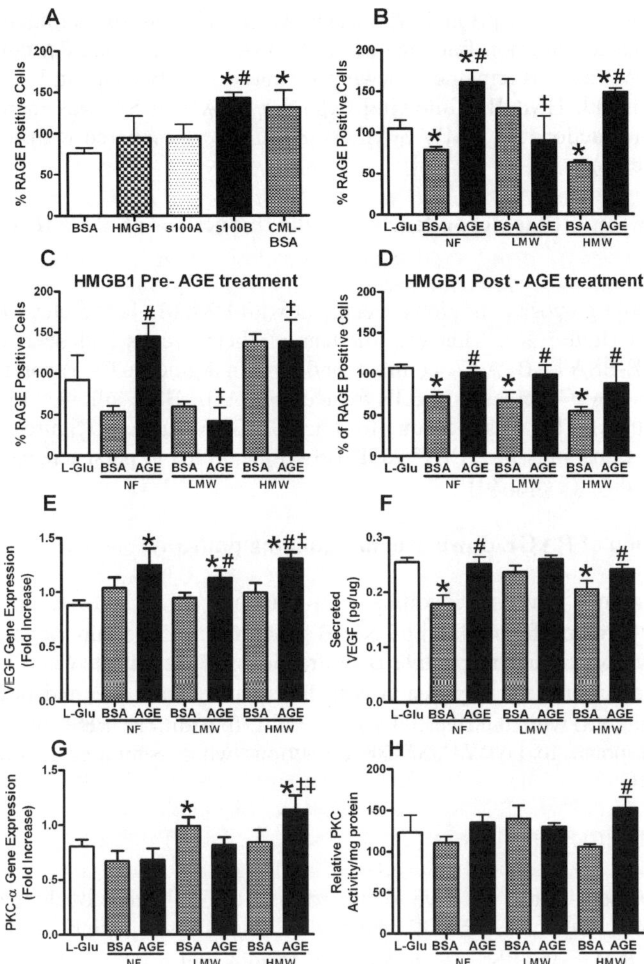

Figure 2. *RAGE modified molecules of high molecular weight contain higher CML levels and show greater binding affinity to RAGE. A) Unfractionated (NF) and fractionated AGE-BSA and control BSA were separated by 10% SDS-PAGE and stained with coomassie blue. CML-BSA was used as a positive control. The left lane is the molecular weight marker. B) CML levels in AGE-BSA fractions were determined by ELISA. C-E Spectra following Ion Spray MSMS of C) BSA control and D) AGE modified BSA E) CML-modified bovine serum albumin as a positive control. F) Binding of ^{125}I-radiolabelled MMW and HMW AGE-BSA fractions to primary rat mesangial cells overexpressing human full length RAGE or soluble RAGE adenovirus. G) Intracellular calcium levels determined by Florescence. n=6/group. *p<0.05 vs ADGO, #p<0.05 vs adRAGE within the same fraction group, ‡p<0.05 vs AdsRAGE different fraction group*

RAGE expression on cells primed with HMGB1 for 2h was further increased in cells subsequently treated with non-fractionated AGE-BSA compared to BSA control, whereas LMW and HMW fractions remained unchanged (figure 2c). By contrast, RAGE expression on cells treated with HMGB1 following exposure to AGE-BSA was increased in cells treated with non-fractionated, LMW or HMW-BSA, when compared with respective BSA fractions (figure 2d).

3.3   Glucose and time-dependence for the induction of AGER expression by ligands

Under high glucose conditions, treatment with HMGB1 led to elevations in AGER gene expression (figure 3b). This was the same, whether cells had been concomitantly exposed to AGE-BSA or BSA By contrast, under normal glucose (NG) conditions (5mM), AGER expression was decreased in cells treated with AGE-BSA followed by priming with HMGB1 (figure 3a). This was the same for all AGE-BSA fractions (figure 3c). However, under HG conditions, only the HMW AGE-BSA fraction led to persistence of AGER gene expression by 7 days (figure 3d).

3.4   Activation of RAGE downstream signalling pathways

Under normal glucose conditions, all fractions of AGE-BSA up-regulated the expression of P47 phox (figure 4a), p65 NFκB gene expression (figure 4c) and IκB (figure 4e) when compared to their respective BSA fractions. By contrast, when cultured under high glucose conditions, the expression both P47 phox (figure 4b) and p65 NFκB gene expression (figure 4d) were unaffected by AGE-BSA. In addition, IκB gene expression was decreased in response to LMW-AGE-BSA fractions when compared to the LMW-BSA group (Figure 4f).

4   DISCUSSION

In the present study, we have identified that RAGE ligands of different molecular weight, not only modulate RAGE expression but cause distinct signaling effects downstream of RAGE via PKC-α , p65 and secreted VEGF. Indeed, we have previously identified that within the circulation of diabetic patients, only immuno-reactive high molecular weight RAGE ligands exist [15]. We have shown that CML modified BSA, S100B, and HMW (>50kDa) AGE-BSA each increased cell surface RAGE expression, but also that RAGE expression in fraction groups was inhibited by priming with HMGB1. HMGB1 challenge, after treatment with AGE-BSA fractions enhanced cell surface RAGE expression irrespective of the size of the fraction. LMW AGE-BSA appears to signal via another receptor other than RAGE, with our results showing they did not bind to RAGE nor activate downstream signalling pathways such as secretion of VEGF. Conversely, our results have shown that HMW AGE-BSA does induce RAGE activation, bind RAGE in cells and signal downstream of RAGE increasing VEGF and PKC activity.

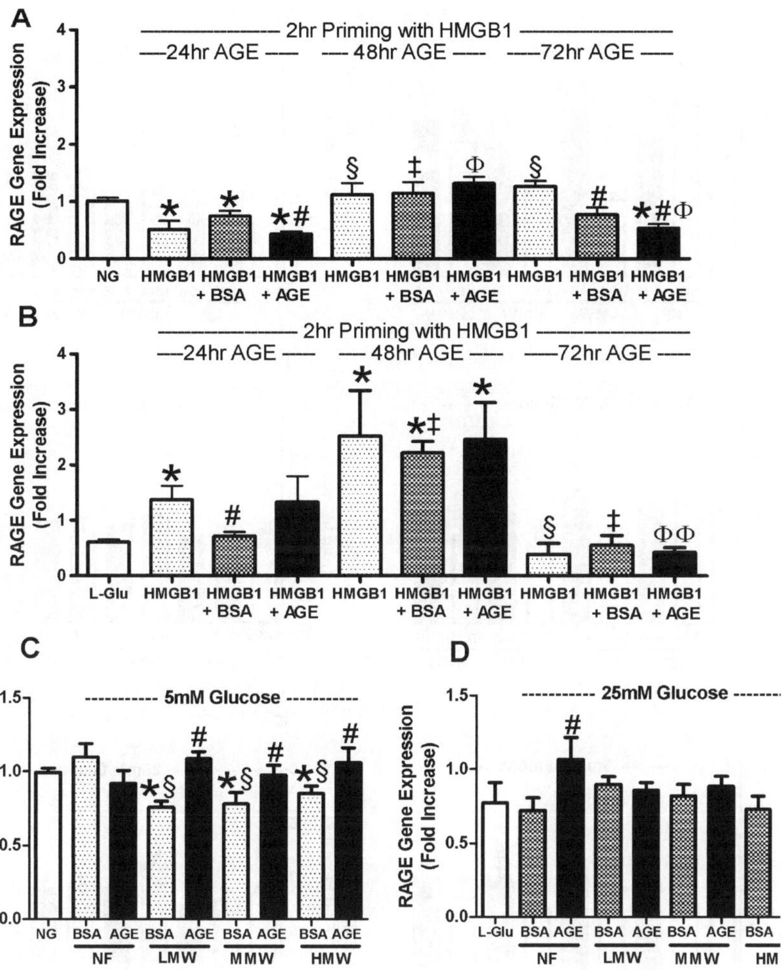

Figure 3: *Modulation of RAGE gene expression by RAGE ligands is time dependent Cells were primed for 2 h with 5ug HMGB1 and treated for 24, 48, 72 h in the presence of BSA or AGE-BSA in either A) 5mM glucose conditions B) 25mM conditions C,D) RAGE gene expression was measure by RT-PCR. n=6/group. *p<0.05 vs L-Glucose, #p<0.05 vs BSA within the same fraction group, §p<0.05 vs HMGB1 + AGE-BSA at different time course.*

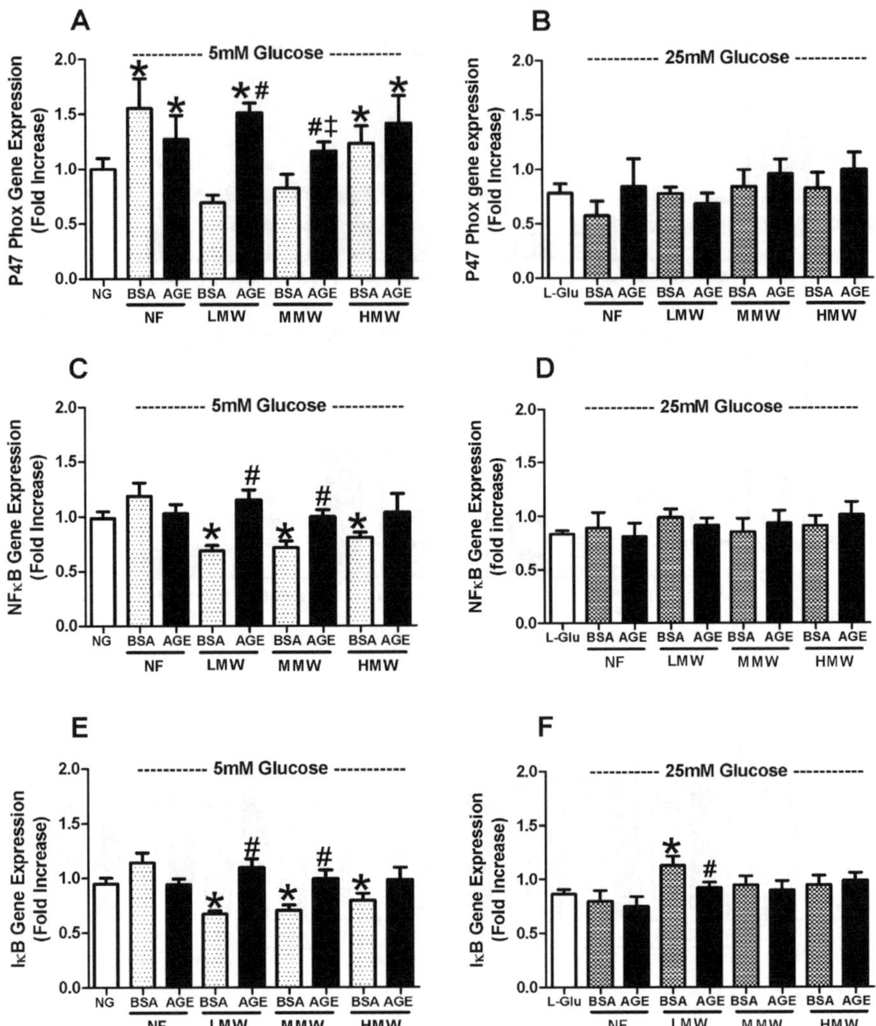

Figure 4: *Activation of RAGE downstream signalling pathways by different sized AGE molecules Primary mesangial cells were exposed to AGE-BSA fractions under eith NG or HG condition for seven days. A-F) Gene expression was measured by RT-PCR. A,B) P47 phox C,D) NFKB E,F) IKB n=6/group. *p<0.05 vs L-Glucose, #p<0.05 vs BSA within the same fraction group, ‡p<0.05 vs AGE-BSA different fraction group*

Interestingly, HMW AGE-BSA fraction clearly had the highest CML concentrations when compared to lower molecular weight fractions. Indeed, immuno-reactive (protein bound) CML was only present in fractions greater than 30kDa. This supported by mass spectrometry where the albumin peak in AGE-BSA sample (72kDa) was larger than the BSA albumin peak (66kDa), identifying that the AGE-BSA had more modifications present and also contained a dimeric peak at approx 133 kDa reminiscent of CML-BSA. The HMW AGE-BSA fraction also had greater binding affinity to cells overexpressing soluble RAGE compared to the MMW fraction. Intracellular calcium levels were elevated in response to the RAGE adenovirus but this effect was not amplified when challenged with AGE-BSA indicating that increases in RAGE are associated with elevations in intracellular calcium but this appears to not involve RAGE signalling.

Our studies have highlighted some major differences in the modulation of RAGE expression by differing molecular weight AGE fractions under normal and high glucose conditions. Indeed, under NG conditions, cells primed with HMGB1 and treated with AGE-BSA for 72 h showed the highest activation of RAGE. Conversely, under high glucose conditions this effect was increased after a 48 h treatment period with AGE-BSA. Interestingly, all AGE-BSA fraction groups increased RAGE gene expression when under NG conditions. However, RAGE gene and protein levels were only increased by HMW AGE-BSA fractions in high glucose conditions, again supporting our previous studies using diabetic patient serum.

Furthermore, our findings show that under normal glucose conditions AGE-BSA fractions are increasing not only RAGE expression but also but the gene expression of P47 phox, NFκB, and IκB through an AGE-mediated effect. However, under high glucose conditions RAGE gene expression is through AGE-mediated response with the HMW AGE-BSA fraction but there were no down stream changes in gene expression seen under these conditions. In conclusion, these data demonstrate that ligands which activate RAGE and cause downstream signalling events are predominantly of high molecular weight.

Acknowledgments:

This research was funded by the Juvenile Diabetes Research Foundation (JDRF) (JMF, MEC), the NIH (MCT) and Diabetes Australia. JMF is a recipient of a JDRF career development award and MEC holds an NHRMC Australia Fellowship and a JDRF Scholars Award.

References

1.   L. C. Maillard, *CR Acad Sci*, 1912, **154**, 66-68.
2.   J. W. Baynes and S. R. Thorpe, *Diabetes*, 1999, **48**, 1-9.
3.   E. D. Schleicher, E. Wagner and A. G. Nerlich, *J Clin Invest*, 1997, **99**, 457-468.
4.   J. M. Forbes, T. Soulis, V. Thallas, S. Panagiotopoulos, D. M. Long, S. Vasan, D. Wagle, G. Jerums and M. E. Cooper, *Diabetologia*, 2001, **44**, 108-114.
5.   J. M. Forbes, V. Thallas, M. C. Thomas, H. W. Founds, W. C. Burns, G. Jerums and M. E. Cooper, *Faseb J*, 2003, **17**, 1762-1764.
6.   J. M. Forbes, M. E. Cooper, V. Thallas, W. C. Burns, M. C. Thomas, G. C. Brammar, F. Lee, S. L. Grant, L. A. Burrell, G. Jerums and T. M. Osicka, *Diabetes*, 2002, **51**, 3274-3282.
7.   A. Bierhaus, P. M. Humpert, M. Morcos, T. Wendt, T. Chavakis, B. Arnold, D. M. Stern and P. P. Nawroth, *J Mol Med*, 2005, **83**, 876-886.
8.   T. Inoguchi, T. Sonta, H. Tsubouchi, T. Etoh, M. Kakimoto, N. Sonoda, N. Sato, N. Sekiguchi, K. Kobayashi, H. Sumimoto, H. Utsumi and H. Nawata, *J Am Soc Nephrol*, 2003, **14**, S227-232.

9.   D. J. Kelly, R. E. Gilbert, A. J. Cox, T. Soulis, G. Jerums and M. E. Cooper, *J Am Soc Nephrol*, 2001, **12**, 2098-2107.

10.   M. A. Hofmann, S. Drury, C. Fu, W. Qu, A. Taguchi, Y. Lu, C. Avila, N. Kambham, A. Bierhaus, P. Nawroth, M. F. Neurath, T. Slattery, D. Beach, J. McClary, M. Nagashima, J. Morser, D. Stern and A. M. Schmidt, *Cell*, 1999, **97**, 889-901.

11.   O. Hori, J. Brett, T. Slattery, R. Cao, J. Zhang, J. X. Chen, M. Nagashima, E. R. Lundh, S. Vijay, D. Nitecki and et al., *J Biol Chem*, 1995, **270**, 25752-25761.

12.   H. Rauvala and A. Rouhiainen, *Biochim Biophys Acta*, **1799**, 164-170.

13.   R. Deane, S. Du Yan, R. K. Submamaryan, B. LaRue, S. Jovanovic, E. Hogg, D. Welch, L. Manness, C. Lin, J. Yu, H. Zhu, J. Ghiso, B. Frangione, A. Stern, A. M. Schmidt, D. L. Armstrong, B. Arnold, B. Liliensiek, P. Nawroth, F. Hofman, M. Kindy, D. Stern and B. Zlokovic, *Nat Med*, 2003, **9**, 907-913.

14.   M. T. Coughlan, D. R. Thorburn, S. A. Penfold, A. Laskowski, B. E. Harcourt, K. C. Sourris, A. L. Tan, K. Fukami, V. Thallas-Bonke, P. P. Nawroth, M. Brownlee, A. Bierhaus, M. E. Cooper and J. M. Forbes, *J Am Soc Nephrol*, 2009, **20**, 742-752.

15.   S. A. Penfold, M. T. Coughlan, S. K. Patel, P. M. Srivastava, K. C. Sourris, S. D., D. E. Webster, M. C. Thomas, R. J. MacIsaac, J. G., L. M. Burrell, M. E. Cooper and J. M. Forbes, *Kidney Int*, 2010, **Accepted for publication (March 10).**

16.   M. T. Coughlan, V. Thallas-Bonke, J. Pete, D. M. Long, A. Gasser, D. C. Tong, M. Arnstein, S. R. Thorpe, M. E. Cooper and J. M. Forbes, *Endocrinology*, 2007, **148**, 886-895.

17.   B. G. Drew, S. J. Duffy, M. F. Formosa, A. K. Natoli, D. C. Henstridge, S. A. Penfold, W. G. Thomas, N. Mukhamedova, B. de Courten, J. M. Forbes, F. Y. Yap, D. M. Kaye, G. van Hall, M. A. Febbraio, B. E. Kemp, D. Sviridov, G. R. Steinberg and B. A. Kingwell, *Circulation*, 2009, **119**, 2103-2111.

GLYCATED PROTEINS BIND TO ERM PROTEINS AND MODULATE THEIR ACTIONS

LA Bach[1,2], A Young[1], MA Gallicchio[1] and EA McRobert[1]

[1]Monash University, Department of Medicine (Alfred), Melbourne, Vic 3004, Australia.
[2]Department of Endocrinology and Diabetes, The Alfred, Melbourne, Vic 3004, Australia.

1 INTRODUCTION

The incidence of diabetes mellitus has reached epidemic proportions and continues to increase rapidly. Patients with diabetes have increased mortality and morbidity due to the development of microvascular complications including nephropathy, retinopathy, neuropathy, as well as macrovascular disease [1]. A number of mechanisms have been suggested for the development of these complications, including accelerated formation of advanced glycation endproducts (AGEs) due to increased non-enzymatic glycation of proteins, lipids and nucleic acids as a result of chronic hyperglycaemia and oxidative stress [2-5]. However, there are many outstanding questions regarding the precise mechanisms whereby AGEs contribute to diabetic complications.

AGEs act via receptor-dependent and receptor-independent mechanisms. With respect to the latter, glycation modifies the physicochemical properties of proteins, thereby changing their functional characteristics. This is especially apparent for long-lived proteins, such as those found within basement membranes. These proteins become relatively resistant to proteolysis, contributing to the characteristic basement membrane thickening found in tissues from patients with diabetes.

A number of AGE receptors have been identified. Of these, the receptor for AGEs (RAGE) is present at sites of diabetes-associated vascular injury, including the kidney, retina, nerve and blood vessels [6,7]. Other AGE binding proteins include AGE-R1, R2 and R3, lysozyme and various scavenger receptors. While some of these receptors mediate clearance of AGEs and activate intracellular messengers, understanding of their full significance is still incomplete [4]. A number of years ago, Youssef et al. showed that diabetes increase binding of AGEs to rat kidney sections, and that this binding was not due to one of the known AGE receptors [8]. We subsequently showed that AGEs bound to the ERM (ezrin, radixin and moesin) family of proteins in diabetic rat kidney extracts [9].

2 THE ERM PROTEIN FAMILY

ERM proteins belong to the band 4.1 superfamily, which also includes the tumor suppressor merlin [10]. ERM proteins share three highly conserved structural domains. The N-terminal FERM (*f*our-point one, *e*zrin, *r*adixin, *m*oesin) domain binds directly or indirectly to membrane proteins, whereas the C-terminal domain binds F-actin, resulting in actin filament rearrangement. ERM proteins therefore act as cross-linkers between cell membrane proteins and the actin cytoskeleton. The N- and C-terminal domains are linked by an α-helical coiled domain that modulates ERM actions [11]. ERM proteins are located in the cytoplasm in a dormant state due to inhibitory interactions between their N- and C-domains. They are activated by phosphorylation and/or interactions with phospholipids, whereupon they move to a submembranous location.

ERM proteins are components of actin-containing cell surface structures including microvilli and membrane ruffles and they regulate the organization and function of these structures [10, 12]. Regulated linkage of membrane proteins to the actin cytoskeleton is necessary for many critical cellular processes such as determination of cell shape, vesicle trafficking, cell adhesion and motility. At the tissue level, ERM proteins stabilise adherens junctions and are critical for morphogenesis in sites such as the intestine and eye [12].

Actions of ERM proteins are mediated by the assembly of multi-protein complexes including receptors, adaptors or scaffolding proteins, Rho GTPase proteins and actin [11]. ERM proteins bind directly or indirectly to a wide range of cell surface proteins. For example, they directly bind to adhesion receptors such as CD44 and ICAMs 1-3 [10, 11]. The ezrin-CD44 interaction underlies directional cell motility, whereas activation of natural killer cells by interleukin-2 relies on the cell surface expression of ICAM-2, which is modulated by the latter's interaction with ezrin. Direct binding of ezrin to the Na^+-H^+ exchanger 1 (NHE-1) ion transporter plays a role in cytoskeletal organisation [10, 11], whereas radixin activates the $\alpha_M\beta_2$-integrin by directly binding to its cytoplasmic tail [13].

ERM proteins also associate with membrane proteins indirectly via binding of their FERM domains to adaptor or scaffolding proteins such as the family of Na^+-H^+ exchanger regulatory factors (NHERFs). For example, ezrin binds to the PDZ domain of NHERF-1, which binds to ion channels such as Na^+-H^+ exchanger 3 (NHE-3). Protein kinase A localises to these complexes via binding of its regulatory subunit to ezrin and regulates the activity of this and other ion channels [14]. ERM/NHERF1 complexes also regulate endocytosis of NHE-3 and receptors, including the β_2 adrenergic receptor and the platelet-derived growth factor receptor. Finally, ERM/NHERF-containing complexes restrict proteins such as podocalyxin to specific plasma membrane domains, which is essential for their function.

ERM proteins modulate a number of intracellular signalling pathways [10]. ERM proteins are substrates for protein kinase $C\alpha$ and θ and are believed to be involved in signalling related to these kinases. The inter-relationship between Rho pathways and ERM proteins is more complex. On the one hand, Rho kinase phosphorylates ERM proteins and may play a role in their activation. On the other hand, ERM proteins modulate Rho activity, in part by binding to RhoGDI and preventing its inhibitory effects. ERM proteins may contribute to locally coordinating membrane receptor and Rho signalling to enhance cell migration [11]. Ezrin interacts with focal adhesion kinase and enhances its phosphorylation and activation [15]. Ezrin has also been implicated in cell survival, via interactions with CD95 (Fas/APO-1) [16], and/or phosphatidylinositol 3-kinase [17]. Ezrin is necessary for metastasis of some tumours, and the activation of mitogen-activated kinase pathways may be involved [18]. Changes in ezrin expression in some adult tumours are associated with poor prognosis [19].

As alluded to above, ezrin plays an important role in renal physiology, including maintenance of the glomerular barrier to proteins, and regulating proximal tubule sodium and phosphate reabsorption and hydrogen ion excretion [14]. Many of these functions are mediated via adaptors including NHERFs. For example, a podocalyxin/NHERF-2/ezrin/actin complex maintains the foot process structure and filtration slit patency of podocytes; disruption of this complex leads to marked changes in morphology and proteinuria such as that seen in diabetic nephropathy. NHERF-1 and ezrin are also important for regulating Na^+-K^+ ATPase activity by parathyroid hormone in renal tubule cells [14].

3 THE INTERACTION OF AGES WITH ERM PROTEINS

3.1 Localization of the ERM binding site for AGEs

Our initial studies showed that N-terminal fragments of ERM proteins sized ~40 kDa bound AGEs [9]. Recombinant ezrin and its fragments were used to further define the AGE binding site [20]. Using surface plasmon resonance analysis, human N-ezrin(1–324) bound to immobilized AGE-BSA with a Kd of ~5 × 10^{-7} M, which is similar to the binding affinity of AGEs for other receptors. Importantly, glycated albumin isolated from patients with diabetes bound to N-ezrin, confirming that the interaction may be physiologically relevant [9]. Similarly to many other proteins that bind to the N-terminal FERM domain [10], inactive full-length ERM proteins did not bind AGEs.

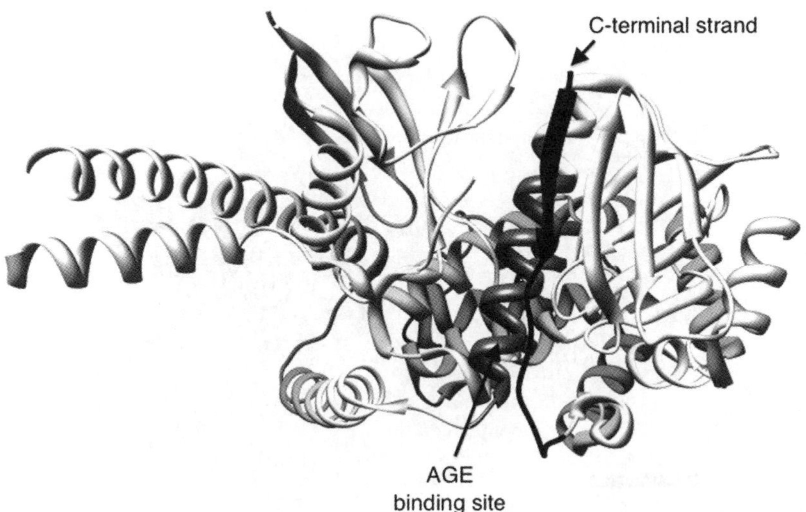

Figure 1   *Three-dimensional structure of dormant full-length insect moesin (PDB accession no 2I1J [21]), showing a C-terminal β-strand (black) overlying the AGE binding site (grey). The figure was prepared using UCSF Chimera, alpha version 1.4 [22].*

AGE binding to a number of deletion constructs (aa 1–280, 1–170 and 1–144) was greatly reduced, whereas binding to glutathione-*S*-transferase-N-ezrin fusion proteins (aa 200–324 and 270–324) was retained, suggesting that amino acids 280-324 are involved in AGE binding [20]. A peptide based on residues 277–299 of ezrin bound AGEs maximally (Kd ~3 × 10^{-8} M), whereas a peptide based on residues 281-296 had only slightly lower affinity. Detailed studies revealed that that Cys284 is necessary but not sufficient for AGE binding and residues 288–299 of ezrin are critical for optimal AGE binding. Within the latter, the R293RRK cluster of basic residues may facilitate binding, which is consistent with the negative charge of AGEs contributing to binding. This may be a generalized property of AGE binding since AGE binding to RAGE is competitively inhibited by heparin, a negatively charged glycosaminoglycan [23].

The AGE binding site is part of an α-helix found at the C-terminal end of the FERM domain. Recently, a crystallographic three-dimensional structure of full-length dormant insect moesin was solved [21]. In this structure, it can be seen that a C-terminal β-strand overlies the AGE binding site (Figure 1), potentially limiting access to this site, which might explain the inability of dormant ERM proteins to bind glycated proteins. This raises the question of how the AGE binding site becomes accessible to AGEs. One possibility is that activation of ERM proteins by phosphorylation disrupts the interaction between the ERM N- and C-domains, resulting in exposure of the N-terminal AGE binding site. Alternatively, proteolytic cleavage could release N-terminal ERM fragments that bind AGEs. In particular, ezrin is susceptible to cleavage by calpain 1, a calcium-dependent protease [24]. High glucose increases calpain activity [25], and preliminary *in vitro* experiments suggest that the calpain-generated N-terminal ezrin fragment binds AGEs (results not shown). Proteolytic cleavage and AGE binding may therefore act in concert to inhibit ezrin actions.

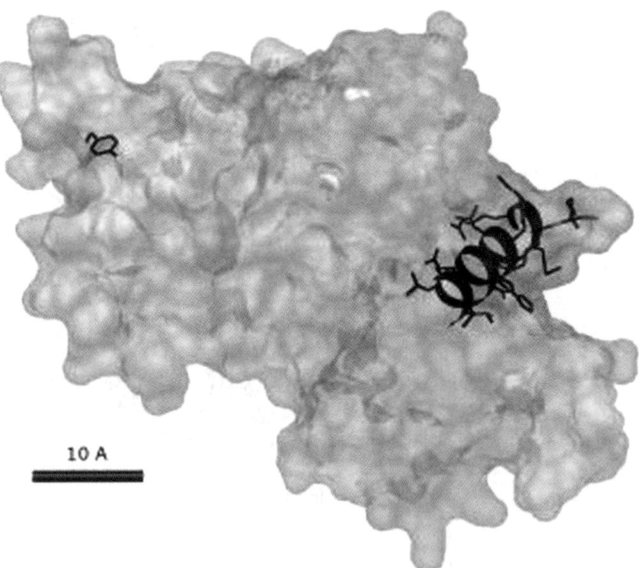

Figure 2   *The surface of activated human N-ezrin is shown (PDB accession no 1NI2 [26]). The AGE binding site of ezrin (on the right) is distant from Tyr145, the EGF receptor phosphorylation site (on the left). The figure was prepared using UCSF Chimera, alpha version 1.4 [22].*

3.2. Effects of AGE binding on ezrin function

ERM proteins can be activated by phosphorylation of a number of tyrosine and threonine kinases [10]. The epidermal growth factor receptor is a tyrosine kinase that phosphorylates Tyr145 of ezrin. Using an *in vitro* assay, we showed that AGE-BSA prevents phosphorylation of this residue in N-ezrin and substantially decreases its phosphorylation in full-length ezrin [9]. We speculated that the latter may have been due to phosphorylation causing a conformational change in full-length ezrin, allowing AGE binding that then prevented further phosphorylation. We further found that a 24 amino acid peptide based on the AGE binding site of ezrin dose-dependently inhibited the effect of AGE-BSA on EGF receptor-induced ezrin phosphorylation [20]. Since the AGE binding site is distant from Tyr145 in the three-dimensional ERM protein structure (Figure 2), it is likely that the inhibitory effect of AGEs on phosphorylation of this residue is due to a conformational change rather than direct steric hindrance.

We next assessed whether AGEs inhibit ezrin-dependent cellular actions in renal proximal tubule epithelial cells *in vitro*. In these cells, hepatocyte growth factor (HGF) mediates tubulogenesis in an ezrin-dependent manner *in vitro* [27]. Co-incubation of glycated proteins with HGF abrogated this effect [20]. Furthermore, over-expression of ezrin overcame the inhibitory effect of AGEs, strongly suggesting that the AGE-ezrin interaction is involved in this effect [28]. Although HGF stimulated phosphorylation of Akt and ERK, both of which contribute to tubulogenesis, AGEs had no effect on this phosphorylation, suggesting that they act via distinct pathways.

As mentioned above, ERM proteins are involved in migration of many cell types, including proximal tubule cells *in vitro* [27]. We confirmed that renal proximal tubule epithelial cell migration is ezrin-dependent [28]. Glycated proteins inhibited migration of these cells, but this effect was also abrogated by ezrin overexpression, providing further evidence that AGEs interfere with the cellular actions of ezrin [28].

4 CONCLUSIONS

The ERM family of proteins is a fascinating one with an expanding range of cellular actions. Our studies show that AGEs bind to the N-domains of these proteins and inhibit their phosphorylation and some of their cellular actions. Several lines of evidence suggest a possible link between altered ERM protein function and diabetic complications. Firstly, ERM proteins are expressed at sites of diabetic complications [29]. Secondly, disrupting ERM-containing complexes in podocytes leads to phenotypic changes such as loss of foot processes and proteinuria, which are features of nephropathy [30]. Thirdly, some of the proteins that associate with ERM proteins are also implicated in diabetic complications; for example, moesin interacts with Na^+-K^+ ATPase and may regulate its cell membrane localization and activity [31]; activity of this enzyme is decreased in diabetes [32]. Further studies are required to determine the extent to which AGE-induced cellular dysfunction is due to the AGE-ERM interaction. These studies, if successful, could lead to a novel therapeutic target to alleviate the burden of diabetic complications.

Acknowledgements

These studies were supported by grants from the National Health and Medical Research Council of Australia, the Juvenile Diabetes Research Foundation, Diabetes Australia Research Trust, the Alfred Research Trust, Eli Lilly Endocrinology Research Grant Scheme and Servier Laboratories.

References

1.   G. Jerums, S. Panagiotopoulos, J. M. Forbes, T. M. Osicka and M. E. Cooper, *Archives of biochemistry and biophysics*, 2003, **419**, 55-62.
2.   M. Brownlee, *Nature*, 2001, **414**, 813-820.
3.   P. J. Thornalley, S. Battah, N. Ahmed, N. Karachalias, S. Agalou, R. Babaei-Jadidi and A. Dawnay, *Biochem J*, 2003, **375**, 581-592.
4.   J. M. Forbes, V. Thallas-Bonke, M. E. Cooper and M. C. Thomas, *Curr Pharmaceut Design*, 2004, **10**, 3361-3372.
5.   S.-Y. Goh and M. E. Cooper, *J Clin Endocrinol Metab*, 2008, **93**, 1143-1152.
6.   M. Neeper, A.-M. Schmidt, J. Brett, S. D. Yan, F. Wang, Y.-C. E. Pan, K. Elliston, D. Stern and A. Shaw, *J Biol Chem*, 1992, **267**, 14998-15004.
7.   T. Soulis, V. Thallas, S. Youssef, R. E. Gilbert, B. McWilliam, R. P. Murray-McIntosh and M. E. Cooper, *Diabetologia*, 1997, **40**, 619-628.
8.   S. Youssef, D. T. Nguyen, T. Soulis, S. Panagiotopoulos, G. Jerums and M. E. Cooper, *Kidney Int*, 1999, **55**, 907-916.
9.   E. A. McRobert, M. Gallicchio, G. Jerums, M. E. Cooper and L. A. Bach, *J Biol Chem*, 2003, **278**, 25783-25789.
10. A. Bretscher, K. Edwards and R. G. Fehon, *Nature Rev Mol Cell Biol*, 2002, **3**, 586-599.
11. A. I. McClatchey and R. G. Fehon, *Trends Cell Biol*, 2009, **19**, 198-206.
12. S. C. Hughes and R. G. Fehon, *Curr Opin Cell Biol*, 2007, **19**, 51-56.
13. P. Tang, C. Cao, M. Xu and L. Zhang, *FEBS Lett*, 2007, **581**, 1103-1108.
14. M. Levi, *J Am Soc Nephrol*, 2003, **14**, 1949-1951.
15. P. Poullet, A. Gautreau, G. Kadare, J. A. Girault, D. Louvard and M. Arpin, *J Biol Chem*, 2001, **276**, 37686-37691.
16. F. Lozupone, L. Lugini, P. Matarrese, F. Luciani, C. Federici, E. Iessi, P. Margutti, G. Stassi, W. Malorni and S. Fais, *J Biol Chem*, 2004, **279**, 9199-9207.
17. A. Gautreau, P. Poullet, D. Louvard and M. Arpin, *Proc Natl Acad Sci USA*, 1999, **96**, 7300-7305.
18. C. Khanna, X. Wan, S. Bose, R. Cassaday, O. Olomu, A. Mendoza, C. Yeung, R. Gorlick, S. M. Hewitt and L. J. Helman, *Nature medicine*, 2004, **10**, 182-186.
19. V. Niggli and J. Rossy, *Int J Biochem Cell Biol*, 2008, **40**, 344-349.
20. E. A. McRobert, A. Tikoo, M. E. Cooper and L. A. Bach, *Int J Biochem Cell Biol*, 2008, **40**, 1570-1580.
21. Q. Li, M. R. Nance, R. Kulikauskas, K. Nyberg, R. Fehon, P. A. Karplus, A. Bretscher and J. J. Tesmer, *J Mol Biol*, 2007, **365**, 1446-1459.
22. E. F. Pettersen, T. D. Goddard, C. C. Huang, G. S. Couch, D. M. Greenblatt, E. C. Meng and T. E. Ferrin, *J Comput Chem*, 2004, **25**, 1605-1612.
23. K. M. Myint, Y. Yamamoto, T. Doi, I. Kato, A. Harashima, H. Yonekura, T. Watanabe, H. Shinohara, M. Takeuchi, K. Tsuneyama, N. Hashimoto, M. Asano, S. Takasawa, H. Okamoto and H. Yamamoto, *Diabetes*, 2006, **55**, 2510-2522.
24. X. Yao, A. Thibodeau and J. G. Forte, *Am J Physiol*, 1993, **265**, C36-C46.

25. S. M. Harwood, D. A. Allen, M. J. Raftery and M. M. Yaqoob, *Kidney Int*, 2007, **71**, 655-663.

26. W. J. Smith, N. Nassar, A. Bretscher, R. A. Cerione and P. A. Karplus, *J Biol Chem*, 2003, **278**, 4949-4956.

27. T. Crepaldi, A. Gautreau, P. M. Comoglio, D. Louvard and M. Arpin, *J Cell Biol*, 1997, **138**, 423-434.

28. M. A. Gallicchio, E. A. McRobert, A. Tikoo, M. E. Cooper and L. A. Bach, *J Am Soc Nephrol*, 2006, **17**, 414-421.

29. L. A. Bach, M. A. Gallicchio, E. A. McRobert, A. Tikoo and M. E. Cooper, *Ann N Y Acad Sci*, 2005, **1043**, 609-616.

30. T. Takeda, T. McQuistan, R. A. Orlando and M. G. Farquhar, *J Clin Invest*, 2001, **108**, 289 -301.

31. D. M. Kraemer, B. Strizek, H. E. Meyer, K. Marcus and D. Drenckhahn, *Eur J Cell Biol*, 2003, **82**, 87-92.

32. P. Finotti and P. Palatini, *Diabetologia*, 1986, **29**, 623-628.

MODIFICATION OF HUMAN SERUM ALBUMIN WITH REACTIVE ALDEHYDES ALTERS THE ANTIOXIDANT ACTIVITY

K Mera[1], K Takeo[1], D Honda[1], TMaruyama[1], M Otagiri[1,2], and R Nagai[3]

[1] Department of Biopharmaceutics, Graduate School of Pharmaceutical Sciences, Kumamoto University, Kumamoto, Japan
[2] Faculty of Pharmaceutical Sciences, Sojo University, Kumamoto, Japan
[3] Department of Food and Nutrition, Laboratory of Nutritional Science and Biochemistry, Japan Women's University, Tokyo, Japan

1 INTRODUCTION

Reducing sugars such as glucose non-enzymatically react with amino residues of proteins to form Schiff base and Amadori products. Further incubation converts these early products into irreversible derivatives termed advanced glycation end-products (AGEs). Protein modification by glucose results in the induction of the functional disruption of proteins such as human serum albumin (HSA)[1, 2], alpha-crystalline[3] and copper-zing-superoxide dismutase[4]. We previously reported that HSA is the major target for glycation and oxidation among plasma proteins[5, 6] and Miyata *et al.* also reported that over 90 % of N^ε-(carboxymethyl)lysine (CML) and pentosidine, well-characterized AGE structures, are generated on HSA[7, 8]. It is well known that HSA exhibits numerous biological functions such as maintenance of colloid osmotic pressure and transport of the endogeneous and exogenous ligands. Recent studies have focused on its antioxidant properties[9], and an inverse relationship between albumin level and cardiovascular diseases has been established[10].

AGEs are generated not only from glucose but also from aldehydes such as glyoxal, methylglyoxal[11], glucosone[12] and glycolaldehyde[13], and those aldehydes rapidly modify proteins. Thiamine and its derivative, benfotiamine, are known to decrease methylglyoxal level *in vivo* and inhibit the development of incipient nephropathy[14] and retinopathy[15] in streptozotocin-induced diabetic rats. Glycolaldehyde is generated by the reaction of hypochloric acid with serine, and reacts with proteins to form AGEs such as CML[16, 17] and GA-pyridine[13] which accumulate in human atherosclerotic lesions. These reports indicate that the modification of proteins with aldehydes *in vivo* may contribute to the pathogenesis of diabetic complications by enhancing post-translational modification in various human tissues. However, little is known about the difference of functional properties and formed AGE structures among aldehydes-modified proteins. To elucidate the effects of reactive aldehydes on change in the structure and function of proteins, we prepared aldehyde-modified HSA and then assessed the effect of these aldehydes on the structure and function of HSA. Our results demonstrated that formed AGE structures are dependent on each aldehyde and that modification by aldehydes may play in a role in the functional deterioration of HSA.

2 MATERIALS AND METHODS

2.1 Chemicals

Human serum albumin (HSA) was donated by the Chemo-Sera-Therapeutic Research Institute (Kumamoto, Japan) and was defatted using charcoal treatment as described by Chen[18]. Xanthine, xanthine oxidase, methylglyoxal, glyceraldehyde and glycolaldehyde were purchased from Sigma (St. Louis, MO). Glyoxal was obtained from Nacalai Tesque (Kyoto, Japan). Horseradish peroxidase (HRP)–conjugated goat anti–mouse IgG antibody was purchased from Kirkegaard Perry Laboratories (Gaitherburg, MD, USA). All other chemicals were of the best grade available from commercial sources.

2.2 Preparation of aldehyde-modified HSA

To prepare the methylglyoxal-modified HSA (methylglyoxal-HSA), glyoxal-modified HSA (glyoxal-HSA), glyceraldehyde-modified HSA (glyceraldehyde-HSA), glycolaldehyde-modified HSA (glycolaldehyde-HSA) and 3DG-modified HSA (3DG-HSA), 2 mg/ml HSA was incubated with 10 mM of methylglyoxal, glyoxal, glyceraldehyde glycolaldehydes or 3DG at 37°C for up to 7 days in 100 mM sodium phosphate buffer (pH 7.4). To prepare the glucosone-modified HSA (glucosone-HSA), 2 mg/ml HSA was incubated with 10 mM of glucosone at 37°C for up to 7 days in 100 mM HPES buffer (ph 7.4). The aliquots were obtained from each reaction mixture and dialyzed against PBS.

2.3 SDS-PAGE

The aldehyde-modified HSA was analyzed via SDS-PAGE, using 12.5% polyacrylamide gel, and detected by staining with Coomassie Brilliant Blue (CBB).

2.4 Measurement of Circular Dichroism Spectra

Circular dichroism spectra were recorded with a Jasco J-720 spectropolarimeter (Jasco Co., Tokyo, Japan), using 10 μM aldehyde-modified HSA in PBS. The α-helix contents (f_H) were estimated from the ellipticity value at 222 nm ($[\theta]_{222}$), according to the method of Chen *et al.*,[19] using the following equation: $f_H = -([\theta]_{222} + 2340) / 30300$

2.5 Enzyme-linked immunosorbent assay

An enzyme-linked immunosorbent assay (ELISA) was performed as described previously[20]. Briefly, each well of a 96-well microtiter plate was coated with 100 μl of the indicated concentration of sample in PBS, and incubated for 2 hr. The wells were washed three times with PBS containing 0.05% Tween 20 (washing buffer), then blocked with 0.5% gelatin in PBS for 1 hr. After washing 3 times, the wells were incubated for 1 hr with 100 μl of the primary antibody including monoclonal anti-CML antibody (2G11: 1 μg/ml), monoclonal anti-CEL antibody (CEL-SP: 5 μg/ml) and monoclonal anti-GA-pyridine antibody (2A2: 0.5 µg/ml). After triplicate washing, the wells were incubated with HRP-conjugated anti-mouse IgG, followed by reaction with 1,2-phenylenediamine dihydrochloride. The reaction was terminated with 100 µl of 1.0 M sulfuric acid, and monitored using absorbance at 492 nm with a micro-ELISA plate reader (TECAN Spectra Fluor Plus).

2.6 Chemiluminescent assay for HOCl-scavenging activity of aldehyde-modified HSA

The concentration of HOCl was measured by using a molar extinction coefficient of 350 M^{-1} cm^{-1} at 290 nm at pH 12. One milliliter solution contained 10 µM HOCl, 250 µM DTPA and 500 µM luminol in the absence or presence of 10 nM of aldehyde-modifed HSA. Chemiluminescent intensity was recorded continuously for 5 min using Mini Lumat LB 9506 luminometer (EG&G Berthold, Germany).

2.7 Chemiluminescent assay for superoxide-scavenging activity of aldehyde-modified HSA

Xanthine/xanthine oxidase (X/XO) system was used to generate $O_2^{\bullet-}$ to measure the superoxide-scavenging activity of aldehyde-modified HSA. Briefly, 1 ml solution contained 100 µg/ml X, 0.01 U/ml XO, 250 µM DTPA and 500 µM luminol in the absence or presence of 10 nM of aldehyde-modifed HSA. Chemiluminescent intensity was recorded as described above.

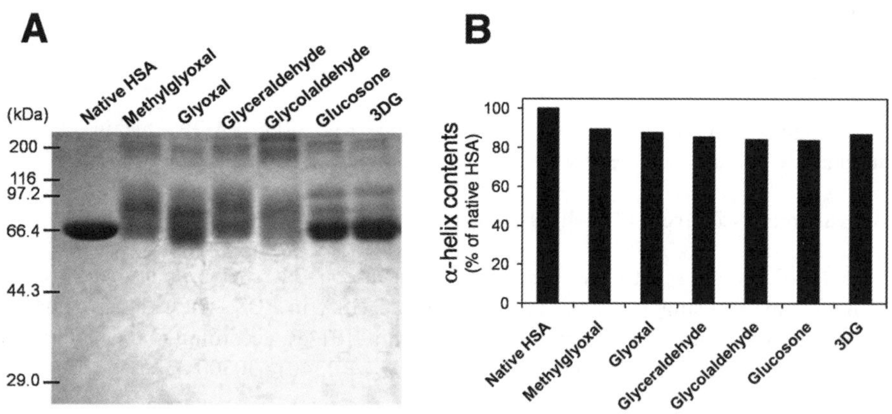

Figure 1.   *Physicochemical properties of aldehyde-modified HSA. (A) SDS-PAGE of aldehyde-modified HSA. Molecular masses in kDa are indicated on the left. Proteins were separated by SDS–PAGE [12.5% (w/v) gel] and revealed by CBB staining. (B) The α-helix contents of aldehyde-modified HSA estimated from circular dichroism spectra.*

3 RESULTS

3.1 Physicochemical properties of aldehyde-modified HSA

To evaluate the effect of reactive aldehydes on the physicochemical properties of HSA, aldehyde-modified HSA was subjected to SDS-PAGE. After incubation with these aldehydes, protein aggregation and cross-linking was clearly indicated by the presence of a smear in the high molecular weight range **(Fig. 1A)**. The α-helix contents of HSA, which estimated from circular dichroism spectra, were slightly decreased by modification with reactive aldehydes **(Fig. 1B)**. These results indicated that modification with reactive aldehydes alters the biological structure of HSA.

3.2 Immunoreactivity of aldehyde-modified HSA with monoclonal antibodies against several AGE structures.

To determine the effect of reactive aldehydes on AGE formation, HSA was incubated with reactive aldehydes, followed by the determination of several AGE structures by noncompetitive ELISA. As shown in **Figure 2A**, consistent with previous reports, CML formation was observed in glyoxal-, glyceraldehyde- and glycolaldehyde-HSA. CEL was generated by incubation with methylglyoxal and glyceraldehydes **(Fig. 2B)**. Furthermore, GA-pyridine was specifically detected in glycolaldehyde-HSA **(Fig. 2C)**. These results indicated that formed AGE structures are dependent on each aldehyde.

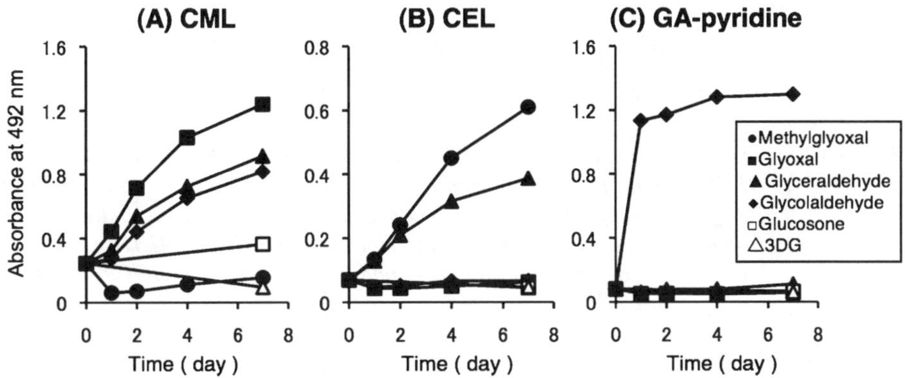

Figure 2. *Immunoreactivity of aldehyde-modified HSA with monoclonal antibodies against several AGE structures. The reactivity of aldehyde-modified HSA with monoclonal anti-CML antibody (2G11; A), monoclonal anti-CEL antibody (CEL-SP; B) and monoclonal anti-GA-pyridine antibody (2A2; C) was determined by non-competitive ELISA.*

3.3 Antioxidant activity of aldehyde-modified HAS

As described in the introduction, HSA play an important role in the antioxidant activity in the extra cellular environment. We next measured the antioxidant activity of aldehyde-modified HSA against HOCl and superoxide anion radical using chemiluminescent assay. As shown in **Figure 3**, a decrease in the HOCl scavenging activity was observed in aldehyde-modified HSA, especially methylglyoxal-HSA. Furthermore, the antioxidant activity of HSA against superoxide anion was decreased by modification with all these aldehydes, and more significantly decreased was observed in methylglyoxal-, glyoxal-, glyceraldehyde-HSA and glycolaldehyde-HSA **(Fig. 4)**. These results indicated that modification by aldehydes may play in a role in the functional deterioration of HSA.

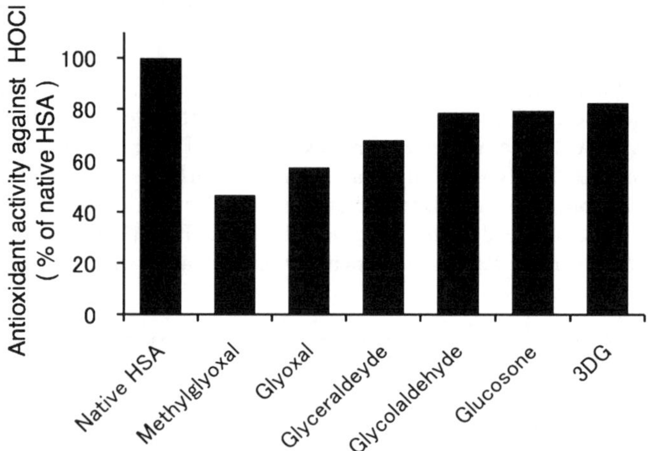

Figure 3.   *Antioxidant activity of aldehyde-modified HSA against hypochlorous acid (HOCl). One milliliter solution contained 10 µM HOCl, 250 µM DTPA, 500 µM luminol and 10 nM of aldehyde-modifed HSA. Chemiluminescent intensity was recorded as described MATERIALS AND METHODS. Antioxidant activities of aldehyde-modified HSA against HOCl were shown as % of native HSA.*

4 DISCUSSION

Recent studies have demonstrated that AGEs are generated not only from glucose but also from reactive aldehydes, which are generated *in vivo* through several pathways such as glycolysis and inflammation. In streptozotocin-induced diabetic rats, the administration of the trapping reagents for reactive aldehydes such as aminoguanidine, pyridoxaimine, thiamine and benfotiamine, reduces AGEs accumulation *in vivo* and inhibits the development of diabetic nephropathy[14, 21, 22] and retinopathy[15, 23], whereas these reagents do not affect the blood glucose level. The findings of these reports indicate that the modification of proteins with aldehydes in vivo may contribute to the pathogenesis of diabetic complications by enhancing the post-translational modification of various tissues. The present study compared the physicochemical properties and AGE structures formed by incubation of HSA with reactive aldehydes. Our result showed that modification with aldehydes changes the secondary structure and antioxidant activity of HSA.

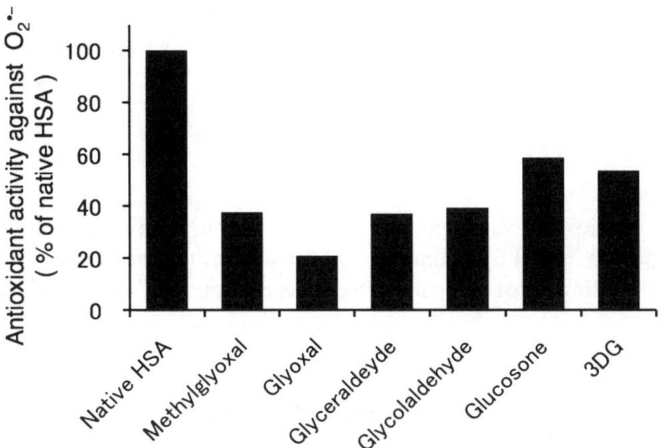

Figure 4. *Antioxidant activity of aldehyde-modified HSA against superoxide anion radical ($O_2^{\bullet-}$). One milliliter solution contained 100 µg/ml X, 0.01 U/ml XO, 250 µM DTPA, 500 µM luminol and 10 nM of aldehyde-modifed HSA. Chemiluminescent intensity was recorded as described MATERIALS AND METHODS. Antioxidant activities of aldehyde-modified HSA against $O_2^{\bullet-}$ were shown as % of native HSA.*

It is well established that free radical and reactive oxygen species contribute to the development of several age-related disorders by inducing oxidative damage. Activated neutrophils and macrophages generate superoxide anion radical and its derivatives such as hydrogen peroxide, hydroxyl radical and singlet oxygen via the nicotinamide adenine dinucleotide phosphate (NADPH) oxidase complex, and HOCl from the myeloperoxidase (MPO) system in the presence of chloride and hydrogen peroxide. MPO is detected in lipid-laden macrophages in human atherosclerotic lesions[24]. Furthermore, Hazell *et al.*[25] demonstrated that HOCl-modified proteins are present in monocyte/macrophages in human atherosclerotic lesions, indicating that MPO is activated and plays a role in protein modification in these lesions. Albumin is the most important extra cellular antioxidant[9], and a lot of potential associations between its plasma concentration or modifications and mortality have been observed[10]. In this study, the antioxidant activity of HSA against HOCl and superoxide anion radical was decreased by modification with reactive aldehydes **(Fig. 3 and 4)**. Since HOCl is believed to play an important role in tissue damage and superoxide is known as the precursor of the potent oxidant peroxynitrite (ONOO-), the loss of antioxidant activityof HSA against HOCl and superoxide may affect extra cellular oxidative stress. Faure *et al.* previously reported that the H_2O_2 scavenging activity of bovine serum albumin was decreased by methylglyoxal modification[26]. However, our results demonstrated that the antioxidant activity of HSA was decreased by modification not only with methylglyoxal but also with other aldehydes **(Fig. 3 and 4)**. Further study will be required to clarify which aldehyde play a role in the oxidative stress *in vivo*.

Furthermore, little is known about the difference of physicochemical properties and formed AGE structures among aldehydes-modified proteins since most of previous immunochemical studies using anti-AGE antibodies were used conventional antibodies which were cross-reactive with those analogues[27] or those epitopes are unidentified[28]. A production system has been developed to make monoclonal antibodies against several AGE structures[29-31]. The current study also provides evidence that formed AGEs, detected by antibodies against CML, CEL and GA-pyridine, on HSA are highly dependent on the kind of aldehydes.

Acknowledgments

We are grateful to Yukio Fujiwara and Mime Nagai for their collaborative endeavours. This work was supported in part by Grants-in-Aid for scientific Research (No. 18790619 to Ryoji Nagai) from the Ministry of Education, Science, Sports and Cultures of Japan.

References

1.   K. Nakajou, H. Watanabe, U. Kragh-Hansen, T. Maruyama and M. Otagiri, *Biochim Biophys Acta*, 2003, **1623**, 88-97.
2.   E. Bourdon, N. Loreau and D. Blache, *Faseb J*, 1999, **13**, 233-244.
3.   P. A. Kumar, M. S. Kumar and G. B. Reddy, *Biochem J*, 2007, **408**, 251-258.
4.   T. Ookawara, N. Kawamura, Y. Kitagawa and N. Taniguchi, *J Biol Chem*, 1992, **267**, 18505-18510.
5.   M. Anraku, K. Kitamura, A. Shinohara, M. Adachi, A. Suenga, T. Maruyama, K. Miyanaka, T. Miyoshi, N. Shiraishi, H. Nonoguchi, M. Otagiri and K. Tomita, *Kidney Int*, 2004, **66**, 841-848.
6.   K. Mera, M. Anraku, K. Kitamura, K. Nakajou, T. Maruyama, K. Tomita and M. Otagiri, *Hypertens Res*, 2005, **28**, 973-980.
7.   T. Miyata, Y. Ueda, T. Shinzato, Y. Iida, S. Tanaka, K. Kurokawa, C. van Ypersele de Strihou and K. Maeda, *J Am Soc Nephrol*, 1996, **7**, 1198-1206.
8.   T. Miyata, M. X. Fu, K. Kurokawa, C. van Ypersele de Strihou, S. R. Thorpe and J. W. Baynes, *Kidney Int*, 1998, **54**, 1290-1295.
9.   B. Halliwell, *Annu Rev Nutr*, 1996, **16**, 33-50.
10. A. Phillips, A. G. Shaper and P. H. Whincup, *Lancet*, 1989, **2**, 1434-1436.
11. P. J. Thornalley, A. Langborg and H. S. Minhas, *Biochem J*, 1999, **344 Pt 1**, 109-116.
12. R. Nagai, Y. Unno, M. C. Hayashi, S. Masuda, F. Hayase, N. Kinae and S. Horiuchi, *Diabetes*, 2002, **51**, 2833-2839.
13. R. Nagai, C. M. Hayashi, L. Xia, M. Takeya and S. Horiuchi, *J Biol Chem*, 2002, **277**, 48905-48912.
14. R. Babaei-Jadidi, N. Karachalias, N. Ahmed, S. Battah and P. J. Thornalley, *Diabetes*, 2003, **52**, 2110-2120.
15. H. P. Hammes, X. Du, D. Edelstein, T. Taguchi, T. Matsumura, Q. Ju, J. Lin, A. Bierhaus, P. Nawroth, D. Hannak, M. Neumaier, R. Bergfeld, I. Giardino and M. Brownlee, *Nat Med*, 2003, **9**, 294-299.
16. M. M. Anderson, S. L. Hazen, F. F. Hsu and J. W. Heinecke, *J Clin Invest*, 1997, **99**, 424-432.
17. M. M. Anderson, J. R. Requena, J. R. Crowley, S. R. Thorpe and J. W. Heinecke, *J Clin Invest*, 1999, **104**, 103-113.
18. R. F. Chen, *J Biol Chem*, 1967, **242**, 173-181.
19. Y. H. Chen, J. T. Yang and H. M. Martinez, *Biochemistry*, 1972, **11**, 4120-4131.
20. K. Mera, R. Nagai, N. Haraguchi, Y. Fujiwara, T. Araki, N. Sakata and M. Otagiri, *Free Radic Res*, 2007, **41**, 713-718.
21. J. M. Forbes, T. Soulis, V. Thallas, S. Panagiotopoulos, D. M. Long, S. Vasan, D. Wagle, G. Jerums and M. E. Cooper, *Diabetologia*, 2001, **44**, 108-114.

22. T. Soulis, M. E. Cooper, D. Vranes, R. Bucala and G. Jerums, *Kidney Int*, 1996, **50**, 627-634.

23. H. P. Hammes, S. Martin, K. Federlin, K. Geisen and M. Brownlee, *Proc Natl Acad Sci U S A*, 1991, **88**, 11555-11558.

24. A. Daugherty, J. L. Dunn, D. L. Rateri and J. W. Heinecke, *J Clin Invest*, 1994, **94**, 437-444.

25. L. J. Hazell, L. Arnold, D. Flowers, G. Waeg, E. Malle and R. Stocker, *J Clin Invest*, 1996, **97**, 1535-1544.

26. P. Faure, L. Troncy, M. Lecomte, N. Wiernsperger, M. Lagarde, D. Ruggiero and S. Halimi, *Diabetes Metab*, 2005, **31**, 169-177.

27. W. Koito, T. Araki, S. Horiuchi and R. Nagai, *J Biochem (Tokyo)*, 2004, **136**, 831-837.

28. R. Nagai, K. Matsumoto, X. Ling, H. Suzuki, T. Araki and S. Horiuchi, *Diabetes*, 2000, **49**, 1714-1723.

29. R. Nagai, Y. Fujiwara, K. Mera, K. Yamagata, N. Sakashita and M. Takeya, *J Immunol Methods*, 2008, **332**, 112-120.

30. R. Nagai, Y. Fujiwara, K. Mera, K. Motomura, Y. Iwao, K. Tsurushima, M. Nagai, K. Takeo, M. Yoshitomi, M. Otagiri and T. Ikeda, *Ann N Y Acad Sci*, 2008, **1126**, 38-41.

31. K. Mera, M. Nagai, J. W. Brock, Y. Fujiwara, T. Murata, T. Maruyama, J. W. Baynes, M. Otagiri and R. Nagai, *J Immunol Methods*, 2008, **334**, 82-90.

A NOVEL MECHANISM OF MENTAL ILLNESS: CARBONYL STRESS INDUCED SCHIZOPHRENIA: A GLYOXALASE I DEFICIT PEDIGREE WITH PSYCHOSIS

M. Itokawa[1,2,3], M. Arai[1], T. Yoshikawa[2], Y. Okazaki[3] and T. Miyata[4]

[1]Project for Schizophrenia & Affective disorders Research, Tokyo Institute of Psychiatry Tokyo 156-8585, Japan
[2]Laboratory for Molecular Psychiatry, RIKEN Brain Science Institute, Saitama 351-0198, Japan
[3]Department of Psychiatry, Tokyo Metropolitan Matsuzawa Hospital, Tokyo 156-0057, Japan
[4]Center for Translational and Advanced Research on Human Disease, Tohoku University Graduate School of Medicine, Miyagi 980-8575, Japan

1 INTRODUCTION

Schizophrenia is a debilitating and complex mental disorder with a prevalence of approximately 1% worldwide. Its pathophysiology remains unclear, despite massive and ongoing research.[1,2] Biochemical and pharmacological studies using human samples and animal models suggest that oxidative/carbonyl stress contributes to the pathophysiology of schizophrenia.[3-6] Oxidative stress is a central mediator of advanced glycation end product (AGE) formation, and pyridoxamine [vitamin (vit)B$_6$]] (biosynthesized from pyridoxal *in vivo*) is known to detoxify reactive carbonyl compounds (RCOs) via carbonyl-amine chemistry. Cellular removal of RCOs hinges largely upon the activity of the zinc metalloenzyme glyoxalase I (GLO1).[7] The glyoxalase detoxification system is ubiquitous in human tissues, including the brain. The GLO1 detoxification system interacts with several metabolizing cascades, and some compounds in these cascades have been reported as causative candidates in the etiology of schizophrenia, such as glutathione, homocysteine, and folic acid metabolites.[8-15]

2 GLO-1 DEFICIENCY AND NEUROPSYCHIATRIC DISORDERS

Recent studies have revealed that dysfunction of GLO1 is involved not only in systemic diseases such as diabetes mellitus[16] and vascular injury, [17] but also in neuropsychiatric disorders such as mood disorder,[18] autism,[19,20] anxiety disorders,[21] and alcoholism[2]. In mice, levels of expression of *Glo1* have been associated with anxiety-like behavioral phenotypes.[23-25] *GLO1* has been mapped to chromosome 6p21, a linkage region for schizophrenia.[26-28] A missense polymorphism, Glu111/Ala111, has been reported in two multiplex Caucasian pedigrees with schizophrenia spectrum disorders.[29]

We recently reported an extremely interesting case with a deficiency of glyoxalase I (GLO1). He was a sixty-year-old male who suffered from severe schizophrenia. He had two brothers who were also affected with schizophrenia, the elder sibling who committed suicide and the other is now hospitalized. Two maternal uncles had also suffered from schizophrenia. Symptoms of the case and his brother remained strongly resistant despite receiving intensive treatments. We identified a novel mutation of an adenine insertion at nt 79 in exon 1, causing a frameshift starting from codon 27 and introducing a premature termination codon after aberrant translation of 15 amino acid residues (T27NfsX15). Expressions of mRNA and protein of GLO1 were decreased by 50% in lymphocytes of this case (Fig 1a, b). Enzymatic activity of GLO1 was approximately half in RBC of the patient compared to that of health controls (Fig 1c). The loss-of-function type frame-shift mutation, accompanied with increased AGE accumulation, appeared to be shared by affected individuals in this family, although we could not access them.

Table 1   *Association between GLO1 genotype and circulating AGE levels*

		AGEs	
		High	Normal
GLO1 genotypes	(frameshift, Ala/Ala)	9 (4.5%)	3 (1.5%)
	(Glu/Ala, Glu/Glu)	57 (28.5%)	107 (53.5%)

Chi-square = 7.727; df = 1; P = 0.0054; Odds = 5.632; 95%CI = 1.466 - 2164

3 GLO-1 MUTATIONS IN PATIENTS WITH SCHIZOPHRENIA

To further explore this association, we re-sequenced the *GLO1* gene of 1,761 schizophrenics and 1,921 control subjects and found another novel frame-shift mutation. Individuals with frameshift mutations had an approximately 40 to 50 % reduction in the enzymatic activity of GLO1 in their erythrocytes. The frameshift *GLO1* carriers with schizophrenia also displayed significantly increased amount of plasma AGEs, consistent with the key role of GLO-1 in AGE accumulation. We also detected homozygotes for Ala111 with schizophrenia who displayed 16% reduction of the enzymatic activity accompanied with significantly increased plasma AGEs. Reduced enzymatic activity of GLO1 with Ala111 as compared with that of Glu111 was confirmed by *in vitro* assay by preparing GFP fused GLO1 constructs carrying Ala111 and Glu111. Interestingly, we also found non-psychiatric controls with frameshift or homozygote for Ala111 who all exhibited normal plasma AGEs levels in contrast to the mutation carriers with schizophrenia, suggesting the existence of compensatory mechanisms.

Vitamin B6 levels in the circulation are a useful biomarker of carbonyl stress, as increased production of toxic alfa-oxoaldehydes induced by GLO1 defects and other factors results inthe elevated consumption to of vitamin B6. In our schizophrenia patients there were also concomitant with a prominent decrease in the vitamin B6 levels, paralleling

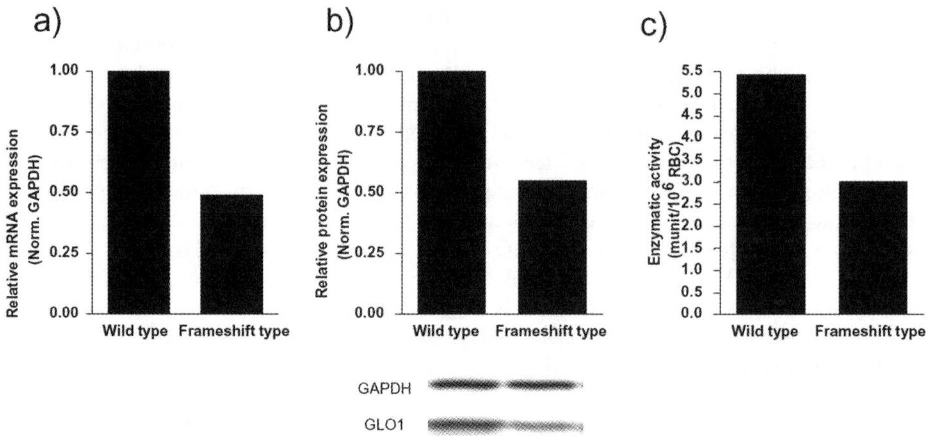

Figure 1   *Glyoxalase I (GLO1) and the impact of frame-shift mutation on (a) mRNA expression (RT-PCR), (b) protein expression (western blotting) and (c) enzymatic activity*

the increases in their AGE levels. In particular, vitamin B6 was markedly reduced in the two frame-shift *GLO1* carriers. Non-psychiatric controls with frame-shift or homozygote for Ala111 exhibited normal vitamin B6 levels, again paralleling their AGE levels.

4 PLASMA AGES IN PATIENTS WITH SCHIZOPHRENIA

To further expand the concept of carbonyl stress-induced schizophrenia to general patients we measured plasma AGEs and vitamin B6 levels using 178 schizophrenia and 76 control subjects that had no diabetes mellitus or renal dysfunction. Concentrations of AGEs were significantly high in patients compared to controls (P < 0.0001) and vitamin B6 levels were significantly lower in schizophrenics than those of control subjects (P < 0.0001) (Fig 2a, b). Sixty six of the patients (37.5%) and three of the controls (3.9%) showed high plasma AGEs (arbitrarily defined as plasma levels greater than standard deviations above the mean for healthy control individuals). In order to examine genetic contribution of *GLO1* to carbonyl stress in schizophrenia, we tested a correlation between the genotype and plasma AGEs levels by performing χ^2 test (Table 1). The frameshift mutations and homozygote of Ala111 are significantly associated with high carbonyl stress ($\chi^2 = 7.72$, $P = 0.005$, odds ratio = 5.63; 95% confidential interval = 1.46-21.64).

5 FUTURE DIRECTIONS

Schizophrenia has been thought to be a heterogeneous syndrome since the end of nineteenth century when E. Kraepelin established the conceptual disease category as *dementia precox*. Our findings suggest that GLO1 deficits and carbonyl stress are linked to the development of some cases of schizophrenia. Moreover, elevated plasma pentosidine and/or low vitamin B6 levels appears to be a cogent and easily measurable 'biomarkers' in

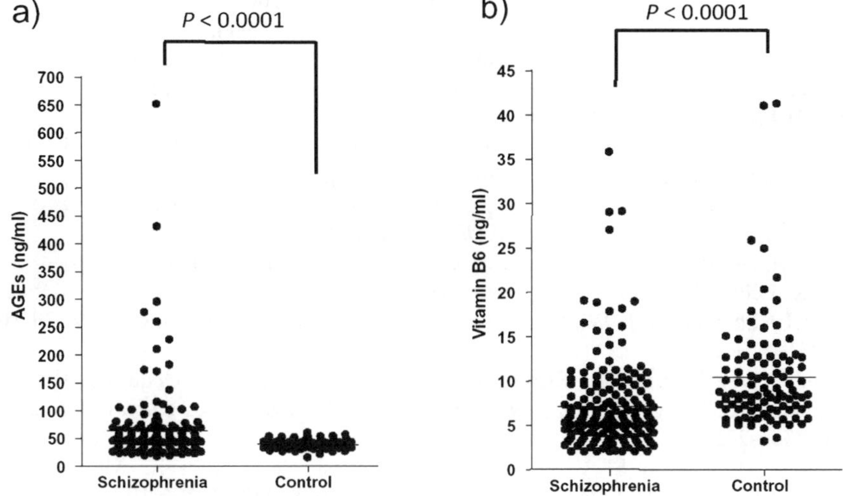

Figure 2   *Plasma AGEs (a) and vitamin B6 levels (b) of schizophrenia and controls.*

schizophrenia, and may be helpful for classifying heterogeneous types of schizophrenia on the basis of their biological causes. If carbonyl stress is directly linked to schizophrenic signs and symptoms, agents able to inhibit AGEs formation or entrap carbonyl compounds may also prove to be of therapeutic value. In particular, the markedly lowered vitamin B6 levels in schizophrenic patients with high pentosidine levels suggest that pyridoxamine, a non-toxic, water-soluble vitamin B6, may prove to be clinically useful means to entrap carbonyl compounds and prevent toxicity.

References

1. Sullivan PF, Kendler KS, Neale MC.*Arch Gen Psychiatry*. 2003, **60(12)**, 1187-1192.
2. Sullivan PF.. *PLoS Med*. 2005,**2(7)**,e212.
3. Schulz JB, Lindenau J, Seyfried J, Dichgans J. *Eur J Biochem*. 2000, **267(16)**,4904-4911.
4. Tosic M, Ott J, Barral S, Bovet P, Deppen P, Gheorghita F, Matthey ML, Parnas J, Preisig M, Saraga M, Solida A, Timm S, Wang AG, Werge T, Cuénod M, Do KQ. *Am J Hum Genet*. 2006, **79(3)**,586-592.
5. Young J, McKinney SB, Ross BM, Wahle KW, Boyle SP. *Prostaglandins Leukot Essent Fatty Acids*. 2007, **76(2)**, 73-85.
6. Ng F, Berk M, Dean O, Bush AI. *Int J Neuropsychopharmacol*. 2008,11(6),851-876.
7. Thornalley PJ. *Mol Aspects Med*. 1993, **14(4)**, 287-371.
8. Brown AS, Bottiglieri T, Schaefer CA, Quesenberry CP Jr, Liu L, Bresnahan M, Susser ES. *Arch Gen Psychiatry*. 2007, **64(1)**, 31-9.

9.   Frankenburg FR. *Harv Rev Psychiatry.* 2007, **15(4),** 146-160.
10. Gilbody S, Lewis S, Lightfoot T. *Am J Epidemiol.* 2007, **165(1),** 1-13.
11. Gysin R, Kraftsik R, Sandell J, Bovet P, Chappuis C, Conus P, Deppen P, Preisig M, Ruiz V, Steullet P, Tosic M, Werge T, Cuénod M, Do KQ. *Proc Natl Acad Sci USA.* 2007, **104(42),** 16621-16626.
12. Haidemenos A, Kontis D, Gazi A, Kallai E, Allin M, Lucia B. *Prog Neuropsychopharmacol Biol Psychiatry.* 2007, **31(6),** 1289-1296.
13. Levine J, Stahl Z, Sela BA, Ruderman V, Shumaico O, Babushkin I, Osher Y, Bersudsky Y, Belmaker RH. *Biol Psychiatry.* 2006, **60(3),** 265-269.
14. Saadat M, Mobayen F, Farrashbandi H. *Psychiatry Res.* 2007,153(1),87-91.
15. Yao JK, Leonard S, Reddy R. *Dis Markers.* 2006, **22(1-2),** 83-93.
16. Kirk RL, Theophilus J, Whitehouse S, Court J, Zimmet P. *Diabetes.* 1979, **28(10),** 949-951.
17. Miyata T, van Ypersele de Strihou C, Imasawa T, Yoshino A, Ueda Y, Ogura H, Kominami K, Onogi H, Inagi R, Nangaku M, Kurokawa K. *Kidney Int.* 2001, **60(6),**2351-2359.
18. Fujimoto M, Uchida S, Watanuki T, Wakabayashi Y, Otsuki K, Matsubara T, Suetsugi M, Funato H, Watanabe Y. *Neurosci Lett.* 2008, **438(2),** 196-199.
19. Junaid MA, Kowal D, Barua M, Pullarkat PS, Sklower Brooks S, Pullarkat RK. *Am J Med Genet A.* 2004, **131(1),** 11-17.
20. Sacco R, Papaleo V, Hager J, Rousseau F, Moessner R, Militerni R, Bravaccio C, Trillo S, Schneider C, Melmed R, Elia M, Curatolo P, Manzi B, Pascucci T, Puglisi-Allegra S, Reichelt KL, Persico AM. *BMC Med Genet.* 2007, **8,** 11.
21. Politi P, Minoretti P, Falcone C, Martinelli V, Emanuele E. *Neurosci Lett.* 2006, **396(2),** 163-166.
22. Ledig M, Doffoel M, Ziessel M, Kopp P, Charrault A, Tongio MM, Mayer S, Bockel R, Mandel P. *Alcohol.* 1986, **3(1),** 11-14.
23. Ditzen C, Jastorff AM, Kessler MS, Bunck M, Teplytska L, Erhardt A, Krömer SA, Varadarajulu J, Targosz BS, Sayan-Ayata EF, Holsboer F, Landgraf R, Turck CW. *Mol Cell Proteomics.* 2006, **5(10),** 1914-1920.
24. Hovatta I, Tennant RS, Helton R, Marr RA, Singer O, Redwine JM, Ellison JA, Schadt EE, Verma IM, Lockhart DJ, Barlow C. *Nature.* 2005, **438(7068),** 662-666.
25. Krömer SA, Kessler MS, Milfay D, Birg IN, Bunck M, Czibere L, Panhuysen M, Pütz B, Deussing JM, Holsboer F, Landgraf R, Turck CW. *J Neurosci.* 2005, **25(17),** 4375-4384.
26. Arolt V, Lencer R, Nolte A, Müller-Myhsok B, Purmann S, Schürmann M, Leutelt J, Pinnow M, Schwinger E. *Am J Med Genet.* 1996, **67(6),** 564-579.
27. Brzustowicz LM, Honer WG, Chow EW, Hogan J, Hodgkinson K, Bassett AS. *Am J Hum Genet.* 1997, **61(6),** 1388-1396.
28. Nurnberger JI Jr, Foroud T. *Am J Med Genet.* 1999, **88(3),** 233-238.
29. Turner WJ. *Biol Psychiatry.* 1979, **14(1),**177-206.

GLYCERALDEHYDE-DERIVED ADVANCED GLYCATION END PRODUCTS DECREASE WHITE ADIPOSE TISSUE WEIGHT AND DOWNREGULATE LEPTIN, ADIPONECTIN, AND MACROPHAGE MARKER.

H. Watanabe[1], Y. Yoshida[2] and F. Hayase[2]

[1]Department of Life Science, Faculty of Agriculture, Meiji University, 1-1-1 Higashimita, Tama-ku, Kawasaki, Kanagawa 214-8571, Japan
[2]Department of Agricultural Chemistry, Faculty of Agriculture, Meiji University, 1-1-1 Higashimita, Tama-ku, Kawasaki, Kanagawa 214-8571, Japan

1 INTRODUCTION

Metabolic syndrome is the coexistence of several risk factors for arteriosclerosis, including visceral obesity, hyperglycemia, dyslipidemia, and hypertension.[1] Among these factors, obesity has been thought the most important determinant for metabolic syndrome.[2] Obesity is marked by the expansion of white adipose tissue which induces the defective secretion of adipocytokines,[3,4] inflammation,[5-7] and oxidative stress,[3,8,9] which cumulatively lead to development and progression of insulin resistance. Recently, it was reported that glucose-, glyceraldehyde(GLA)-, and glycolaldehyde-derived advanced glycation end products (AGEs) induce generation of intracellular ROS, and then contribute to inhibition the differentiation to adipocytes, attenuation of insulin sensitivity, and upregulation of monocyte chemoattractant protein-1 (MCP-1) in 3T3-L1 cells.[10] However, it is not clear whether AGEs alter function of adipose tissue in vivo. In this study, we examine the effects of GLA-BSA on white adipose tissuess accumulation as well as the expression of adipocytokines and markers of inflammation.

2 MATERIALS AND METHODS

2.1 Preparation of GLA-modified bovine serum albumin (GLA-BSA)

GLA-BSA was prepared by the method previously described.[11-13] Briefly, 100mM GLA and 10 mg/ml fatty acid-free bovine serum albumin (BSA) were dissolved in 200mM sodium phosphate buffer (pH 7.4) and incubated at 37 °C for 7 days. The solution was dialyzed against PBS for 1 day and against water for 3 days, and then lyophilized and dissolved in PBS at 10mg/ml. Control BSA was prepared in the same conditions without glyceraldehyde.

2.2 Animals

Male Sprague-Dawley (SD) rats at 8-week-old (260-290 g of body weight) were purchased from Clea Japan. The commercial chow (MF, Oriental Yeast, Tokyo, JAPAN) and tap water were provided *ad libitum*.

2.3 Administration of GLA-BSA

The administration of GLA-BSA was conducted as previously reported.[14,15] After the acclimation (6 days), rats were divided into two groups (n=7 for GLA-BSA-infused group and n=6 for BSA-infused group), so that the average body weights were equivalent. The rats were intravenously infused with GLA-BSA or BSA at 1mg/day under anaesthesia with diethylether. GLA-BSA or BSA was infused through right femoral vein at first day, and it was performed for right and left alternation every other day for up to ten days. Blood was obtained from abdominal aorta 45-60 min after infusion at final day. Plasma was harvested after centrifugation (3,000 × g, 20 min).
Subcutaneous white adipose tissue, epididymal white adipose tissue retroperitoneal white adipose tissue and mesenteric white adipose tissue were collected and washed in PBS twice. Plasma and white adipose tissues were stored at -80 °C until analysis. The concentration of plasma AGEs was determined by the method previously reported.[16, 17]

2.4 Determination of the concentration of plasma AGEs

Briefly, the fluorescence intensity in diluted plasma was measured at excitation 370 nm and emission 440 nm. The concentration of plasma AGEs was indicated in arbitrary units (AU) after normalized to the concentration of plasma protein.

2.5 Real time RT-PCR

Total RNA from white adipose tissues was purified with DNase I and used for RT-PCR. cDNA synthesis with AMV Reverse Transcriptase XL. cDNA fragments were amplified with a thermal cycler for initial denaturation at 94 °C for 90s, followed by 18-34 cycles of amplification; each cycle consisted of 94 °C for 30s, 60-63 °C for 30s, 72 °C for 45s. Acidic ribosomal phosphoprotein P0 (36B4) 18S was used as the internal standard. After agarose electrophoresis, the results were quantified using Scion Image (Scion Corporation, Frederick, MD, US) and normalized to 36B4 mRNA expression. PCR primers were as follows:

aP2,	5'-atgtgtgatgcctttgtgggg-3'	5'-gtagaagtcacgcctttcatga-3';
leptin,	5'-ctgtcctatgttcaagctgtgc-3'	5'-tgctcagagccaccacctc-3';
adiponectin,	5'-ctgttgcaagcgctcctgttc-3'	5'-tcgtaggtgaagagaacggcc-3';
EMR1,	5'-tcctgggacaaagacttaacgg-3'	5'-aattcctggagcactcatccac-3';
36B4,	5'-gggccacctggagaacaacc-3'	5'-ggcaacagtcgggtagccaatc-3'.

2.6 Statistical analysis

All data were expressed as mean ± SEM. The statistical significant difference was established by Student's t-test using JMP 4.0.5J microcomputer program (SAS Institute, Cary, NC, US) and considered as p < 0.05.

Table 1.   *Total and tissue weights and plasma AGEs in response to exposure to GLA-BSA Final body weight was measured at 9th day before food deprivation. Concentration of plasma AGEs was determined 45 min after the infusion of GLA-BSA or BSA at 10th day. Data are expressed as mean ± SEM. *p < 0.05 vs. BSA.*

	BSA (n=6)	GLA-BSA (n=7)
Initial body weight (g)	319.9 ± 5.1	319.3 ± 4.0
Final body weight (g)	334.3 ± 6.7	337.9 ± 4.4
Body weight gain (g)	14.3 ± 4.5	18.6 ± 2.2
Tissue weight (g/ 100g body weight)		
- subcutaneous white adipose tissues	1.36 ± 0.12	1.26 ± 0.09
- epididymal white adipose tissues	0.77 ± 0.05	0.61 ± 0.03*
- retroperioneal white adipose tissues	0.74 ± 0.16	0.55 ± 0.08
- mesenteric white adipose tissues	0.49 ± 0.09	0.33 ± 0.03
- liver	3.00 ± 0.10	3.04 ± 0.10
- heart	0.30 ± 0.00	0.31 ± 0.01
- kidney	0.78 ± 0.02	0.77 ± 0.02
Plasma AGEs (AU))	230.3 ± 19.2	221.4 ± 18.5

3   RESULTS AND DISCUSSION

Administration of glyceraldehyde-modified bovine serum albumin (GLA-BSA), a model AGE, to rats resulted in a significant decrease in epididymal white adipose tissues weight when compared to BSA alone (Table 2). It is known that major determinants for white adipose tissues weight are adipocyte size and adipose number. Therefore, we measured aP2, adipogenic and adipocyte marker in white adipose tissues, mRNA expression in subcutaneous white adipose tissues, epididymal white adipose tissues, and retroperioneal white adipose tissues. Fig. 1 shows decrease in aP2 mRNA expression in epididymal white adipose tissues by GLA-BSA infusion, which was consistent with decrease in tissue weight. In addition, the other white adipose tissuess weight and aP2 mRNA expression in subcutaneous white adipose tissues also had a tendency to decrease. On the other hand, liver, heart, kidney and total body weight were not changed by the infusion of GLA-BSA. These results suggest that GLA-BSA is able to reduce adipocyte size and/or adipocyte number, especially in epididymal white adipose tissues. This is consistent with a recent study reporting that GLA-BSA inhibits the differentiation of 3T3-L1 cells to adipocytes.[10] It is possible to speculate that GLA-BSA might therefore decrease adipocyte size rather than adipocyte number in white adipose tissues.

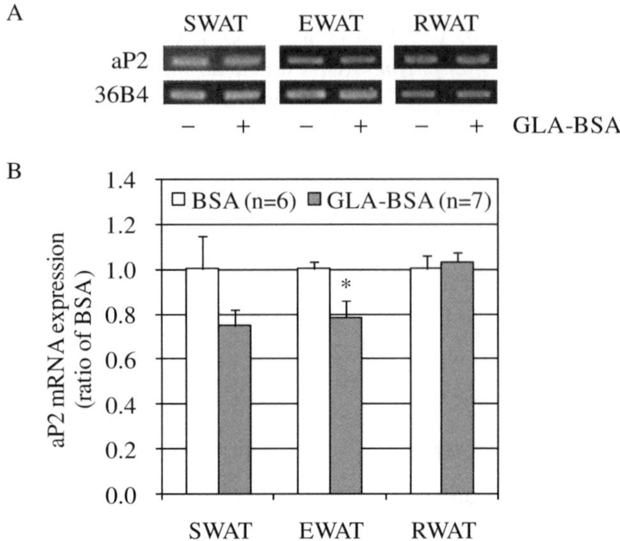

Figure 1   ***Decrease in aP2 mRNA expression in epididymal white adipose tissues.*** *Representative images of agarose gel electrophoresis (A) and aP2 mRNA expression normalized to 36B4 mRNA expression (B). Data are expressed as mean ± SEM. $*p < 0.05$ vs. BSA.*

As shown in Fig. 2, GLA-BSA significantly downregulated the expression of leptin and adiponectin in epididymal white adipose tissues. Interestingly, it also had a tendency to decrease of these genes in subcutaneous white adipose tissues, which constitutes about 80% of all white adipose tissues.[22] On the other hand, no significant changes in resistin mRNA expression were observed (data not shown). Leptin is an adipocytokine that inhibits body weight gain.[23] Therefore, the long-term administration of GLA-BSA may induce obesity, while GLA-BSA decreased the weight of white adipose tissuess. Further study about this discrepancy will be needed. On the other hand, leptin increases insulin sensitivity.[23] Adiponectin has insulin sensitizing, anti-inflammatory, and anti-atherogenic effect.[23] Therefore, the enhancement of AGEs formation under hyperglycemia may result in deterioration of metabolic syndrome and change of systemic homeostasis mediated by the downregulation of these genes.

It was reported that GLA-BSA induces MCP-1 mRNA expression in 3T3-L1 adipocytes.[10] Therefore GLA-BSA may induce macrophage infiltration into white adipose tissues. However, in the present study MCP-1 mRNA expression in white adipose tissues did not increase in GLA-BSA-infused rats (data not shown). On the other hand, expression of EMR1, a macrophage marker, was reduced by the infusion of GLA-BSA in subcutaneous white adipose tissues and retroperioneal white adipose tissues (Fig. 3). It is known that macrophages in white adipose tissues are classified into two types, anti-inflammatory M2 macrophages and inflammatory M1 macrophages.[24] EMR1 is a marker of both M1 and M2 macrophages. In this study, mRNA expression of IL-6, a M1 macrophage marker, was not changed by GLA-BSA-infusion (data not shown). These results indicate that GLA-BSA may decrease M2 macrophages in white adipose tissues, although further studies will be required.

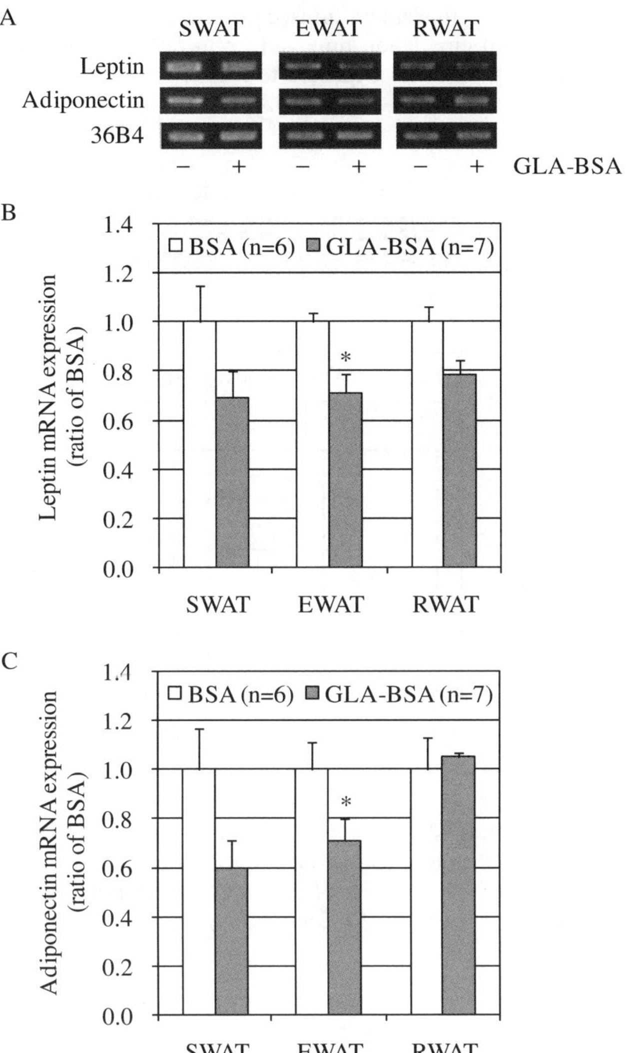

Figure 2.   ***Decrease in leptin and adiponectin mRNA expression in epididymal white adipose tissues.*** *Representative images of agarose gel electrophoresis (A) and leptin (B) and adiponectin (C) mRNA expression normalized to 36B4 mRNA expression. Data are expressed as mean ± SEM. *p < 0.05 vs. BSA.*

Notably, administration of GLA-BSA to rats did not significantly change circulating plasma AGEs (GLA-BSA, 221.4 ± 18.5; BSA-infused rats 30.3 ± 19.2. However, previous reports suggest that glucose-modified BSA administered to a mouse is rapidly (within several minutes) decreased by scavenger receptor-mediated endocytosis in hepatic sinusoidal Kupffer and endothelial cells.[19, 20] Because plasma was harvested 45-60 min after infusion of GLA-BSA in this study, infused GLA-BSA might be rapidly scavenged. However, because the blood circulation time is 15 s in rat (250g body weight)[21] it may nonetheless provide enough time and exposure to trigger receptor mediated siganlling in white adipose tissues.

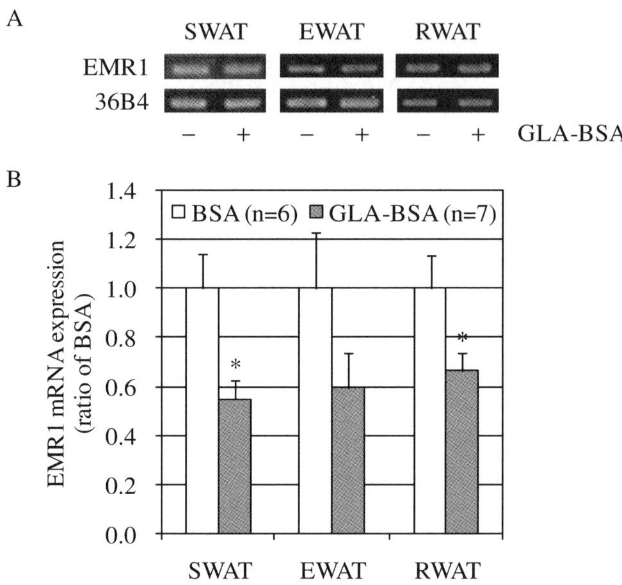

Figure 3   ***Decrease in EMR1 mRNA expression in subcutaneous white adipose tissues and retroperioneal white adipose tissues.*** *Representative images of agarose gel electrophoresis (A) and EMR1 mRNA expression normalized to 36B4 mRNA expression (B). Data are expressed as mean ± SEM. *p < 0.05 vs. BSA.*

4 SUMMARY AND CONCLUSION

Although recent studies have suggested AGEs alter various functions of adipocytes, it is unclear whether AGEs are involved in adipose tissue dysfunction in vivo. Our present study suggests that AGEs may alter pathology of obesity and diabetes in vivo. In this study, we examined the effect of GLA-BSA, a model AGE, on white adipose tissue weight and adipocytokine mRNA expression in vivo. Epididymal white adipose tissue weight decreased in Sprague-Dawley rats infused with GLA-BSA. In addition, GLA-BSA suppressed leptin and adiponectin mRNA expression in epididymal white adipose tissues. Furthermore, macrophage marker mRNA expression in subcutaneous white adipose tissue and retroperitoneal white adipose tissue decreased by GLA-BSA administration. These suggest that AGEs may alter the pathology of obesity and diabetes and may provide preventive and therapeutic strategies for obesity- and insulin resistance-related disorders.

References

1. Alberti KG, Zimmet P, and Shaw J; IDF Epidemiology Task Force Consensus Group, *Lancet* 2005, **366**, 1059-1062.
2. Fève B. *Best Pract Res Clin Endocrinol Metab* 2005, **19**, 483-499.
3. Furukawa S, Fujita T, Shimabukuro M, Iwaki M, Yamada Y, Nakajima Y, Nakayama O, Makishima M, Matsuda M, and Shimomura I. *J Clin Invest,* 2004, **114**, 1752-1761.
4. Guzik TJ, Mangalat D, and Korbut R. *J Physiol Pharmacol*, **57**, 505-528.
5. Wellen KE and Hotamisligil GS, *J Clin Invest*, 2005, **112**, 1785-1788.
6. Weisberg SP, McCann D, Desai M, Rosenbaum M, Leibel RL, and Ferrante AW Jr,. *J Clin Invest*, 2003, **112**, 1796-1808.
7. Xu H, Barnes GT, Yang Q, Tan G, Yang D, Chou CJ, Sole J, Nichols A, Ross JS, Tartaglia LA, and Chen H, *J Clin Invest*, 2003, **112,** 1821-1830.
8. Houstis N, Rosen ED, and Lander ES, *Nature*, 2006, *440,* 944-948.
9. Fujita K, Nishizawa H, Funahashi T, Shimomura I, and Shimabukuro M. *Circ. J.,* 2006, **70**, 1437-1442.
10. Unoki H, Bujo H, Yamagishi S, Takeuchi M, Imaizumi T, and Saito Y, *Diabetes Res. Clin. Pract*, 2007, **76**, 236-244.
11. Usui T, Shimohira K, Watanabe H, and Hayase F. *Biofactors*, 2004, **21**, 391-394
12. Usui T, Shizuuchi S, Watanabe H, and Hayase F, *Biosci Biotechnol Biochem*, 2004, **68**, 333-340.
13. Usui T, Shimohira K, Watanabe H, and Hayase F, *Biosci Biotechnol Biochem*, 2007, **71**, 442-448.
14. Yamagishi S, Matsui T, Nakamura K, Takeuchi M, and Imaizumi T, *Microvasc Res*, 2006, **72**, 86-90.
15. Yamagishi S, Nakamura K, Matsui T, Inagaki Y, Takenaka K, Jinnouchi Y, Yoshida Y, Matsuura T, Narama I, Motomiya Y, Takeuchi M, Inoue H, Yoshimura A, Bucala R, and Imaizumi T, *J Biol Chem*, 2006, **281**, 20213-20220.
16. Yanagisawa K, Makita Z, Shiroshita K, Ueda T, Fusegawa T, Kuwajima S, Takeuchi M, and Koike T. *Metabolism*, 1998, **47**, 1348-1353.
17. Nagai R, Matsumoto K, Ling X, Suzuki H, Araki T and Horiuchi S. *Diabetes*, 2000, **49**, 1714-1723.
18. Akamine R, Yamamoto T, Watanabe M, Yamazaki N, Kataoka M, Ishikawa M, Ooie T, Baba Y, and Shinohara Y, *J Biochem Biophys Methods*, 2007, **70**, 481-486.
19. Smedsrød B, Melkko J, Araki N, Sano H, and Horiuchi S, *Biochem J*, 1997, **22**, 567-573.
20. Nagai R, Mera K, Nakajou K, Fujiwara Y, Iwao Y, Imai H, Murata T, and Otagiri M, *Biochim Biophys Acta*, 2007, **1772**, 1192-1198.
21. Stahl WR, *J Appl Physiol,* 1967, **22**, 453-460.
22. Arner P, *Biochem Soc Trans*, 2001, **29**, 72-75.
23. Havel PJ, *Curr Opin Lipidol*, 2002, **13**, 51-59.
24. Lumeng CN, Bodzin JL, and Saltiel AR, *J Clin Invest*, 2007, **117**, 175-184.

METHYLGLYOXAL MODIFICATION OF HEAT-SHOCK PROTEIN 27 IN COLON MUCOSA OF HUMAN ULCERATIVE COLITIS

T. Oya-Ito[1], Y. Naito[1], T. Takagi[2], O. Handa[2], H. Matsui[3], K. Uchida[4], M.Yamada[5], K. Shima[5], T. Yoshikawa[1,2]

[1]Department of Medical Proteomics and [2]Department of Biomedical Safety Science, Kyoto Prefectural University of Medicine, 465Kaji-i, Kyoto 602-8566, JAPAN
[3]Graduate School of Comprehensive Human Sciences, University of Tsukuba,
[4]Graduate School of Bioagricultural Sciences, Nagoya University,
[5]Shimadzu Corporation

1 INTRODUCTION

Patients with chronic inflammatory bowel disease (IBD) including ulcerative colitis (UC) are at increased risk of developing colorectal cancer (CRC). Indeed, IBD ranks among the top three high-risk conditions for CRC, together with familial adenomatous polyposis [1-3] and hereditary nonpolyposis colorectal cancer [3]. CRC development in association with IBD appears to be closely relate to chronic inflammation of the large bowel mucosa [3]. Also, IBD-associated colon carcinogenesis can be summarized as an inflammation-dysplasia-carcinoma sequence: hyperplastic lesions in the inflamed mucosa develop CRC through flat dysplasia [4,5].

Enhanced glycolysis is observed in most of cancerous cells and tissues, called as the Warburg effect [6]. In colon and stomach cancer, tumor tissues contained nearly equal or higher amounts of glycolytic intermediates than their corresponding normal counterparts, and this trend was clearer in colon tissues [7]. Methylglyoxal (MGO), a metabolite of the glycolysis pathway, reacts with certain proteins to yield irreversible advanced glycation end products (AGEs), which can alter protein structure and functions. In human non-small cell lung cancer tissues AGEs, Nε-(carboxymethyl)lysine (CML) and argpyrimidine have been detected by immunohistochemistry [8]. The receptor for AGEs is associated with an increased risk of gastric cancer in a Chinese population [9]. Pentosidine expression was up-regulated in the inflamed tissue of IBD [10]. The urinary concentration of pentosidine in active ulcerative colitis is significantly greater than that in inactive ulcerative colitis, and is greater in active Crohn's disease than in inactive Crohn's disease [10]. In this study, we identified the methylglyoxal-modified proteins as a target molecule in colitis.

2   METHODS

2.1   Reagents

Human Hsp27 monoclonal antibody and mouse Hsp25 polyclonal antibody were obtained from StressGen Biotechnologies Corp. (Victoria, British Columbia, Canada). Horseradish peroxidase (HRP)-linked anti-rabbit IgG was obtained from Cell Signaling Technology, Inc. (Beverly, MA). HRP-linked anti-mouse IgG, Deep Purple Total Protein Stain and enhanced chemiluminescence (ECL) Plus Western blotting detection reagents were obtained from GE Healthcare UK Ltd. (Buckinghamshire, England). AOM was purchased from Sigma-Aldrich Co. (St. Louis, MO, USA). DSS with a molecular weight of 36,000–50,000 was obtained from MP Biomedicals, LLC (Aurora, OH, USA).

All cells were incubated at 37°C in a humidified incubator with 5% CO_2. Rat intestinal epithelial (RIE) cells were grown in a 1:1 mixture of Dulbecco's modified Eagle medium and Ham's F-12 medium (DMEM/F12; Cosmo Bio, Tokyo, Japan) supplemented with 10% fetal calf serum (FCS; Gibco, Grand Island, NY). Young adult mouse colon (YAMC) epithelial cells were grown in RPMI with 5% fetal calf serum, 100 ng/μL IFN-γ, 2 mM glutamine, 50 μg/ml gentamicin, 100 units/ml penicillin, 100 μg/ml streptomycin, and 1x insulin/transferrin/selenium.

2.2   Animals and diets

Male Crj: CD-1 (ICR) mice (Charles River Japan, Inc., Tokyo) aged five weeks were used in this study. DSS for the induction of colitis was dissolved in distilled water at a concentration of 2% (w/v). Charles River Formula (CRF)-1 (Oriental Yeast Co., Ltd., Tokyo, Japan) was used as a basal diet throughout the study.

2.3   DSS-treated mice

After arriving, mice were acclimated for seven days with tap water and a pelleted basal diet of CRF-1, ad libitum. They received a single intraperitoneal (i.p.) injection of 10 MGO/kg body weight azoxymethane (AOM). Starting one week after the AOM injection, the animals were exposed to 2% dextran sodium sulfate (DSS) in the drinking water for seven days, and then were followed without any further treatment until the experiment was done. They were sacrificed by CO_2 euthanasia at week 20 for the analysis. All mice were maintained at the Kyoto Prefectural University of Medicine Animal Facility according to the Institutional Animal Care Guidelines and were maintained under controlled conditions of humidity (50±10%), light (12/12 hr light/dark cycle), and temperature (23±2°C).

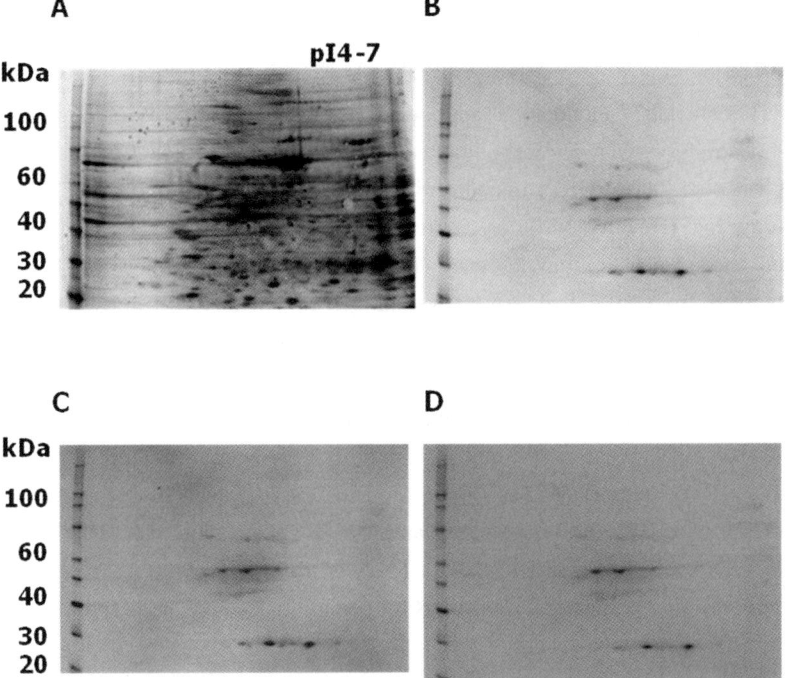

Figure 1   *Two-dimensional electrophoresis and Western blot analysis of proteins obtained from UC. Protein extracts from colonic biopsy samples were separated by two-dimensional electrophoresis and stained with Deep Purple Total Protein Stain (A). MGO-modified proteins were identified by immunoblot analysis using anti-argpyrimidine antibody (B). Hsp 27 and pHsp27 (Ser15) were identified by immunoblot analysis using anti-Hsp27 antibody (C) and anti- pHsp27 (Ser15) antibody (D).*

2.4   Sample preparation and proteomic assays

Cells were harvested by centrifugation, rinsed in phosphate-buffered saline, and resuspended in homogenization buffer (8M Urea, 4% CHAPS, 40 mM Tris) containing nuclease and protein inhibitors (GE Healthcare UK Ltd.). A buffer volume approximately equal to the packed cell volume was used. For proteomic assays using two-dimensional gel-electrophoresis, colonic biopsy samples were taken from the inflamed mucosa in four UC patients (all males; age range 25-42 years, average age 33 years). The frozen tissue samples (200 MGO) were homogenized in 2 mL homogenization buffer containing nuclease and protease inhibitors using a homogenizer at 25 000 r/min. Colonic tumors (histologically confirmed as well-differentiated tubular adenocarcinomas) and nontumorous mucosa tissues were collected from the mice that received AOM and DSS and were stored at −80°C prior to use. The frozen tissues were homogenized with five volumes of homogenization buffer. Homogenized samples were transferred to a ultracentrifuge tube, and the nucleic acids were removed by centrifugation (20 min at 20,000 × g, 25 °C). Lysates were precipitated using the Plus One 2D Clean-up kit as recommended by the manufacturer (GE Healthcare UK Ltd.). The protein concentration in the supernatant fraction was determined by the Bradford assay, using bovine serum albumin as the standard. Samples were solubilized in 8 M urea, 2% CHAPS, 20 mM dithiothreitol (DTT), 0.5% ZOOM Carrier Ampholytes 3-10 (Invitrogen Japan K.K., Tokyo, Japan), and 0.002% bromophenol blue. Protein lysates (50 μg) were separated by two-dimensional PAGE. IPG strips, pH 3–10NL or pH 4-7 (Invitrogen Japan K.K.), were rehydrated overnight with protein samples. Proteins were separated on the basis of their isoelectric point by IEF using the ZOOM IPG Runner (Invitrogen Japan K.K.) with a maximal voltage of 2000 V and 50 μA per gel. Following IEF, IPG strips were incubated in equilibration buffer I (6 M urea, 130 mM dithiothreitol, 30% glycerol, 45 mM Tris base, 1.6% LDS, 0.002% bromophenol blue; Genomic Solutions) and once in equilibration buffer II (6 M urea, 135 mM iodoacetamide, 30% glycerol, 45 mM Tris base, 1.6% LDS, 0.002% bromphenol blue; Genomic Solutions) for 15 min. Equilibrated IPG strips were applied to 4–12% Bis-Tris gradient gels (Invitrogen Japan K.K.), and proteins were separated in the second dimension based on their molecular size using NuPAGE MOPS buffer (Invitrogen Japan K.K.) at 200 V for 55 min. Following electrophoresis, gels were transferred onto nitrocellulose and immunoblotted with antibody.

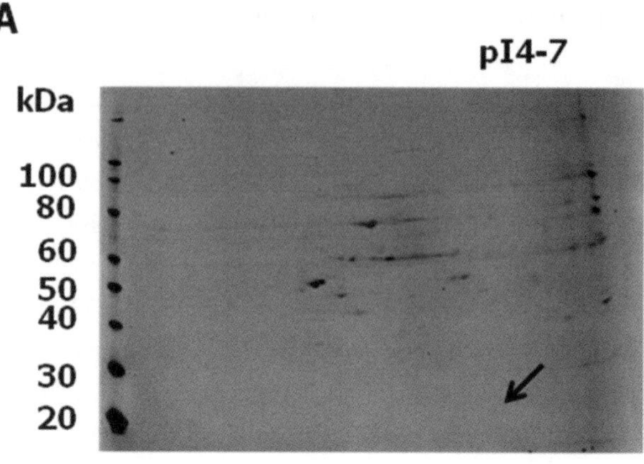

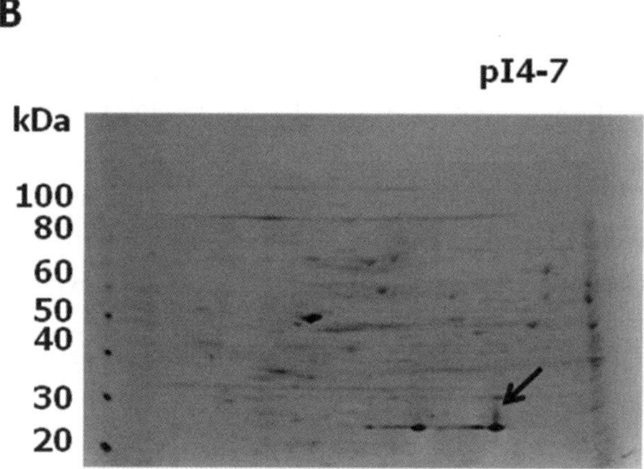

Figure 2   *Two-dimensional electrophoresis and Western blot analysis. Whole cell extracts from YAMC cells (A and B) were separated by two-dimensional electrophoresis. Forced expression of Hsp25 was induced by treatment with butyrate. MGO-modified proteins were identified by immunoblot analysis using anti-argpyrimidine antibody (A). Hsp 25 was identified by immunoblot analysis using anti-Hsp25 antibody (B).*

3	RESULTS

In a search for protein modification by methylglyoxal, we used a proteomic approach on colonic mucosal tissue obtained from UC. Whole protein extracts from the tissues were separated by two-dimensional electrophoresis, followed by immunoblot analysis with anti-methylglyoxal-modified protein monoclonal antibody, which specifically recognized argpyrimidine [11]. Total proteins were detected by using Deep Purple Protein Stain (Fig. 1A). As shown in Fig. 1B, two-dimensional electrophoresis on a strip that focuses proteins between pI4.0 and 7.0 followed by immunoblot analysis revealed MGO-modified proteins with molecular weights between 20 and 30 kDa, and between 50 and 60 kDa. Most proteins that reacted with antibody to Hsp27 showed immunoreactivity to argpyrimidine antibody (Fig. 1C). In addition, these proteins were immunoreacted with an antibody to phosphorylated Hsp27 (Ser.15).

The modifications of human Hsp27 and rat Hsp25 with MGO were also observed in human colon carcinoma HT-29 cells and rat gastric carcinoma mucosal (RGK-1) cells, respectively (data not shown). The Hsp25 in YAMC cells, which was forcedly expressed by treatment with butyrate, was not modified by MGO (Fig. 2A and B). Moreover, the rat intestinal epithelial (RIE) cells treated with butyrate showed forced expression of Hsp25 and no modification of Hsp25 by MGO (data not shown). These results suggested that MGO modifications of human Hsp27 and rat Hs25 were specific to inflammatory cells. In order to investigate the extent of MGO-modification upon Hsp25 in colonic mucosa of colitic mice, protein extracts were analyzed by Western Blot. Colitis in mice was induced by the presence of 2% dextran sulfate sodium (DSS) in the drinking water. An azoxymethane (AOM)/DSS mouse model[6] has been used to investigate the changes of proteins in the background of inflammation-related colon cancer[7]. A number of proteins have been analyzed to isolate and identify tumor specific proteins that might be involved in the development of colitis-related CRC in AOM/DSS model mice[6] by two-dimensional gel electrophoresis to further investigate the protein expression during colitis-associated carcinogenesis (8). In the current study, whole protein extracts from the tissue were separated by SDS-PAGE, followed by immunoblot analysis with anti-methylglyoxal modified protein monoclonal antibody. Hsp25 in protein extracts obtained from each colonic mucosa were confirmed by immunoblotting using anti-hsp25 antibody, as shown in Fig. 3. The immunoreactivity of a ~25 kDa protein in colonic mucosa from DSS-treated mice was identified by an antibody to argpyrimidine (Fig. 3A). The expression levels of Hsp25 were similar between in control and in DSS-treated groups (Fig. 3B). These results indicated that MGO-modified mouse Hsp25 abounds in DSS-induced colitis.

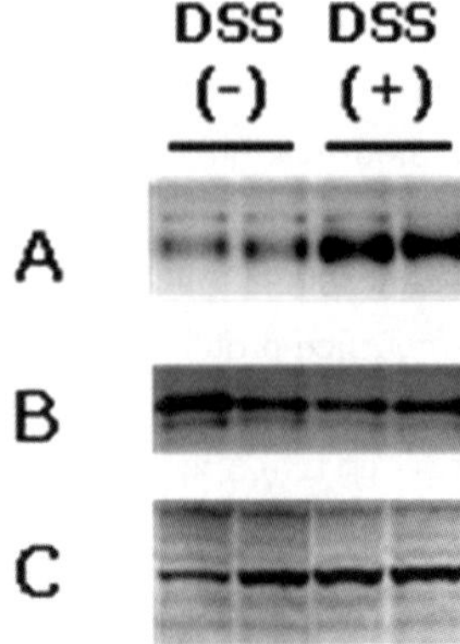

Figure 3   *Detections of argpyrimidine and Hsp27 in the colonic mucosa of mice.*
Colitis   was   induced   by   2%   DSS   in   the   drinking   water.
Methylglyoxal-modified proteins were identified by immunoblot analysis
using anti-argpyrimidine monoclonal antibody (A). Hsp25 and β-actin
were identified by immunoblot analysis using anti- Hsp25 antibody (B)
and anti- β-actin antibody (C), respectively.

4   DISCUSSION

Chronic inflammation is a well recognized risk factor for cancer and patients with long-standing UC are at an increased risk for colorectal carcinoma development. In this study, we uncovered the posttranslational modification linked with UC and carcinogenesis. Mouse Hsp25 and human Hsp27 were identified as major MGO-modified proteins in colon mucosa from colitis. Recently, argpyrimidine formation in Hsp27 has also been identified in human non-small lung carcinoma cells [8]. It is uncertain whether the concentration of MGO is higher in inflammatory cells than in normal cells.

Blood plasma profiles in chronic inflammation have provided evidence of loss of energy homeostasis, impaired metabolism of lipoproteins and glycosylated proteins [12]. In particular, IL-10-/-mice were characterized by decreased levels of VLDL and increased concentrations of LDL and polyunsaturated fatty acids, which are related to the etiology of IBD. Moreover, higher levels of lactate, pyruvate, citrate and lowered glucose suggested increased fatty acid oxidation and glycolysis, while higher levels of free amino acids reflected muscle atrophy, breakdown of proteins and interconversions of amino acids to produce energy [12]. It has reported that the concentration of AGEs, such as pentosidine was increased in the inflamed tissue of IBD [10].

Murine Hsp25 and human Hsp27 act as molecular chaperones and are constitutively expressed in several mammalian cells, particularly in pathological conditions [13]. Short-chain fatty acids, such as butyrate, induce a time- and concentration-dependent increase in Hsp25 protein expression in rat intestinal epithelial cells [14]. Small heat-shock proteins share functions as diverse as protection against toxicity mediated by aberrantly folded proteins or oxidative-inflammation conditions [13, 15]. These proteins can have pernicious effects through their ability to protect cancer cells against the immune system- or drug-mediated death. High levels of Hsp27 constitutive expression have been detected in several cancer cells, particularly those of carcinoma origin [16, 17]. In addition, these proteins share anti-apoptotic properties and are tumorigenic when expressed in cancer cells [13]. Hsp27 is phosphorylated at serines 15, 78 and 82 by mitogen-activated protein kinases associated protein kinases (MAPKAP kinases 2,3) which are themselves activated by phosphorylation by MAP p38 protein kinase [18, 19]. However, the regulation mechanism of Hsp27 action in IBD is still unknown.

The argpyrimidine formation in Hsp27 has also been found in endothelial cells [20], lens epithelial cells [21] and rat kidney mesangial cells [22], indicating that Hsp27 is a protein highly susceptible for MGO modification. In the previous study, we found that Hsp27 becomes a better anti-apoptotic protein after modification by MGO in lens epithelial cells, which may be due to multiple mechanisms that include enhancement of chaperone function and inhibition of activity of caspases [21, 23, 24]. MGO-modification of Arg-188 in Hsp27 process is essential to its repressing activity for cytochrome *c*-mediated caspase activation by using 293T cells transfected with the mutant of Hsp27 at Arg-188 to glycine [25]. Indeed, the modification of Hsp25 by MGO was not seen in RIE cells, YAMC cells or colonic mucosa of normal mice, whereas Hsp27 in HT-29 cells and colonic mucosa of UC and colitic mice were highly modified by MGO.

5 CONCLUSION

In conclusion, murine Hsp25 and human Hsp27 were identified as a major MGO-modified protein in IBD. Furthermore, almost all MGO-modified Hsp27 was phosphorylated in colon mucosa of UC. These data suggest that methylglyoxal-modified Hsp27 in colonic mucosa may play a crucial role in for epithelial injury in colitis. Hsp27 may also be a candidate for target protein in UC-associated carcinogenesis. Hsp27 is thought to be protective against apoptotic cell death [13, 26], however, the detailed mechanisms responsible for the protective effect of Hsp27 were unclear. It will be extremely necessary to investigate the changes in the function of Hsp27 protein after this modification.

Acknowledgements

We would like to thank Dr. Ram H. Nagaraj, Case Western Reserve University, whose comments made enormous contribution to our work. We would also like to thank Dr. Etsuo Niki, National Institute of Advanced Industrial Science and Technology, who gave us constructive comments and warm encouragement. This research was partially supported by FY 2009 Grant-in-Aid for Scientific Research for young scientists (B) to T. O.–I. from Japan Society for the Promotion of Science, by FY 2009 Research for Promotiong Technological Seeds A (discovery type) to T. O.–I. from Independent Administrative Corporation Japan Science and Technology Agency, and by the 170th Redox Life Sciences Committee to T. Y. from Japan Society for the Promotion of Science.

References

1.   S. P. Hussain, L. J. Hofseth and C. C. Harris, *Nat Rev Cancer*, 2003, **3**, 276-285.

2.   B. A. Lashner, K. S. Provencher, J. M. Bozdech and A. Brzezinski, *Am J Gastroenterol*, 1995, **90**, 377-380.

3.   S. H. Itzkowitz and X. Yio, *Am J Physiol Gastrointest Liver Physiol*, 2004, **287**, G7-17.

4.   J. Xie and S. H. Itzkowitz, *World J Gastroenterol*, 2008, **14**, 378-389.

5.   T. L. Zisman and D. T. Rubin, *World J Gastroenterol*, 2008, **14**, 2662-2669.

6.   O. Warburg, *Science*, 1956, **123**, 309-314.

7.   A. Hirayama, K. Kami, M. Sugimoto, M. Sugawara, N. Toki, H. Onozuka, T. Kinoshita, N. Saito, A. Ochiai, M. Tomita, H. Esumi and T. Soga, *Cancer Res*, 2009, **69**, 4918-4925.

8.   J. W. van Heijst, H. W. Niessen, R. J. Musters, V. W. van Hinsbergh, K. Hoekman and C. G. Schalkwijk, *Cancer Lett*, 2006, **241**, 309-319.

9.   H. Gu, L. Yang, Q. Sun, B. Zhou, N. Tang, R. Cong, Y. Zeng and B. Wang, *Clin Cancer Res*, 2008, **14**, 3627-3632.

10.   S. Kato, K. Itoh, M. Ochiai, A. Iwai, Y. Park, S. Hata, K. Takeuchi, M. Ito, J. Imaki, S. Miura, K. Yakabi and M. Kobayashi, *J Gastroenterol Hepatol*, 2008, **23 Suppl 2**, S140-145.

11.   T. Oya, N. Hattori, Y. Mizuno, S. Miyata, S. Maeda, T. Osawa and K. Uchida, *J Biol Chem*, 1999, **274**, 18492-18502.

12.   F. P. Martin, S. Rezzi, D. Philippe, L. Tornier, A. Messlik, G. Holzlwimmer, P. Baur, L. Quintanilla-Fend, G. Loh, M. Blaut, S. Blum, S. Kochhar and D. Haller, *J Proteome Res*, 2009, **8**, 2376-2387.

13. A. P. Arrigo, S. Simon, B. Gibert, C. Kretz-Remy, M. Nivon, A. Czekalla, D. Guillet, M. Moulin, C. Diaz-Latoud and P. Vicart, *FEBS Lett*, 2007, **581**, 3665-3674.

14. H. Ren, M. W. Musch, K. Kojima, D. Boone, A. Ma and E. B. Chang, *Gastroenterology*, 2001, **121**, 631-639.

15. C. Decroos, Y. Li, G. Bertho, Y. Frapart, D. Mansuy and J. L. Boucher, *Chem Res Toxicol*, 2009, **22**, 1342-1350.

16. D. R. Ciocca and S. K. Calderwood, *Cell Stress Chaperones*, 2005, **10**, 86-103.

17. S. K. Calderwood, M. A. Khaleque, D. B. Sawyer and D. R. Ciocca, *Trends Biochem Sci*, 2006, **31**, 164-172.

18. D. Stokoe, K. Engel, D. G. Campbell, P. Cohen and M. Gaestel, *FEBS Lett*, 1992, **313**, 307-313.

19. J. Rouse, P. Cohen, S. Trigon, M. Morange, A. Alonso-Llamazares, D. Zamanillo, T. Hunt and A. R. Nebreda, *Cell*, 1994, **78**, 1027-1037.

20. C. G. Schalkwijk, J. van Bezu, R. C. van der Schors, K. Uchida, C. D. Stehouwer and V. W. van Hinsbergh, *FEBS Lett*, 2006, **580**, 1565-1570.

21. T. Oya-Ito, B. F. Liu and R. H. Nagaraj, *J Cell Biochem*, 2006, **99**, 279-291.

22. A. K. Padival, J. W. Crabb and R. H. Nagaraj, *FEBS Lett*, 2003, **551**, 113-118.

23. R. H. Nagaraj, T. Oya-Ito, P. S. Padayatti, R. Kumar, S. Mehta, K. West, B. Levison, J. Sun, J. W. Crabb and A. K. Padival, *Biochemistry*, 2003, **42**, 10746-10755.

24. A. Biswas, A. Miller, T. Oya-Ito, P. Santhoshkumar, M. Bhat and R. H. Nagaraj, *Biochemistry*, 2006, **45**, 4569-4577.

25. H. Sakamoto, T. Mashima, K. Yamamoto and T. Tsuruo, *J Biol Chem*, 2002, **277**, 45770-45775.

26. M. T. Akbar, A. M. Lundberg, K. Liu, S. Vidyadaran, K. E. Wells, H. Dolatshad, S. Wynn, D. J. Wells, D. S. Latchman and J. de Belleroche, *J Biol Chem*, 2003, **278**, 19956-19965.

EXTRACELLULAR MATRIX GLYCATION AND PATHOGENESIS OF DIABETIC COMPLICATIONS

P.A. Voziyan[1*] and Billy G. Hudson[1]

[1]Department of Medicine, Division of Nephrology and Center for Matrix Biology, Vanderbilt University Medical Center, Nashville, Tennessee, 37232, USA

1   INTRODUCTION

Extracellular matrix (ECM) proteins provide critical scaffolding for tissue and organ formation. They also serve as ligands for cellular receptors such as integrins and dystroglycans and are involved in signaling cascades that regulate many cellular functions including adhesion, proliferation, migration, and apoptosis. Diverse structural and regulatory roles of ECM proteins make them critical players in normal and pathophysiology. Along with genetic mutations such as those causing the Alport syndrome[1] , ECM proteins are also affected by post-translational modifications including non-enzymatic glycation, which has been suggested to play a major role in the pathogenesis of diabetic complications[2]. This review will look at the impact of modifications caused by the formation of Advanced Glycation End products (AGEs) on the structure and function of matrix proteins.

2   THE MODIFICATION OF ECM PROTEINS BY AGES

AGEs can form via several major reactions, which are accelerated in diabetes. In one these reaction, the aldehyde group of glucose interacts directly with the protein amino group forming a Schiff base followed by essentially irreversible rearrangement to an Amadori intermediate. The Amadori intermediate then undergoes further rearrangements to yield heterogeneous AGEs[3]. They range from relatively simple, such as Nϵ-carboxymethyllysine (CML), to more complex products such as pyralline[4, 5]. Post-Amadori glycation and glycoxidation reactions can also result in intra- and inter-molecular cross-links, among them lysine-lysine cross-links crosslines [6] and vesperlysines [7] and the lysine-arginine crosslinks pentosidine and glucosepane [8, 9].

Another source of AGEs is reactive carbonyl species (RCSs). RCSs such as methylglyoxal (MGO) and 3-deoxyglucosone (3-DG) can form enzymatically as byproducts of glycolysis [10], during oxidation of acetone [11] , and in the reaction catalyzed by fructosamine-3-kinase during the repair of glycated protein intermediates [12]. Glyoxal (GO) and MGO can also form via non-enzymatic oxidative degradation of circulating glucose and glycated protein intermediates [13, 14]. While the aldehyde moiety of glucose reacts with the ϵ-amino groups of lysine and N-terminal α-amino groups of proteins and can modify arginine only via crosslink formation, RCS can react directly with both lysine and arginine side chains with high reaction rates [15]. Protein modifications specific to RCS are MOLD and GOLD lysine-lysine crosslinks derived from MGO and glyoxal (GO) [3]; MGO-derived modifications of arginine – tetrahydropyrimidine, hydromethylimidazolone, and argpyrimidine [15-17]; and lysine-arginine crosslinks MODIC, GODIC, and DOGDIC derived from MGO, GO, and 3-DG, respectively [8]. All described AGE modifications can accumulate in ECM proteins.

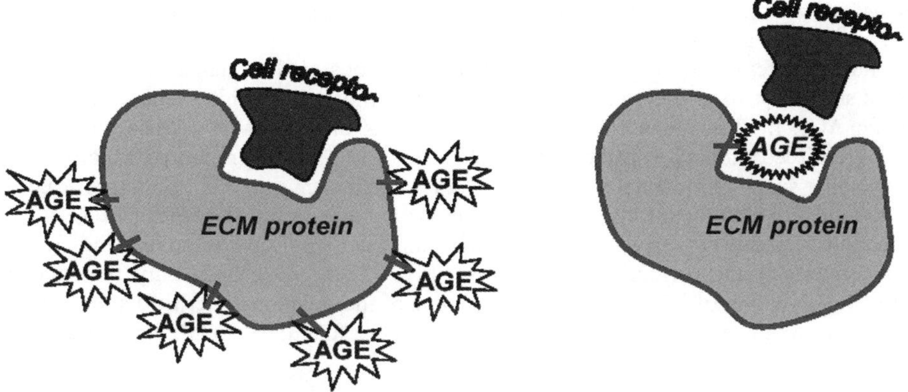

Figure 1   *Schematic illustration of importance of site-specific ECM protein glycation in pathogenesis.*

The ECM proteins are significantly more susceptible to glycoxidative damage compared to both circulating and intracellular proteins. The ECM proteins have relatively long life times. For example, human skin and cartilage collagens have half-lives of ~117 and ~15 years, respectively [18] while half-life of aggrecan in human inter vertebral discs is ~12 years [19]. Even ECM proteins of renal glomerular basement membrane are relatively long-lived with a half-life of > 100 days [20]. Slow ECM turnover rates allow for glycation modification to accumulate at significant levels and they are more likely to interfere with protein function. ECM proteins are also significantly less protected from glycoxidative damage compared to intracellular proteins, which reside in generally reducing environment under protection of robust enzymatic defenses including carbonyl detoxification and deglycation [12, 21]. Consequently, AGE modifications are significantly increased in ECM subjected to chronic hyperglycemia and carbonyl stress of diabetes. For example, CML adducts were detected in experimental models and clinical renal specimens taken from diabetics [22-25]. High level of CML was found in diabetic renal GBM and in ECM of human retina [26, 27]. Similarly, elevated MGO- and GO-derived arginine adducts were found in sciatic endoneurium and aortal collagen of diabetic rats and in renal mesangial matrix of diabetic patients.[28-31]. AGE cross-links, including pentosidine and glucosepane, have been shown to significantly increase in ECM proteins of diabetic animal models and diabetic patients [32-34].

The specific role of ECM glycation/glycoxidation in pathogenesis of diabetic complications is poorly understood. It is yet unclear whether ECM modifications that accumulate in diabetes have mechanistic significance in pathogenesis or simply indicate secondary glycoxidative damage. The measurements of total levels of AGE modifications in ECM cannot provide significant insights into the pathogenic mechanisms since only a small fraction of susceptible amino acids is located within functionally important protein sites. Therefore, depending on microenvironment and reactivity of specific amino acid side chains, the levels of AGE modifications may not always reflect functional protein damage as schematically illustrated in Fig. 1. Indeed, the total levels of specific AGEs are often not reliable as disease biomarkers [35, 36].

In contrast, determination of site-specific modifications within functionally important protein sites may offer direct insight into early (triggering) pathogenic events. These site-specific AGE modifications could also be utilized as more reliable disease biomarkers which are directly linked to pathogenic mechanisms. In ECM, there are potentially three types of such functionally critical sites that could be damaged upon glycation/glycoxidation: 1) the cellular receptor binding sites conferring cell- ECM interactions; 2) the proteolytic cleavage sites important in ECM turnover, and 3) the subunit interface sites critical for ECM assembly. There have been only few studies addressing glycation of ECM proteins within some of the functionally critical sites. These findings are reviewed below along with proposed other functional ECM sites which may be targeted by glycation reactions.

3   AGES AND CELL-ECM INTERACTIONS.

Integrin binding motifs in ECM proteins have multiple potential glycation sites ([37-50] and Table 1). Mechanistic *in vitro* studies have suggested that modification of arginine residues within some of these sites using MGO can perturb integrin-dependent cell-ECM adhesion [31, 51-53]. Specifically, modification of RGD and GFOGER integrin binding motifs caused disruption of glomerular and dermal endothelial cell adhesion to MGO-modified collagen IV [31, 51]. The disruption of these integrin-ECM interactions is important in the pathogenesis as mice lacking integrin α1β1, a principal collagen IV receptor, developed more severe expansion of mesangial matrix than their wild type counterparts [24]. On the other hand, glycoxidation of lysine residues in collagen IV or collagen I had no effect on cell adhesion to these ECM proteins [51, 52], but specifically inhibited cell proliferation [54]. This suggests that specific AGE modifications of ECM may affect distinct cell functions by interfering with binding of different cellular receptors [53].

Besides integrins, receptor for advanced glycation end products (RAGE) may also participate in cell-ECM interactions. While glycation of arginine residues within ECM binding sites inhibited integrin binding, glycoxidation of ECM lysine residues to CML is required for RAGE binding. Even though RAGE can bind circulating CML-albumin [55], it also has many other ligands including amphoterin, S100 proteins, lypopolysaccharide, and amyloid â peptide, which may compete with circulating AGEs for binding [56]. The importance of circulating AGEs as RAGE ligands has been questioned based on relatively low degree of modification of circulating proteins under physiological conditions [56-58]. Compared to circulating proteins, long-lived ECM proteins accumulate much higher levels of AGE modifications, including CML, under pathophysiological conditions in vivo [30, 59]. This could make glycoxidized ECM a better RAGE ligand capable of providing multiple binding sites necessary for RAGE dimerization and activation [60]. There are very few reports addressing RAGE-ECM interactions. Glucose-modified collagen can modulate neutrophil adhesion and migration via RAGE-dependent PI3K signaling pathway [61].

RAGE also appears to mediate dermal fibroblast apoptosis induced by glycoxidized ECM [62]. On the other hand, RAGE was not involved in podocyte apoptosis induced by AGE-modified collagen [63]. Identification of CML clusters with potential of binding RAGE in glycoxidized ECM proteins will help to uncover the mechanisms of RAGE-ECM interactions. It is yet unknown whether other AGE receptors, i.e. scavenger receptor A, CD36, and AGE receptors 1-3 [64-66], can be specifically engaged by glycoxidized ECM.

Table1.   *Potential glycation sites in the integrin binding motifs of ECM proteins (compiled from [37-50])*

Binding motif	ECM protein	Cellular integrin receptors
RGD	fibronectin, vitronectin,	$\alpha v \beta 3, \alpha v \beta 5, \alpha v \beta 1$
	laminin ($\alpha 5$), entactin, osteopontin	$\alpha v \beta 1, \alpha IIb \beta 3, \alpha 5 \beta 1$
KQNCLSSRASFRGCVRNLRLSR	Laminin ($\alpha 1$)	$\alpha 3 \beta 1$
YIGSR	laminin ($\beta 1$)	$\alpha 2 \beta 1$
YGYYGDALR	laminin ($\beta 1$)	$\alpha 2 \beta 1$
GFOGER	collagens I and IV	$\alpha 2 \beta 1$
FYFDLR	collagen IV	$\alpha 2 \beta 1$
GEFYFDLRLKGDK	collagen IV ($\alpha 1$),	$\alpha 3 \beta 1$
D461($\alpha 1$)R461($\alpha 2$)	collagen IV	$\alpha 1 \beta 1$

4   AGES AND MATRIX TURNOVER.

The rates of ECM turnover are significantly decreased in diabetes [2]. The role of matrix glycation in this phenomenon is supported by detection of AGE cross-links in tissues and by a decreased activity of specific matrix metalloproteases (MMPs) towards glycated ECM proteins [2, 32, 34]. Consistent with these findings, are reports of attenuation of ECM accumulation in diabetic tissue by AGE cross-link breakers [67], even though the exact mechanism of AGE cross-link breakers *in vivo* is still controversial [68, 69]. Besides AGE cross-linking, there other processes that contribute to the increase in ECM turnover rates. Engagement of cellular receptors by AGE-ECM may also inhibit matrix turnover rates. Consistent with this notion, cells grown on glucose-modified collagen express less of MMP-2 and more of endogenous collagen IV and MMP inhibitor TIMP-1 compared to cells grown on unmodified collagen [54, 70]. Exposure of tenocytes to CML-collagen I increased intracellular activity of transglutaminase, enzyme that catalyzes formation of non-AGE cross-links in ECM [71]. In addition, non cross-link AGEs may also contribute to the loss of ECM digestibility because the consensus cleavage sites of many MMPs contain lysine and arginine residues [72]. Identification of site-specific ECM crosslinks and AGE modifications within ECM cellular receptor binging and protease cleavage sites may provide important clues about the role of cross-link and non cross-link AGEs in matrix turnover.

Table 2. *Potential glycation sites in the chain recognition region (VR3) in collagen IV NC1 domains [76].*

Collagen IV chain	Species	VR3sequence
	Bovine	ATIERSEMFKKPTPS
	Chimpanzee	ATIERSEMFKKPTPS
α1	Human	ATIERSEMFKKPTPS
	Mouse	ATIERSEMFKKPTPS
	Bovine	ASLDPKRMFRKPIPS
	Chimpanzee	ASLNPERMFRKPIPS
α3	Human	ASLNPERMFRKPIPS
	Mouse	ASLNPERMFRKPIPS
	Chimpanzee	ATVDVSDMFSKPQSE
α5	Human	ATVDVSDMFSKPQSE
	Mouse	ATVDMSDMFNKPQSE

5 AGES AND ECM ASSEMBLY.

There are several reports indicating that AGE modifications can affect matrix conformation and assembly. AGE-modified ECM had increased rigidity as well as greater mean fiber radius and pore size compared to unmodified ECM [73-75]. However, the mechanisms of these effects are unknown and the studies of potential ECM sites susceptible to AGE modifications may prove critical for understanding pathogenic mechanisms involving ECM protein conformation and network assembly. For example, the VR3 regions of NC1 domains of collagen IV, the conserved regions critical in chain recognition during protomer assembly, are enriched in arginine and lysine residues ([76] and Table 2). AGE modifications of arginine and/or lysine side chains within these regions in diabetes are likely to interfere with collagen IV network assembly and stability.

6 CONCLUSIONS

ECM proteins are the most susceptible to non-enzymatic AGE modifications compared to circulating or cellular proteins. These modifications can affect ECM assembly and turnover as well as multiple cell-ECM interactions, thus contributing to pathogenesis of diabetic complications. However, the exact mechanisms are unknown. Determination of AGE modifications within functionally critical sites of ECM proteins under diabetic conditions in vivo will provide important clues for uncovering pathogenic mechanisms and identification of new disease biomarkers.

Acknowledgements

This work was supported by the NIH grant DK65138.

References

1. Hudson, B.G., Tryggvason, K., Sundaramoorthy, M.Neilson, E.G., *N Engl J Med.* 2003, **348**, 2543-56.
2. Monnier, V.M., Mustata, G.T., Biemel, K.L., Reihl, O., Lederer, M.O., Zhenyu, D.Sell, D.R., *Ann N Y Acad Sci.* **1043**, 533-44, 2005.
3. Thorpe, S.R. Baynes, J.W., *Amino Acids.* 2003, **25**, 275-81.
4. Ahmed, M.U., Thorpe, S.R.Baynes, J.W., *J Biol Chem.* 1986, **261**, 4889-94.
5. Hayase, F., Nagaraj, R.H., Miyata, S., Njoroge, F.G.Monnier, V.M., *J Biol Chem.* 1989, **264**, 3758-64.
6. Ienaga, K., Nakamura, K., Hochi, T., Nakazawa, Y., Fukunaga, Y., Kakita, H.Nakano, K., C. *Contrib Nephrol.* 1995, **112**, 42-51.
7. Nakamura, K., Nakazawa, Y.Ienaga, K.,. *Biochem Biophys Res Commun.* 1997, .**232**, 227-30.
8. Biemel, K.M., Friedl, D.A.Lederer, M.O., *J Biol Chem.* 2002, **277**, 24907-15.
9. Sell, D.R. Monnier, V.M., *J Biol Chem.* 1989, **264**, 21597-602.
10. Ohmori, S., Mori, M., Shiraha, K.Kawase, M., *Prog Clin Biol Res.* 1989, **290**, 397-412.
11. Casazza, J.P., Felver, M.E.Veech, R.L., *Biol Chem.* 1984, **259**, 231-6.
12. Szwergold, B.S., Howell, S.Beisswenger, P.J.,. *Diabetes.* 2001, **50**, 2139-47.
13. Thornalley, P.J., Langborg, A.Minhas, H.S., *Biochem J.* 1999, **344 Pt 1**, 109-16.
14. Wells-Knecht, K.J., Zyzak, D.V., Litchfield, J.E., Thorpe, S.R.Baynes, J.W., *Biochemistry.* 1995, **34**, 3702-9.
15. Lo, T.W., Westwood, M.E., McLellan, A.C., Selwood, T.Thornalley, P.J., *J Biol Chem.* 1994, **269**, 32299-305.
16. Oya, T., Hattori, N., Mizuno, Y., Miyata, S., Maeda, S., Osawa, T.Uchida, K., *J Biol Chem.* 1999, **274**, 18492-502.
17. Shipanova, I.N., Glomb, M.A.Nagaraj, R.H., *Arch Biochem Biophys.* 1997, **344**, 29-36.
18. Verzijl, N., DeGroot, J., Thorpe, S.R., Bank, R.A., Shaw, J.N., Lyons, T.J., Bijlsma, J.W., Lafeber, F.P., Baynes, J.W.TeKoppele, J.M.,. *J Biol Chem.* 2000, **275**, 39027-31.
19. Sivan, S.S., Tsitron, E., Wachtel, E., Roughley, P.J., Sakkee, N., van der Ham, F., DeGroot, J., Roberts, S.Maroudas, A., *J Biol Chem.* 2006, **281**, 13009-14.
20. Price, R.G. Spiro, R.G., *J Biol Chem.* 1977, **252**, 8597- 602.
21. Thornalley, P.J.,. *Biochem Soc Trans.* 2003, **31**, 1343-8.
22. Schleicher, E.D., Wagner, E. Nerlich, A.G.,. *J Clin Invest.* 1997, **99**, 457-68.
23. Nakamura, S., Tachikawa, T., Tobita, K., Aoyama, I., Takayama, F., Enomoto, A.Niwa, T., *Am J Kidney Dis.* 2003.**41**, S68-71.
24. Zent, R., Yan, X., Su, Y., Hudson, B.G., Borza, D.B., Moeckel, G.W., Qi, Z., Sado, Y., Breyer, M.D., Voziyan, P.Pozzi, A.,. *Kidney Int.* 2006, **70**, 460-70.
25. Duran-Jimenez, B., Dobler, D., Moffatt, S., Rabbani, N., Streuli, C.H., Thornalley, P.J., Tomlinson, D.R.Gardiner, N.J., *Diabetes*2009.
26. Wendt, T., Tanji, N., Guo, J., Hudson, B.I., Bierhaus, A., Ramasamy, R., Arnold, B., Nawroth, P.P., Yan, S.F., D'Agati, V.Schmidt, A.M., *J Am Soc Nephrol.* 2003, **14**, 1383-95.

27. Glenn, J.V. Stitt, A.W., *Biochim Biophys Acta.* 2009, **1790**, 1109-16.
28. Niwa, T., Katsuzaki, T., Miyazaki, S., Miyazaki, T., Ishizaki, Y., Hayase, F., Tatemichi, N.Takei, Y.,. *J Clin Invest.* 1997, **99**, 1272-80.
29. Niwa, T., Katsuzaki, T., Ishizaki, Y., Hayase, F., Miyazaki, T., Uematsu, T., Tatemichi, N.Takei, Y., *FEBS Lett.* 1997, **407**, 297-302.
30. Thornalley, P.J., Battah, S., Ahmed, N., Karachalias, N., Agalou, S., Babaei-Jadidi, R.Dawnay, A., *Biochem J.* 2003, **375**, 581-92.
31. Dobler, D., Ahmed, N., Song, L., Eboigbodin, K.E.Thornalley, P.J., *Diabetes.* 2006, **55**, 1961-9.
32. Sell, D.R., Biemel, K.M., Reihl, O., Lederer, M.O., Strauch, C.M.Monnier, V.M.,. *J Biol Chem.* 2005, **280**, 12310-5.
33. Monnier, V.M., Bautista, O., Kenny, D., Sell, D.R., Fogarty, J., Dahms, W., Cleary, P.A., Lachin, J.Genuth, S., *Diabetes.* 1999, **48**, 870-80.
34. Mott, J.D., Khalifah, R.G., Nagase, H., Shield, C.F., 3rd, Hudson, J.K.Hudson, B.G., *Kidney Int.* 1997, **52**, 1302-12.
35. Senolt, L., Braun, M., Vencovsky, J., Sedova, L.Pavelka, K., *Physiol Res.* 2007, **56**, 771-7.
36. Busch, M., Franke, S., Wolf, G., Brandstadt, A., Ott, U., Gerth, J., Hunsicker, L.G.Stein, G., *Am J Kidney Dis.* 2006, **48**, 571-9.
37. Ruoslahti, E., *Annu Rev Cell Dev Biol.* 1996, **12**, 697- 715.
38. Plow, E.F., Haas, T.A., Zhang, L., Loftus, J.Smith, J.W., Ligand binding to integrins. *J Biol Chem.* 2000, **275**, 21785-8.
39. Aumailley, M., Gerl, M., Sonnenberg, A., Deutzmann, R.Timpl, R., *FEBS Lett.* 1990, **262**, 82-6.
40. Sasaki, T. Timpl, R.,. *FEBS Lett.* 2001, **509**, 181-5.
41. Pierschbacher, M.D. Ruoslahti, E., *Nature.* 1984, **309**, 30-3.
42. Cherny, R.C., Honan, M.A.Thiagarajan, P.,. *J Biol Chem.* 1993, **268**, 9725-9.
43. Vandenberg, P., Kern, A., Ries, A., Luckenbill-Edds, L., Mann, K.Kuhn, K., *J Cell Biol.* 1991, **113**, 1475-83.
44. Eble, J.A., Golbik, R., Mann, K.Kuhn, K., *Embo J.* 1993, **12**, 4795-802.
45. Kramer, R.H. Marks, N., *J Biol Chem.* 1989, **264**, 4684-8.
46. Knight, C.G., Morton, L.F., Peachey, A.R., Tuckwell, D.S., Farndale, R.W.Barnes, M.J., *J Biol Chem.* 2000, **275**, 35-40.
47. Graf, J., Ogle, R.C., Robey, F.A., Sasaki, M., Martin, G.R., Yamada, Y.Kleinman, H.K., *Biochemistry.* 1987.**26**, 6896-900.
48. Yi, X.Y., Wayner, E.A., Kim, Y.Fish, A.J., *Cell Adhes Commun.* 1998, **5**, 237-48.
49. Gehlsen, K.R., Sriramarao, P., Furcht, L.T.Skubitz, A.P., *J Cell Biol.* 1992, **117**, 449-59.
50. Maeda, T., Titani, K.Sekiguchi, K., *J Biochem (Tokyo).* 1994, **115**, 182-9.
51. Pedchenko, V.K., Chetyrkin, S.V., Chuang, P., Ham, A.J., Saleem, M.A., Mathieson, P.W., Hudson, B.G.Voziyan, P.A., *Diabetes.* 2005, **54**, 2952-60.
52. Chong, S.A., Lee, W., Arora, P.D., Laschinger, C., Young, E.W., Simmons, C.A., Manolson, M., Sodek, J.McCulloch, C.A *J Biol Chem.* 2007, **282**, 8510-20.
53. Paul, R.G. Bailey, A.J., *Int J Biochem Cell Biol.* 1999, **31**, 653-60.
54. Pozzi, A., Zent, R., Chetyrkin, S., Borza, C., Bulus, N., Chuang, P., Chen, D., Hudson, B.Voziyan, P., *J Am Soc Nephrol.* 2009, **20**, 2119-25.
55. Kislinger, T., Fu, C., Huber, B., Qu, W., Taguchi, A., Du Yan, S., Hofmann, M., Yan, S.F., Pischetsrieder, M., Stern, D.Schmidt, A.M., *J Biol Chem.* 1999, **274**, 31740-9.
56. Heizmann, C.W., *Mol Nutr Food Res.* 2007, **51**, 1116-9.

57. Valencia, J.V., Mone, M., Koehne, C., Rediske, J.Hughes, T.E., *Diabetologia.* 2004, **47**, 844- 52.
58. Buetler, T.M., Leclerc, E., Baumeyer, A., Latado, H., Newell, J., Adolfsson, O., Parisod, V., Richoz, J., Maurer, S., Foata, F., Piguet, D., Junod, S., Heizmann, C.W.Delatour, T., *Mol Nutr Food Res.* 2008, **52**, 370-8.
59. Hamelin, M., Borot-Laloi, C., Friguet, B.Bakala, H., *Arch Biochem Biophys.* 2003.**411**, 215-22.
60. Ostendorp, T., Leclerc, E., Galichet, A., Koch, M., Demling, N., Weigle, B., Heizmann, C.W., Kroneck, P.M.Fritz, G.,. *Embo J.* 2007.**26**, 3868-78.
61. Toure, F., Zahm, J.M., Garnotel, R., Lambert, E., Bonnet, N., Schmidt, A.M., Vitry, F., Chanard, J., Gillery, P.Rieu, P.,. *Biochem J.* 2008, **416**, 255-61.
62. Niu, Y., Xie, T., Ge, K., Lin, Y.Lu, S., *Am J Dermatopathol.* 2008, **30**, 344-51.
63. Chuang, P.Y., Yu, Q., Fang, W., Uribarri, J.He, J.C., *Kidney Int.* 2007, **72**, 965-76.
64. Li, Y.M., Mitsuhashi, T., Wojciechowicz, D., Shimizu, N., Li, J., Stitt, A., He, C., Banerjee, D.Vlassara, H., *Proc Natl Acad Sci U S A.* 1996.**93**, 11047-52.
65. Miyazaki, A., Nakayama, H.Horiuchi, S., S*Trends Cardiovasc Med.* 2002.**12**, 258-62.
66. Ohgami, N., Nagai, R., Ikemoto, M., Arai, H., Miyazaki, A., Hakamata, H., Horiuchi, S.Nakayama, H.,). *J Diabetes Complications.* 2002, **16**, 56-9.
67. Thallas-Bonke, V., Lindschau, C., Rizkalla, B., Bach, L.A., Boner, G., Meier, M., Haller, H., Cooper, M.E.Forbes, J.M., *Diabetes.* 2004, **53**, 2921-30.
68. Yang, S., Litchfield, J.E.Baynes, J.W., *Arch Biochem Biophys.* 2003, **412**, 42-6.
69. Price, D.L., Rhett, P.M., Thorpe, S.R.Baynes, J.W., *J Biol Chem.* 2001, **276**, 48967-72.
70. Anderson, S.S., Wu, K., Nagase, H., Stettler-Stevenson, W.G., Kim, Y.Tsilibary, E.C., *Cell Adhes Commun.* 1996, **4**, 89-101.
71. Rosenthal, A.K., Gohr, C.M., Mitton, E., Monnier, V.Burner, T.,. *J Investig Med.* 2009, **57**, 460-6.
72. Turk, B.E., Huang, L.L., Piro, E.T.Cantley, L.C. *Nat Biotechnol.* 2001, **19**, 661-7.
73. Howard, E.W., Benton, R., Ahern-Moore, J.Tomasek, J.J., *Exp Cell Res.* 1996, **228**, 132-7.
74. Boyd-White, J. Williams, J.C., Jr., *Diabetes.* 1996, **45**, 348-53.
75. Makino, H., Shikata, K., Hironaka, K., Kushiro, M., Yamasaki, Y., Sugimoto, H., Ota, Z., Araki, N.Horiuchi, S.,. *Kidney Int.* 1995, **48**, 517-26.
76. Khoshnoodi, J., Sigmundsson, K., Cartailler, J.P., Bondar, O., Sundaramoorthy, M.Hudson, B.G., *J Biol Chem.* 2006, **281**, 6058-69.

STRUCTURAL ANALYSIS OF A SKIN COLLAGEN-LINKED FLUOROPHORE THAT INCREASES IN DIABETES AND END-STAGE RENAL DISEASE

D.R. Sell[1,2], I.Nemet[1,2] and V.Monnier[1,2]

[1]Department of Pathology Case Western Reserve University, Cleveland, Ohio, 44106 USA
[2]Department of Biochemistry, Case Western Reserve University, Cleveland, Ohio 44106, USA

1   INTRODUCTION

Protein-linked autofluorescence at 440 nm upon excitation at 370 nm; i.e., henceforth long-wavelength fluorescence or LW, was introduced many years ago by us as a putative and surrogate marker for the Maillard reaction *in vivo* [1,2]. Ever since, this marker has been used in a large number of studies with the assumption that it is an advanced glycation endproduct (AGE) representative of the broader Maillard reaction in vivo. However, our recent data with EDIC patients [3,4] together with studies by Smit's group [5-7], as well as the introduction onto the commercial market of diagnostic devices capable of noninvasively measuring LW in human forearm skin for screening individuals for diabetes and its complications [7,8] strongly suggests that the biochemical origin needs to be precisely understood. Most noteworthy is that long-wave fluorscence can be generated *in vivo* through the oral administration of methylglyoxal to rodents [9]. Possible precursors for LW include glucose [10-13], ribose [11], glycolaldehyde [14], ascorbic acid [15], Ehrlich-positive Chromogen-related product [16], lipid peroxidation [17], but not lysyl oxidase-dependent hydroxypyridinium crosslinks [18]. Undoubtedly, both oxidative and glycoxidative processes can generate LW [19]. In short, there is a long list of potential precursors for LW, and while many authors have assumed that LW is a marker for AGEs, there is no evidence that LW formed *in vivo* is biochemically related to AGE formation. Therefore, much of the existing data regarding the origins of LW in tissue proteins is confusing and controversial. Thus, clarification of these structures is urgently needed. Below, we describe the isolation, purification, and partial molecular characterization of a single major protein-bound fluorophore from enzymatic digests of human skin collagen referred to as **LW1** whose levels are elevated by aging, diabetes and end-stage renal disease (ESRD).

2   METHODS

Procedures have been published in detail elsewhere[20]. Briefly, human skin was obtained at autopsy from individuals with diabetes and chronic renal failure or ESRD. After removal of the epidermis, a total of 75 g wet-weight skin tissue was defatted (2:1 chloroform-methanol) and extracted (1 M sodium chloride, 0.5 M acetic acid, 1mg/ml pepsin in 0.5 M acetic acid) which yielded ~12 g of insoluble collagen. This collagen was digested sequentially with collagenase, peptidase, pronase, and aminopeptidase M [20]. LW-1 was purified by reverse-phase C18 HPLC and subjected to UV-fluorescence spectrometry, 1H-NMR, 1H-13C HSQC, 1H-1H TOCSY and mass spectrometry as described [20]. HPLC methods are described in Figs. 1-2.

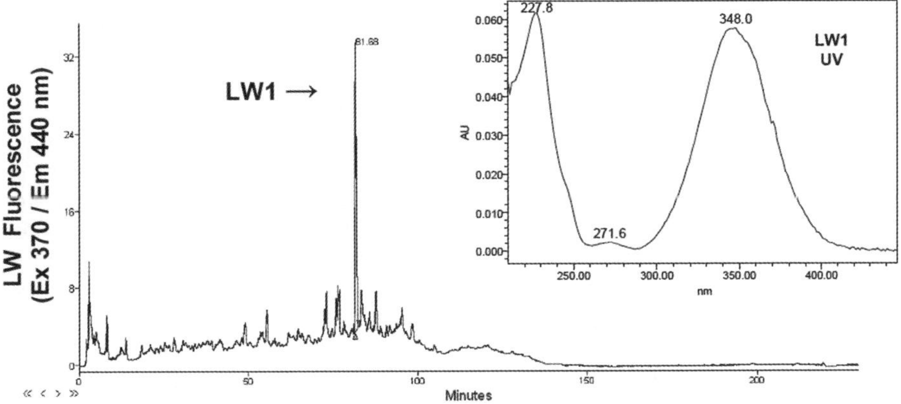

Figure 1   *HPLC- fluorescence (ex/em 370/440 nm) chromatogram of enzymatic digestion of human insoluble skin collagen sample from a 62 year-old patient with type 2 diabetes & ESRD. A total of 64 μg of collagen in 20 μl was injected onto a Waters HPLC system20 containing a 15 cm X 2.1, 3 μm Discovery HS C18 column (Sigma-Aldrich) used at flow rate 0.2 ml/min and equilibrated and eluted with the following gradient program: 0 - 5 min, 0%B; 5.1 - 100 min, 5 - 36%B (linear); 100 – 200 min, 36 - 100%B (linear); 200 – 212 min, 100%B; 213 – 230 min, 0%B. Solvent A: 98% water - 2% acetonitrile with 0.13% heptafluorobutyric acid (HFBA); solvent B: 60% acetonitrile - 40% water. The eluate was monitored using a JASCO spectrofluorometer (20. Insert: UV spectrum of purified LW1 determined by a HPLC-online Waters 2996 PDA Detector. LW1: long-wavelength fluorophore.*

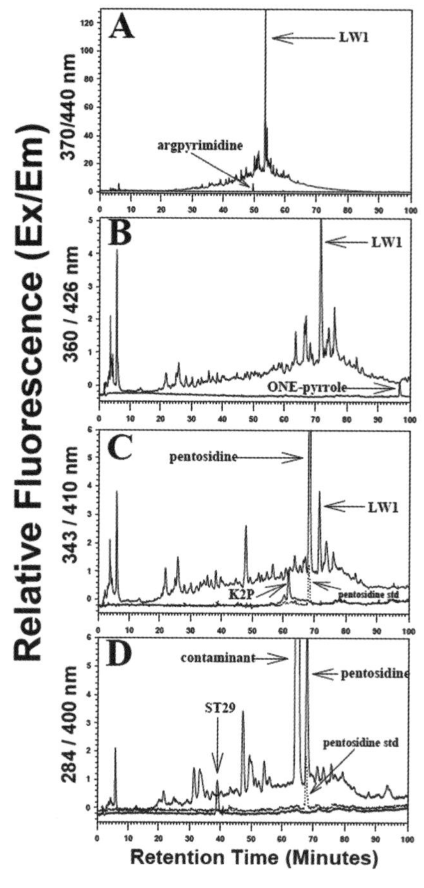

Figure 2
HPLC chromatograms of enzymatic digests of human insoluble skin collagen samples injected onto a HPLC monitored by a fluorescence detector set at different wavelengths as shown. (A) argpyimidine standard (std), peak ~283 pmol injected; (B) ONE-pyrrole std ~42 fmol; (C) K2P std ~676 pmol; (D) ST29 std ~1.28 pmol; (C&D) pentosidine std ~196 fmol. The HPLC, column, flow rate, and solvents were the same as Fig. 1. However, the gradient used in (A) was different from that used in (B,C,D) as follows: (A) 0 - 10 min, 0%B; 10 - 100 min 0 - 82%B (linear); 101 min, 100%B; (B,C,D) 0 - 5 min, 0%B; 5.1 min, 5%B; 5.1 - 100 min, 5 – 45% (linear); 101 min, 100%B. The retention times of the fluorophores were as follows: LW1(A), LW1(B,C,D), argpyrimidine, ONE-pyrrole, pentosidine, K2P, ST29 at ~ 53.6, 71.5, 49.7, 96.9, 68.3, 61.8, 39 min, respectively. The amount of collagen injected: (A) 640 µg; (B,C,D) 64 µg. Patient donors: (A) 47 year-old with type 1 diabetes & ESRD; (B&C) 68 year-old with type 2 diabetes & ESRD; (D) 62 year-old with type 2 diabetes & ESRD. The contaminant in (D) is from the buffer used in the enzymatic digestion procedure (20).

3   RESULTS

Injection of an enzymatic digest of human insoluble skin collagen onto a HPLC showed a single major peak in the chromatogram when monitored by a LW fluorescence detector (Fig. 1). It was noted that the size of the peak from collagen digests prepared from age 10 and 47 year-old individuals with type 1 diabetes were comparatively more intense *versus* those from nondiabetic individuals of similar age [20]. Furthermore, peak area increased with chronological age and was considerably more intense in individuals with ESRD. Subsequently the peak was named LW1.

LW1 was unstable to acid hydrolysis; i.e., 110°C in 6 N HCl for 24 hrs. Additionally, it was non-borohydride reducible; i.e., fluorescence was not quenched by reduction with sodium borohydride (data not shown). LW1 did not co-chromatograph on HPLC with other described fluorophores; namely, pentosidine [21], argpyrimidine [22], ONE-pyrrole [23], K2P [24], and ST29 [25] (see Fig. 2). LW1 was the largest peak at fluorescent wavelengths ex/em 370/440 (Fig. 2A) and 360/426 nm (Fig. 2B), became a minor peak at 343/410 nm (Fig. 2C), and was not observed at 284/ 400 nm (Fig. 2) where pentosidine predominated (Fig. 2C&D).

Because of lingering questions concerning the possible presence of unresolved peaks in the HPLC chromatogram and the extent of total autofluorescence in collagen explained by LW1, the HPLC gradient was modified to increase its shallowness such that the HPLC program and its chromatogram were run and recorded over a 230 min period (Fig. 1). However, even with this shallow gradient, LW1 still emerged as the single most major peak in the chromatogram (Fig. 1). It accounted for up to ~20% of the total fluorescence in these digests with the rest of the autofluorescence being ubiquitously spread-out throughout the chromatogram in many minor peaks characterized by unresolved broadening and tailing effects (Fig. 1). The latter observation may be due to incomplete digestion in these samples which averaged ≈ 60% [20].

Based on the results of Fig. 1, a preparative protocol for the quantitative preparation of LW1 was made [20]. Expecting low yields, a total of 75 g (wet-weight) human skin from autopsy was processed which yielded ≈ 12 g (dry-weight) of insoluble collagen after extraction. After enzymatic digestion and purification procedures, ~ 440 μg of pure LW1 was obtained. The baseline-resolved UV and fluorescence spectra demonstrated that LW1 had a UV maximum of 348 nm (Fig. 1) and fluorescence maxima at ex/em 348/465 nm with a broad emission spectrum maximum between 440 to 470 nm [20]. These spectra were not pH-dependent; i.e., no bathochromic shift. The pure LW1 underwent many different instrumental analyses using the most sensitive state-of-the art NMR and mass spectral (MS) instruments which we had access to since only minute amounts of LW1 could be obtained. A baseline-resolved 800 MHz proton (1H) NMR spectrum was obtained as well as an ESI (Electro-Spray Ionization) full-scan MS and MS/MS.

The 1H-NMR spectrum showed that LW1 had three signals *a, b, c* in the far downfield (aromatic) region with chemical shifts at 8.6 (singlet), 8.02 (doublet) and 6.95 (doublet) with integral ration of 1:1:1 (Fig. 3A). These protons were bound to carbon atoms with chemical shifts at 110, 138, and 96 ppm, respectively, as determined by HSQC experiments (Fig. 3C&D) [20]. A TOCSY experiment showed strong coupling between protons *b* & *c* due to vicinal coupling (1NMR, J 3.2, Fig. 3A) and weak cross-signals between protons *a* & *c* as a result of long-range coupling (Fig. 3A). However, none of these protons correlated with protons in the aliphatic region (Fig. 3B).

The most prominent signals in the aliphatic region (Fig. 3B) were a highly visible singlet labeled *d* at 2.69 ~3 protons; a multiplet complex between 3.6 to 3.8 ppm labeled as box *e* in Fig. 3B ~ px ≈10 protons; and a triplet *f* (J = 7 Hz) at 4.43 ppm ~2 protons. Protons *d* & *f* were bound to carbon atoms at 2.69 and 4.43 ppm, respectively (HSQC, Fig. 3D). Protons *d* belonged to a methyl group attached to an aromatic ring while protons *f* were attributed to the ε-CH2 group of a lysine incorporated into an aromatic ring [20]. Of particular interest were protons of box e (Fig. 3B) which represent the overlap region for proton signals of amino acids and sugars. Proton e' at 3.7 ppm (Fig 3B) was bound to a carbon atom with a chemical shift of 55 ppm (HSQC) and corresponded to the α-proton of an amino acid. In the HSQC (Fig. 3D), five carbon atoms were identified with chemical shifts at 70, 72, 73, 75, 77 ppm corresponding to proton signals ~ 3.63 to 3.75, which were characteristic for the presence of a sugar moiety.

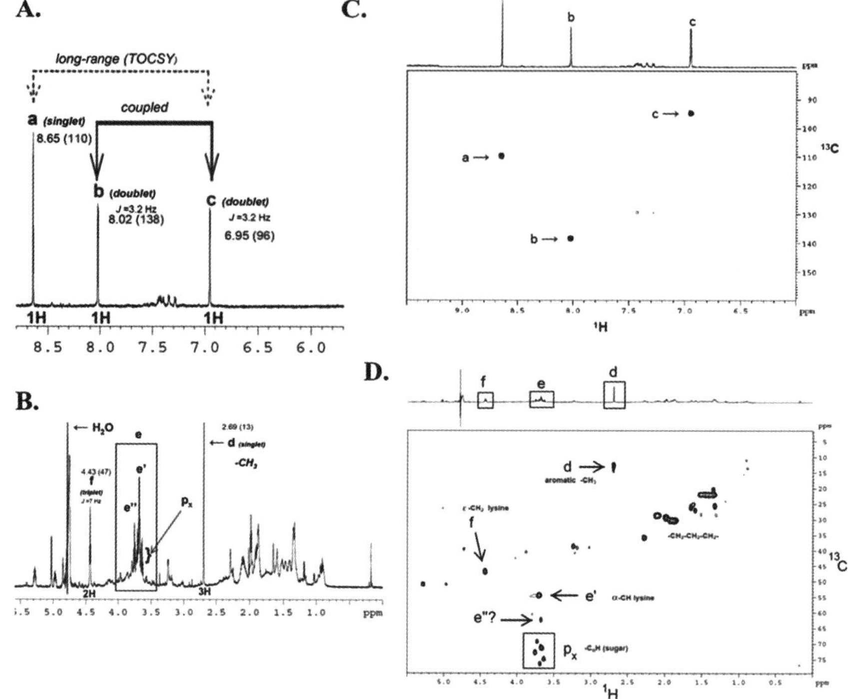

Figure 3   *NMR spectra of purified LW1 (see Methods). (A) 1H-NMR aromatic region; the chemical shifts for (1H, 13C) are indicated for signals a,b,c; chemical coupling is indicated as determined by a TOCSY experiment; (B) 1H-NMR for aliphatic region; (C) 1H-13C HSQC for aromatic region; (D) 1H-13C HSQC for aliphatic region.*

Full-scan MS of LW1 determined using a triple quadrupole MS (Micromass Quattro Ultima) instrument showed that LW1 has a molecular (parent) mass of 623 Da (Fig 4). The MS/MS splitting of the m/z 623 parent ion revealed m/z 447 as the major fragment followed by m/z 318 (Fig. 4). Other fragments were m/z 402 > 578 > 130 > 273 > 430 > 494 [20]. Further studies using high resolution MS (Micromass Q-Tof Ultima API) determined that LW1 had an exact mass of 623.2746 ± 0.0017 Da. The predicted elemental composition of LW1 based on C,H, N, O content suggests one of four empirical formulas: $C_{25}H_{43}N_3O_{14}$ (-3.0 mDa), $C_{20}H_{43}N_6O_{16}$ (+1.0 mDa), $C_{26}H_{39}N_8O_{10}$ (-4.3 mDa), and $C_{21}H_{39}N_{10}O_{12}$ (-0.3 mDa) as the only possible candidates.

Levels of LW1 were quantified by LC/MS/MS in insoluble collagen samples prepared from human donor skin samples obtained by autopsy. The results showed that total LW fluorescence strongly correlated with LW1 MS signals at m/z 623 → m/z 447 (r=0.94, n=14, P<0.0001). LW1 expressed as nmol/mg collagen significantly (P<0.0001) increased with age at an exponential rate in nondiabetic individuals (r=0.92, n=13, P<0.0001). Levels were also significantly elevated by diabetes (P<0.0001), chronic/ESRD (P=0.021), and in non-diabetic patients with severe respiratory problems (P<0.05)[20].

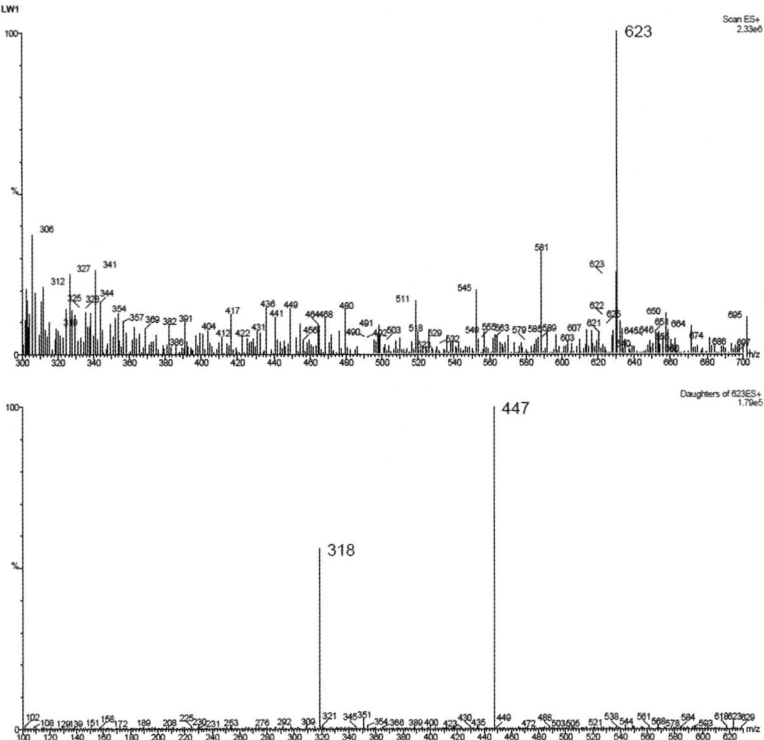

Figure 4   *Mass Spectrum (Full MS - upper) and MS/MS (lower) of purified LW1 determined by a Waters Micromass Quattro Ultima Triple Quadrupole LC/MS/MS*

4   DISCUSSION

In this investigation, studies were conducted into the structural elucidation of a long-wavelength fluorescent molecule (LW1), identified to be the single major fluorescent peak in HPLC chromatograms from enzymatic digest of human skin collagen, particularly from patients with diabetes and ESRD. Unlike pentosidine which fluoresces at lower wavelengths; i.e., ex/em 335/385 nm [21], LW1 is not stable to acid hydrolysis, thereby making its structural determination more difficult and considerably more challenging. Sufficient quantities were isolated and purified for full analysis by 1H-NMR, 1H-13C HSQC, and 1H-1H TOSCY as well as mass spectrometric analysis. In addition, levels were quantified by LC/MS/MS in these samples. However, because LW1 was a large molecule that is present in low yield, it was not possible to obtain a critical 13C NMR spectrum necessary for determination of its complete structure, thus only its partial structure is currently known.

From the results of the 1H-NMR (Fig. 3A&B), we initially thought its structure consisted of a stand-alone pyridinium ring whereas protons *a,b,c,* together with protons *d* (methyl group) were all attached to the same ring. However, the results of the HSQC and TOCSY experiments, in particular, the 1H, 13C chemical shifts for signals *a,b,c* (Fig. 3A&C) were indicative of a heteroaromatic five and six-member ring structure whereas protons *b* & *c* were part of the five-member ring while protons *a* & *d* (methyl group) were part of the six-member ring.

Of particular interest is that some of the NMR signals noted here for LW1, including J coupling constants, shared many similarities in chemical shifts with those previously reported for vesperlysine B (26) as follows: (1H, 13C ppm, J coupling) (1) signal *d* (methyl group) bound to aromatic ring: LW1 (2.69, 13) *vs.* vesperlysine B (2.8, 15.4); (2) LW1 signals *a* (8.65, 110), *b* (8.02, 138, J 3.2 Hz), *c* (6.95, 96, J 3.2 Hz) *vs.* analogous signals for vesperlysine B (7.97, 115), (7.85, 138, J 3 Hz), (6.86, 98, J 3 Hz), respectively; ([3]) triplet proton signal *f* (Fig. 2B), LW1 (4.43, 47, J 7 Hz) *vs.* vesperlysine B (4.32, 48.9, J 7 Hz). Additionally, for the latter signal *f*, an equivalent signal with similar 1H, 13C chemical shifts have been noted for pentosidine as well as many other proteinaceous cross-links[20] and has been attributed to the ε-CH2 of a lysine residue incorporated into an aromatic ring through its ε-amino group. However, although similarity with vesperlysine B suggests ε-amino group of LW1 lysine as part of a pyridinium ring, there is currently no unequivocal data which proves lysine as a part of a pyridinium (six-member) or pyranium (five-member) aromatic ring. Furthermore, vesperlysine B shows a bathochromic shift [26] while LW1 does not. Thus, LW1 has no hydroxyl group attached to the aromatic ring unlike vesperlysine B [26] and observed for many other glycation products.

Evidence for the presence of lysine in the structure of LW1 include the following: (1) amino acid analysis of purified LW1 after acid hydrolysis quantified lysine (data not shown); (2) chemical shifts (1H, 13C) for signal e' in the HSQC spectrum (Fig. 3D) for LW1 (3.7, 55) is in agreement with those published for the α-CH of lysine (4.5, 56) (27); (3) similarly, and as discussed, signal f in the HSQC is also compatible with lysine; (4) fragmentation of LW1, in particular m/z 623 → m/z 578 fragment, indicates the loss of norleucine ≈ loss of a deaminated lysine residue explained by the signal at m/z 130 in the MS/MS spectrum[20, 28], also observed in the fragmentation MS of pentosidine [21]) and vesperlysine A [29].

It has been suggested that minor signals between 7.25 to 7.47 ppm in the 1H-NMR spectrum (Fig. 3A) is indicative of the presence of a peptide. However, we feel that these signals represent a contaminant since they also ubiquitously appear in the 1H NMR spectra of collected fractions adjacent to LW1 during its purification. Additionally, the HSQC (Fig. 3D) of LW1 showed weak (1H ,13C) correlated cross-signals at ≈ (7.43, 130 ppm) characteristic of phenylalanine or possibly tyrosine contaminant based upon published NMR spectra for these amino acids in peptides [30-33].

Lastly, the HSQC (Fig. 3D) showed a group of (1H ,13C) signals at (3.6 – 3.85, 70-75 ppm) characteristic for chemical shifts of sugars. Signals with similar shifts were noted in the high resolution NMR (HMQC) spectroscopic metabolic investigations of rat liver by Bollard et al.[34] and were assigned as CnH groups of glucose. In the present study, it was also suggested that m/z 623 → m/z 447 fragment split (Fig. 3B) ~ loss of a 176 fragment could possibly represent the sugar portion of the LW1 molecule. However, this needs further investigating.

Since our findings suggests that skin collagen LW is associated with the risk in progression of atherosclerosis determined by coronary calcium deposition [4], we feel that elucidating the structure of LW1 and conducting studies on its clinical and biological significance is of critical importance. However, because LW1 could not be synthesized from a combination of numerous amino acids and carbonyl compounds, as we previously did for pentosidine [21], it needs to be prepared from scratch and chemically broken down into subspecies for individual analysis.

Acknowledgements

This research was supported by NIH Grant NIDDK R21 DK-79432 to D.R. Sell. We thank Dr. A. Daniel Jones of the Mass Spectrometry Research Technology Support Facility (Michigan State University, East Lansing, MI) for performing the high resolution mass analysis of LW1. We also thank the Tissue Procurement Facility at University Hospitals Case Medical Center (Cleveland, OH) and the National Disease Research Interchange (Philadelphia, PA) for providing tissue samples.

References

1. Monnier VM, Cerami A. *Science* 1981, **211**, 491-493.
2. Monnier VM, Kohn RR, Cerami A. *Proc. Natl. Acad. Sci. U.S.A.* 1984, **81**, 583-587.
3. Monnier VM, Bautista O, Cleary PA, Sell DR, Genuth S, Group D-ER. *Diabetes* 2004, **53**, supplement 2, A77-A78.
4. Monnier VM, Bautista O, Cleary P, Sell DR, Genuth S, Group TDER. *Ann. N.Y. Acad. Sci.* 2005, **1043**, 917.
5. Meerwaldt R, Hartog JW, Graaff R, Huisman RJ, Links TP, den Hollander NC, et al. *J. Am. Soc. Nephrol.* 2005, **16**, 3687-3693.
6. Meerwaldt R, Lutgers HL, Links TP, Graaff R, Baynes JW, Gans RO, et al. *Diabetes Care* 2007, **30**, 107-112.
7. Lutgers HL, Gerrits EG, Graaff R, Links TP, Sluiter WJ, Gans RO, et al. *Diabetologia* 2009, **52**, 789-797.
8. Maynard JD, Rohrscheib M, Way JF, Nguyen CM, Ediger MN. *Diabetes Care* 2007, **30**, 1120-1124.
9. Golej J, Hoeger H, Radner W, Unfried G, Lubec G. *Life Sci.* 1998, **63**, 801-807.
10. Suarez G, Maturana J, Oronsky AL, Raventos-Suarez C. *Biochim. Biophys. Acta* 1991, **1075**, 12-19.
11. Bailey AJ, Sims TJ, Avery NC, Halligan EP. *Biochem. J.* 1995, **305(2)**, 385-390.
12. Hou FF, Boyce J, Chertow GM, Kay J, Owen WF, Jr. J. Am. Soc. Nephrol.. 1998, **9**, 277-283.
13. Sajithlal GB, Chithra P, Chandrakasan G. *Mol. Cell Biochem.* 1999, **194**, 257-263.
14. Luo ZJ, King RH, Lewin J, Thomas PK. J. Neurol. 2002, **249**, 424-431.
15. Sajithlal GB, Chithra P, Chandrakasan G. *Biochim. Biophys. Acta* 1998, **1407(3)**, 215-224.
16. Stefek M, Gajdosik A, Gajdosikova A, Krizanova L. Biochim. Biophys. Acta 2000, **1502**, 398-404.
17. Meng J, Sakata N, Takebayashi S, Asano T, Futata T, Nagai R, et al. *Atherosclerosis* 1998, 136(2), 355-365.
18. Sady C, Khosrof S, Nagaraj R. *Biochem. Biophys. Res. Commun.* 1995, **214**, 793-797. 19. Baynes JW. *Diabetes* 1991, 40, 405-412.
19. Sell DR, Nemet I, Monnier VM. *Arch. Biochem. Biophys.* 2010, **493**, 192-206.
20. Sell DR, Monnier VM. *J. Biol. Chem.* 1989, **264(36)**, 21597-21602.
21. Shipanova IN, Glomb MA, Nagaraj RH. *Arch. Biochem. Biophys.* 1997, **344**, 29-36.
22. Zhang W-H, Liu J, Xu G, Yuan Q, Sayre LW.. Chem. Res. Toxicol. 2003, **16**, 512-523.

23. Cheng R, Feng Q, Argirov OK, Ortwerth BJ. *J.Biol. Chem.* 2004, **279**, 45441-45449.

24. Argirov OK, Lin B, Olesen P, Ortwerth BJ. *Biochim. Biophys. Acta* 2003, **1620**, 235-244.

25. Nakamura K, Nakazawa Y, Ienaga K. *Biochem. Biophys. Res. Commun.* 1997, **232**, 227-230.

26. Pretsch E, Buhlmann P, Affolter C. New York, Springer-Verlag, 2000.

27. Argirov OK, Leigh ND, Ortwerth BJ. *Ann. N. Y. Acad. Sci.* 2005, **1043**, 903.

28. Tessier F, Obrenovich M, Monnier VM. *J. Biol. Chem.* 1999, **274(30)**, 20796-20804.

29. Barden. JA, Kemp BE. *Biochemistry* 1987, **26**, 1471-1478.

30. Roy M, Lee RW-K, Kaarsholm NC, Thøgersen H, Brange J, Dunn MF. *Biochim. Biophys. Acta* 1990, **1053**, 63-73.

31. de Beer T, van Zuylen CWE, Hård K, Boelens R, Kaptein R, Kamerling JP, et al. *FEBS Lett.* 1994, **348**, 1-6.

32. Breslow E, Sardana V, Deeb R, Barbar E, Peyton DH. *Biochemistry* 1995, **34**, 2137-2147.

33. Bollard ME, Garrod S, Holmes E, Lindon JC, Humpfer E, Spraul M, et al. *Magn. Reson. Med.* 2000, **44**, 201-207.

ADVANCED GLYCATION ENDPRODUCTS: BIOMARKERS FOR AGE-RELATED MACULAR DEGENERATION

J. Ni[1,2], R.H. Nagaraj[3] and J.W. Crabb[1,2,4,5]

[1]Cole Eye Institute and Lerner Research Institute, Cleveland Clinic Foundation;
[2]Department of Chemistry, Cleveland State University;
[3]Deptartment of Ophthalmology and Visual Sciences, Case Western Reserve University, Cleveland, OH
[4]Department of Chemistry, Case Western Reserve University; and
[5]Departments of Ophthalmology and Molecular Medicine, Cleveland Clinic Lerner College of Medicine of Case Western Reserve University, Cleveland, OH 44106

1 INTRODUCTION

Age-related macular degeneration (AMD) is a complex, progressive disease involving multiple genetic and environmental factors and a major cause of severe visual loss in the elderly worldwide [1,2]. Deposition of debris (drusen) along Bruch's membrane in the macula (Figure 1) is the first evidence of early AMD. Advanced AMD occurs in two forms, geographic atrophy and choroidal neovascularization. Geographic atrophy (advanced dry AMD) develops slowly and results in blindness when focal areas of the retinal pigment epithelium (RPE) degenerate in the macula. Choroidal neovascularization (wet AMD) is characterized by the growth of new blood vessels from the choroid through Bruch's membrane and the RPE. When these vessels hemorrhage, a blood clot accumulates between the RPE and the macular photoreceptors, causing immediate central vision loss. Wet AMD accounts for over 80% of debilitating visual loss in AMD yet only 10-15% of AMD cases progress to neovascular AMD.

There is growing consensus that AMD is an inflammatory disease involving dysregulation of the complement system based in part upon the identification of AMD susceptibility genes encoding complement factors and the presence of complement proteins in drusen [1,2]. However the molecular inducers of inflammation in AMD remain largely unknown. Oxidative stress appears to play an important role as smoking significantly increases the risk of AMD [3] and antioxidant vitamins can selectively slow AMD progression [4]. Oxidative protein modifications have been proposed as catalysts of AMD pathology [5] and a host of oxidative modifications have now been detected at elevated levels in AMD ocular tissues and plasma [5-12]. Emphasizing the complexity of the disease, AMD susceptibility genes have been implicated in over 50% of AMD cases [13], but many individuals carrying AMD risk genotypes may never develop the disease [13] and only a fraction of those diagnosed with early AMD progress to advanced disease [4].   Accordingly, strategies are needed to facilitate early intervention that might halt or delay disease progression and prevent significant visual loss. This chapter reviews current evidence supporting two plasma protein advanced glycation endproducts (AGEs) as proteomic biomarkers of AMD, namely carboxymethyllysine (CML) and pentosidine [12].

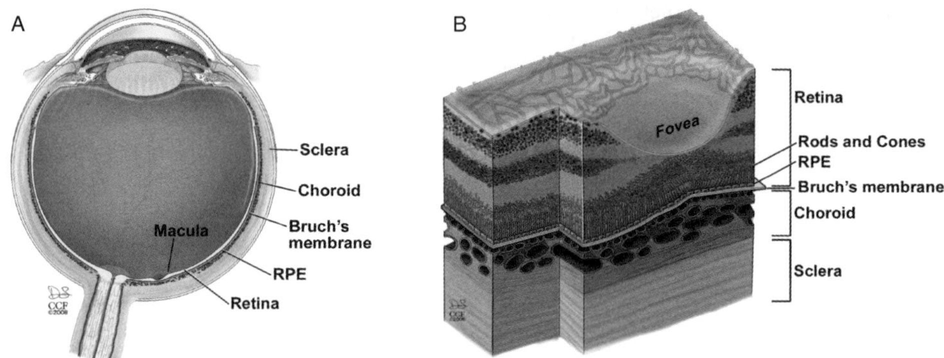

Figure 1. *The macular interface. (A) The macular region of the retina, centered around the fovea, is responsible for high acuity, central vision. (B) Cross-section through the macular region. Bruch's membrane is a permeable extra cellular matrix supporting the RPE, a key single cell layer responsible for many essential processes, including vectorial transport of nutrients and waste products between the blood bearing choroid and the rod and cone photoreceptors. Age-related changes at this critical interface disrupt normal retinal physiology and contribute to AMD pathology. (Reproduced with copyright permission from the Cleveland Clinic Foundation.)*

2   OXIDATIVE MODIFICATIONS IN AMD PATHOLOGY

The finding that carboxyethylpyrrole (CEP) protein adducts are elevated in AMD Bruch's membrane and drusen lead to the hypothesis that oxidative protein modifications may be catalysts or initiators of AMD pathology [5]. CEP modifications are generated by covalent adduction of primary amino groups by a reactive fragment derived uniquely from oxidative cleavage of docosahexaenoate (DHA)-containing phospholipids [10, 14]. CEP adducts and CEP autoantibodies are both elevated in AMD plasma [10, 11] and CEP adducts stimulate neovascularization *in vivo,* suggesting a role in the induction of choroidal neovascularization [15]. Furthermore, CEP immunization has been shown to induce an AMD-like phenotype in mice [16]. Other direct evidence of oxidative damage in AMD tissues in addition to AGEs and

CEP include 4-hydroxynonenal, acrolein, nitrotyrosine and the DNA modification 8-hydroxy-2'-deoxyguanosine in advanced dry AMD retina [9]; and nitrotyrosine in AMD plasma [17].

3   AGES IN OCULAR TISSUES

AMD has long been hypothesized to be a systemic disease [18], based first on the presence of retinal drusen in patients with membranoproliferative glomerulonephritis type II [19] and more recently by systemic complement activation in AMD [20]. Support for this hypothesis also comes from mounting evidence that AGEs play a role in AMD [6-8, 21, 22]. AGEs are a heterogeneous group of mostly oxidative modifications resulting from the Maillard reaction and have been associated with several age-related diseases and diabetic complications [23, 24]. CML, was the first AGE to be found in AMD Bruch's membrane and drusen [6]. Subsequently pentosidine, a fluorescent AGE, was detected in drusen and both CML and pentosidine have been shown to increase with age in Bruch's membrane [21, 25]. Several other reports have documented the presence of CML, pentosidine and fluorescent AGEs in Bruch's membrane, drusen, RPE, and choroidal extracellular matrix from healthy eyes [5, 7, 8, 26, 27]. RPE grown *in vitro* on AGE-modified basement membrane results in increased RPE lipofuscin accumulation[27]; lipofuscin also accumulates in AMD with potential adverse consequences [28]. CML has been shown to stimulate neovascularization *in vivo,* suggesting it may also contribute to choroidal neovascularization [29].

4   RECEPTORS FOR AGES IN OCULAR TISSUES

Receptors for AGE are pattern-recognition receptors like the complement system which plays a role in AMD pathology [1, 2]. Receptors for AGEs bind multiple ligands and RAGE and AGE-R1 have been found elevated on RPE and photoreceptor cells in early and advanced dry AMD [8], especially in RPE overlying drusen-like deposits on Bruch's membrane [22]. RAGE appears to mediate AGE induced damage of RPE cells *in vitro* [8] and of neurons *in vivo* [30]. Using quantitative proteomics, we have recently found another receptor for AGEs, namely AGE-R3 (also known as galectin-3) to be significantly elevated in the macular region of advanced dry AMD Bruch's membrane/choroid [31]. These findings suggest a role for AGEs and receptors for AGEs in the mechanisms of progression to advanced dry AMD.

5   PENTOSIDINE AND CML AS BIOMARKERS OF AMD

We quantified plasma protein-bound pentosidine and CML by liquid chromatography-fluorimetry and liquid chromatography-tandem mass spectrometry, respectively, in a study population of 90 patients [12]. Both control subjects (n = 32) and AMD patients (n = 58) were evaluated, including 27-early/mid stage AMD and 31 advanced AMD patients. Plasma protein and fructosyl-lysine, a marker of early glycation, were quantified by amino acid analysis. We found AMD patients exhibited ~54% higher CML and ~64% higher pentosidine concentrations relative to control plasma (p < 0.0001 for both AGEs) (Table 1). Amounts of plasma protein fructosyl-lysine were the same in control and non-diabetic AMD donors, supporting an association between AMD and increased levels of CML and pentosidine, independent of other diseases like diabetes. Both AGEs were elevated in early/mid stage AMD and over a broad age range in those with AMD. Comparison of CML between early/mid stage AMD and advanced AMD donors revealed no significant difference in concentrations, however the average pentosidine level was ~17% higher in advanced AMD (p ~0.06) (Table 1). While more extensive studies are required for confirmation, plasma protein pentosidine may be useful as a predictor of susceptibility to AMD progression. Autoantibody titers for CML and pentosidine were detectable but were not significantly different between AMD and control donors. Correlation of CML and pentosidine concentrations showed both markers to be elevated above median control levels in 84% of AMD patients compared with only 28% of controls [12]. The areas under the receiver operating characteristic (ROC) curves (c-statistics) suggest that CML can discriminate between AMD and control plasma donors with ~78% accuracy and that pentosidine can discriminate with ~88% accuracy. Odds ratios (OR) for all AMD risk based on plasma protein concentrations of CML (OR 6.3) and pentosidine (OR 10.6) were significantly elevated, however additional studies in a larger study population are required to refine the precision of the associated confidence intervals.

Table 1. *CML and Pentosidine Markers in Control and AMD Plasma*

	n	CML (pmol/mg protein)		CML Elevated Above Control Median		
		Mean ± SD	Median (Q1, Q3)	Odds Ratio	95% CI	*P* value
Control	32	27.9 ± 10.0	26.7 (19.7, 34.7)	1	(Reference)	
Early/Mid Stage AMD	27	41.0 ± 16.1	39.6 (28.4, 57.3)	2.9	1.0, 8.6	*0.068*
Advanced AMD	31	44.8 ± 14.0	43.3 (35.5, 48.6)	30.0	3.6, 247.3	*< 0.0001*
All AMD	58	43.0 ± 15.0	42.0 (34.7, 51.0)	6.3	2.3, 17.3	*< 0.0001*
	n	Pentosidine (pmol/mg protein)		Pentosidine Elevated Above Control Median		
		Mean ± SD	Median (Q1, Q3)	Odds Ratio	95% CI	*P* value
Control	32	1.00 ± 0.25	0.93 (0.84, 1.02)	1	(Reference)	
Early/Mid Stage AMD	27	1.51 ± 0.41	1.48 (1.22, 1.82)	8.0	2.0, 32.0	*0.002*
Advanced AMD	31	1.76 ± 0.59	1.71 (1.34, 1.99)	14.5	3.0, 71.2	*< 0.0001*
All AMD	58	1.64 ± 0.53	1.59 (1.29, 1.92)	10.6	3.4, 33.5	*< 0.0001*

CML and pentosidine concentrations were determined by LC-MS and LC-fluorimetry; protein was quantified by amino acid analysis. Determined mean concentrations of CML expressed in fmol CML / nmol Lys were: control, 33.8 ± 11.9; early/mid-stage AMD, 49.8 ± 19.8; advanced AMD, 54.6 ± 17.4; and all AMD 52.3 ± 18.5. Determined mean concentrations of pentosidine expressed in fmol pentosidine / nmol Lys were: control, 1.2 ± 0.3; early/mid-stage AMD, 1.8 ± 0.5; advanced AMD, 2.2 ± 0.8; and all AMD 2.0 ± 0.7. The odds ratio (OR) reflects the AMD risk for donors exhibiting elevated levels of either CML or pentosidine markers relative to median control levels. P values were determined using the Fischer Exact Test. Odds ratios, 95% CI and p-values are based on log-transformed CML and pentosidine concentrations. (Reproduced with copyright permission from The American Society for Biochemistry and Molecular Biology.)

The AMD discriminatory accuracy of CEP, CML and pentosidine plasma protein concentrations was compared in a study population of 86. The results showed CEP adducts and CML to be elevated in ~69% of the AMD cohort (n = 54) and CEP adducts and pentosidine to be elevated in ~74%.[12] CEP autoantibody titer was not informative, due in part to the variability associated with the autoantibody assay and the relatively small sample size. The determined c-statistics suggests that CML and CEP adducts alone discriminate between AMD and control patients with about equal accuracy (~78% to ~79%). CML in combination with pentosidine provided ~89% accuracy and combining CEP with pentosidine provided ~ 92% discriminatory accuracy (Table 2). These preliminary results suggest the use of all three biomarkers may provide a useful prognostic test for AMD.

Table 2   *Sensitivity and specificity of CML, Pentosidine and CEP*

Markers Alone	CEP adducts	CML	Pentosidine
Sensitivity (%)	67	84	84
Specificity (%)	81	72	88
C-statistic	0.78	0.79	0.88
95% CI	0.68, 0.88	0.70, 0.89	0.81, 0.95
Joint Effect of Markers	**CML + Pentosidine**	**CEP + CML**	**CEP + Pentosidine**
Sensitivity (%)	83	69	89
Specificity (%)	97	91	91
C-statistic	0.89	0.87	0.92
95% CI	0.81, 0.96	0.80, 0.94	0.86, 0.99
P value	0.06 (vs CEP)	0.10 (vs CML) 0.13 (vs CEP)	0.20 (vs Pentosidine) 0.02 (vs CEP)

Sensitivity and specificity were determined from ROC curves to maximize the sum of the two values and constructed from the output of logistic regression analysis fit with either CEP adduct, CML, or pentosidine concentrations alone, or in combination. The concentration of CML was expressed in nmol/ml, and pentosidine and CEP adducts were expressed in pmol/ml. C-statistics, 95% CI and p values derived from single and joint markers were determined with SAS 9.1 based on log-transformed maker concentrations. Verification of c-statistics and 95% CI was performed by bootstrap resampling and 10-fold cross-validation. Pentosidine plus CEP appears to be a better discriminator than CEP alone (p ~0.02) and CML plus pentosidine may be a better discriminator than CEP alone (p~ 0.06).

6   POSSIBLE CONFOUNDING FACTORS

While no confounding influences were detected regarding CML, plasma protein pentosidine was higher in AMD donors with hypertension or cardiovascular disease, as also found for CEP adducts.[11] Plasma protein fructosyl-lysine provided an effective parameter to narrow the cause of increased AGEs. Except for 5 diabetic AMD donors, the AMD and control plasma donors in our study exhibited essentially the same fructosyl-lysine concentrations. When fructosyl-lysine, CML and pentosidine are all elevated, more information will be required for an accurate clinical assessment. In such cases, we suggest that CEP adduct levels may help rule out diabetic complications since plasma CEP adducts are elevated in AMD but not in diabetes [11].

7   CONCLUSIONS

Efforts are now focused on evaluating plasma protein CML and pentosidine in a larger study population and in combination with AMD risk genotypes. CML, pentosidine and CEP together in combination with genomic markers may offer the optimum approach to predicting AMD susceptibility among potential confounding factors. This hypothesis is based on previous studies showing combined CEP proteomic and genomic measurements are more effective in predicting AMD risk than either method alone [11]. Additional larger multicenter prospective and longitudinal studies of plasma protein CEP, CML, pentosidine as AMD biomarkers are now justifiable. We anticipate that plasma protein biomarkers will eventually be used to facilitate early interventions to halt or delay AMD progression and for monitoring the efficacy of AMD therapeutic measures.

Acknowledgements

The contributions of all collaborators, including The Clinical Genomic and Proteomic AMD Study Group, and participating patients are gratefully acknowledged. This work was supported in part by US National Institute of Health grants EY015638, EY014239, EY09912, BRTT 05-29 from the State of Ohio, a Foundation Fighting Blindness Center Grant to the Cole Eye Institute, a Research to Prevent Blindness (RPB) Challenge Grant to the Cole Eye Institute, a RPB Senior Investigator Award to JWC, a Steinbach Award to JWC, and the Cleveland Clinic Foundation. JWC was a consultant for Alcon Research Ltd and Allergan, Inc. during the course of this research and holds a patent for CEP biomarker technology licensed to SKS Ocular LLC.

References

1.   R. D. Jager, W. F. Mieler and J. W. Miller, *N Engl J Med*, 2008, **358**, 2606-2617.
2.   X. Ding, M. Patel and C. C. Chan, *Prog Retin Eye Res*, 2009, **28**, 1-18.
3.   J. M. Seddon, W. C. Willett, F. E. Speizer and S. E. Hankinson, *Jama*, 1996, **276**, 1141-1146.
4.   A.-R. E. D. S. G. (AREDs), *Arch Ophthalmol*, 2001, **119**, 1417-1436.
5.   J. W. Crabb, M. Miyagi, X. Gu, K. Shadrach, K. A. West, H. Sakaguchi, M. Kamei, A. Hasan, L. Yan, M. E. Rayborn, R. G. Salomon and J. G. Hollyfield, *Proc Natl Acad Sci U S A*, 2002, **99**, 14682-14687.
6.   T. Ishibashi, T. Murata, M. Hangai, R. Nagai, S. Horiuchi, P. F. Lopez, D. R. Hinton and S. J. Ryan, *Arch Ophthalmol*, 1998, **116**, 1629-1632.
7.   H. P. Hammes, H. Hoerauf, A. Alt, E. Schleicher, J. T. Clausen, R. G. Bretzel and H. Laqua, *Invest Ophthalmol Vis Sci*, 1999, **40**, 1855-1859.
8.   K. A. Howes, Y. Liu, J. L. Dunaief, A. Milam, J. M. Frederick, A. Marks and W. Baehr, *Invest Ophthalmol Vis Sci*, 2004, **45**, 3713-3720.
9.   J. K. Shen, A. Dong, S. F. Hackett, W. R. Bell, W. R. Green and P. A. Campochiaro, *Histol Histopathol*, 2007, **22**, 1301-1308.
10.   X. Gu, S. G. Meer, M. Miyagi, M. E. Rayborn, J. G. Hollyfield, J. W. Crabb and R. G. Salomon, *J Biol Chem*, 2003, **278**, 42027-42035.
11.   J. Gu, G. J. Paeur, X. Yue, U. Narendra, G. M. Sturgill, J. Bena, X. Gu, N. S. Peachey, R. G. Salomon, S. A. Hagstrom, J. W. Crabb and T. C. G. P. S. Group., *Mol Cell Proteomics*, 2009, **8**, 1338-1349.
12.   J. Ni, X. Yuan, J. Gu, X. Yue, X. Gu, R. H. Nagaraj, J. W. Crabb and T. C. G. a. P. A. S. Group., *Mol Cell Proteomics*, 2009, **8** , 1921-1933
13.   L. G. Fritsche, T. Loenhardt, A. Janssen, S. A. Fisher, A. Rivera, C. N. Keilhauer and B. H. Weber, *Nat Genet*, 2008, **40**, 892-896.
14.   X. Gu, M. Sun, B. Gugiu, S. Hazen, J. W. Crabb and R. G. Salomon, *J Org Chem*, 2003, **68**, 3749-3761.

15. Q. Ebrahem, K. Renganathan, J. Sears, A. Vasanji, X. Gu, L. Lu, R. G. Salomon, J. W. Crabb and B. Anand-Apte, *Proc Natl Acad Sci U S A*, 2006, **103**, 13480-13484.

16. J. G. Hollyfield, V. L. Bonilha, M. E. Rayborn, X. Yang, K. G. Shadrach, L. Lu, R. L. Ufret, R. G. Salomon and V. L. Perez, *Nat Med*, 2008, **14**, 194-198.

17. J. Gu, X. Zhan, J. S. Crabb, E. Bala, K. Renganathan, S. A. Hagstrom, H. Lewis, R. G. Salomon and J. W. Crabb, *Invest Ophthalmol Vis Sci 48, E-abstract 34*, 2007.

18. G. S. Hageman, P. J. Luthert, N. H. Victor Chong, L. V. Johnson, D. H. Anderson and R. F. Mullins, *Prog Retin Eye Res*, 2001, **20**, 705-732.

19. S. J. Huang, D. L. Costa, N. E. Gross and L. A. Yannuzzi, *Retina*, 2003, **23**, 429-431.

20. H. P. Scholl, P. Charbel Issa, M. Walier, S. Janzer, B. Pollok-Kopp, F. Borncke, L. G. Fritsche, N. V. Chong, R. Fimmers, T. Wienker, F. G. Holz, B. H. Weber and M. Oppermann, *PLoS ONE*, 2008, **3**, e2593.

21. J. T. Handa, N. Verzijl, H. Matsunaga, A. Aotaki-Keen, G. A. Lutty, J. M. te Koppele, T. Miyata and L. M. Hjelmeland, *Invest Ophthalmol Vis Sci*, 1999, **40**, 775-779.

22. Y. Yamada, K. Ishibashi, I. A. Bhutto, J. Tian, G. A. Lutty and J. T. Handa, *Exp Eye Res*, 2006, **82**, 840-848.

23. J. W. Baynes, *Exp Gerontol*, 2001, **36**, 1527-1537.

24. S. Y. Goh and M. E. Cooper, *J Clin Endocrinol Metab*, 2008, **93**, 1143-1152.

25. J. V. Glenn, J. R. Beattie, L. Barrett, N. Frizzell, S. R. Thorpe, M. E. Boulton, J. J. McGarvey and A. W. Stitt, *FASEB J*, 2007, **21**, 3542-3552.

26. B. Farboud, A. Aotaki-Keen, T. Miyata, L. M. Hjelmeland and J. T. Handa, *Mol Vis*, 1999, **5**, 11.

27. J. V. Glenn, H. Mahaffy, K. Wu, G. Smith, R. Nagai, D. A. Simpson, M. E. Boulton and A. W. Stitt, *Invest Ophthalmol Vis Sci*, 2009, **50**, 441-451.

28. K. P. Ng, B. Gugiu, K. Renganathan, M. W. Davies, X. Gu, J. S. Crabb, S. R. Kim, M. B. Rozanowska, V. L. Bonilha, M. E. Rayborn, R. G. Salomon, J. R. Sparrow, M. E. Boulton, J. G. Hollyfield and J. W. Crabb, *Mol Cell Proteomics*, 2008, **7**, 1397-1405.

29. T. Okamoto, S. Tanaka, A. C. Stan, T. Koike, M. Kase, Z. Makita, H. Sawa and K. Nagashima, *Microvasc Res*, 2002, **63**, 186-195.

30. B. G. Hassid, M. N. Nair, A. F. Ducruet, M. L. Otten, R. J. Komotar, D. J. Pinsky, A. M. Schmidt, S. F. Yan and E. S. Connolly, *J Clin Neurosci*, 2009, **16**, 302-306.

31. X. Gu, X. Yuan, J. S. Crabb, K. Shadrach, J. G. Hollyfield and J. W. Crabb, *Invest Ophthalmol Vis Sci*, 2009, **50**, E-abstract 2343.

METAL CATALYZED LENS CRYSTALLIN OXIDATION DURING AGING AND IN DIABETES : THE ROLE OF GLUTATHIONE

X. Fan[1], J. Zhang[1,3], I. Nemet[1], V.M. Monnier[1,2]

[1]Departments of Pathology, Case Western Reserve University, Cleveland, USA
[2]Biochemistry Case Western Reserve University, Cleveland, USA
[3]Chemistry, Case Western Reserve University, Cleveland, USA

1   INTRODUCTION

Considerable evidence has accumulated over the years in support of the presence of catalytic metals in the lens [1]. Our own interest in this field was triggered by the discovery that the Maillard reaction protein residue carboxymethyl-lysine (CML) has structural homology with EDTA, and therefore should bind redox active metals, such as Cu^{2+} and Fe^{3+} as well divalent metals, such as Zn^{2+} and Ca^{2+} [2]. Evidence for redox active metals that were associated with CML was then found in human lens, in diabetic rats and in endstage renal disease[2,3]. Similar studies and results were done and obtained by Stadtman and others[4,5], and evidence for hydroxyl radical mediated oxidation of human lens crystallins was reported by Fu et al. [6]. The current study builds on the pioneering work by Stadtman and colleagues who reported that lysine, arginine and proline residues from proteins exposed to catalytic metals were oxidized into adipic semialdehyde ("allysine") residues from lysine, and glutamyl semialdehyde from arginine and proline, respectively[7]. The introduction of the "protein carbonyl" concept and the demonstration that these could be easily assayed as dinitrophenyl hydrazones, either colorimetrically or with anti-DNP antibodies in ELISA [8] that are now commercially available has enormously stimulated the field of protein oxidation. However, we have argued that the formation of protein carbonyls is but an intermediary step in the oxidation that eventually leads to a stable carboxylic acids [9], whereby there potentially are several potential pathways for the formation of the protein carbonyl (PC) (**Figure 1a**). In presence of an oxidant, such as hydrogen peroxide, the stable end product 2-aminoadipic acid and glutamic acid will form. The adipic semialdehyde ("allysine") can be assayed as reduced product, i.e. 6-hydroxynorleucine. In the lens the most likely pathway for allysine formation involves ascorbate (Stadtman[10]) or glycation of lysyl residues by ascorbate, glucose or methylglyoxal to form a metal complex that oxidatively deaminates lysine (**Figure 1b**) [11]. Since both methylglyoxal and ascorbate are present in the lens we may never know which glycation agent is more important for allysine formation. In collagen and arteries however, myeloperoxidase (MPO)[12] may play a role, while lysyl oxidase and semicarbazide sensitive oxidase/vascular adhesion protein might participate in oxidation of plasma proteins[13-15].

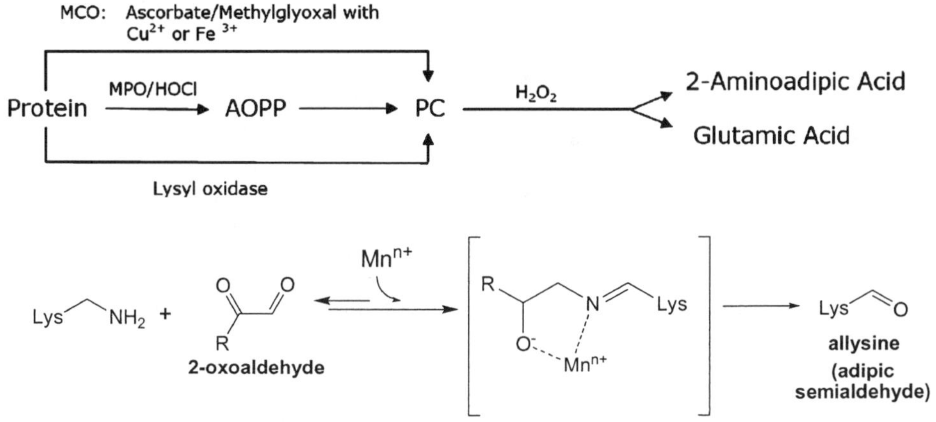

Figure 1   *Protein oxidation into adipic semialdehyde can occur via multiple pathways. Scheme I (Figure 1a) in presence of an oxidant, such as hydrogen peroxide, the stable end product 2-aminoadipic acid and glutamic acid will form. The adipic semialdehyde ("allysine") can be assayed as reduced product, i.e. 6-hydroxynorleucine. However, in the lens the most likely pathway for allysine formation involves ascorbate (Stadtman[10]) or glycation of lysyl residues by ascorbate, glucose or methylglyoxal to form a metal complex that oxidatively deaminates lysine (figure 1b) [11]*

What prompted the present study was our recent discovery that the protein carbonyl allysine (adipic semialdehyde) can become further oxidized into 2-aminoadipic acid (2-AAA) (see Scheme I). In human skin, this thermodynamically stable oxidation endproduct of lysine, accumulated to a much larger extent than its precursor measured as 6-hydroxynorleucine[9]. This finding implied that hydrogen peroxide mediated oxidation was important in the extracellular matrix and raised the question whether intracellular long-lived proteins, such as the crystallins were also subject to lysyl oxidation and 2-AAA formation.

To address these questions, we have used gas chromatography – mass spectrometry (GC/MS) to determine allysine and 2-AAA formation in human water soluble and insoluble human lens crystallins as a function of age, degree of lens pigmentation and presence of diabetes. We have then performed mechanistic studies on the role of oxygen levels for MCO using the hyperbaric oxygen /rabbit lens model and determined its relationship to other pathways of carbonyl and oxidant stress in the lens. Finally, to understand the role of glutathione (GSH) as an endogenous inhibitor of carbonyl and oxidative stress in the lens, we cultured intact mouse lenses in the presence of ascorbic acid but under depletion of GSH using buthionine sulfoximine.

2   MATERIALS AND METHODS

2.1   Preparation of Lens Proteins

Human lenses were decapsulated and homogenized in 5mM Chelex-100 treated sodium phosphate buffer (pH 7.4) with protease inhibitors added (Roche, IN) (2.0ml per lens) in a Con-Torqueglass homogenizer (Eberbach, Ann Arbor, MI) for 3min on ice. The supernatant (WS) and pellet (WI) were separated. The pellet was suspended in 2.0ml of ice-cold buffer and recentrifuged. The WS fraction was further dialyzed against water for 48h with change of water after 24h. The dialysate was lyophilized. Both the WS and WI fractions were delipidated by treatment with 5.0ml of chloroform/methanol, 2:1 (vol/vol).

2.2   AGEs Determination by GS/MS and LC-MS/MS

Amounts of 6-hydroxynorleucine, 2-aminoadipic acid, carboxymethyl-lysine (CML), carboxyethyl-lysine (CEL) and fructoselysine (FL) were determined in acid hydrolysates of processed lens samples, and   derivatized and subjected to GC/MS analysis as previously described[16]. Amounts of FL, MetSOX, G-H1 and MG-H1 in enzymatically digested samples were measured by LC-MS/MS system following the previously published procedure by Thornalley et al[17].

2.3   Other procedures

All other procedures were described in detail in Fan et al. [16].

2.4   Statistical methods

Regression analysis, Spearman's correlations and the Mann–Whitney Test were computed using SPSS software REF Testing for homogeneity of variance was done using either the F-test or the Burr–Foster Q-Test as previously described. Data were transformed with either the square-root or log y transformations. Significance was considered at $P<0.05$.

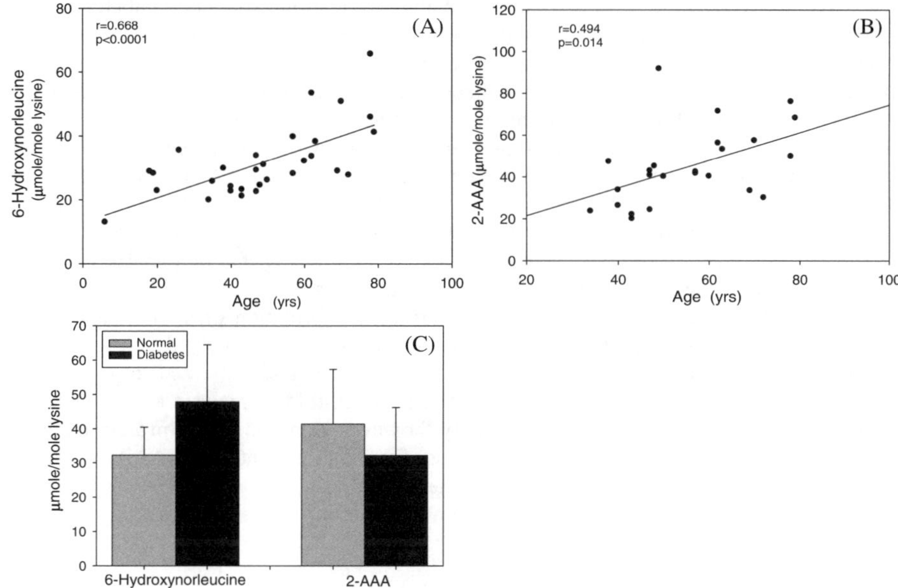

Figure 2 *Levels of 6-hydroxynorleucine (p<0.0001) and 2-aminoadipic acid (p<0.014) versus chronological age in water insoluble lens protein fractions (WI) from non-diabetic individuals, and effects of diabetes on levels of 6-hydroxynorleucine (p<0.0001) and 2-aminoadipic acid (n.s) in water insoluble fraction (WI) upon adjustment to age 50yrs. (A) 6-hydroxynorleucine vs. chronological age; Y=0.38x+12.81, n=30, r=0.67, p<0.0001; (B) 2-aminoadipic acid vs. chronological age. y=0.66x+8.25, n=23, r=0.49, p< 0.01. Regression line equations where x=age and y=parameter. (C) Effect of diabetes. Student t test was used to compare two groups, p value below 0.05 was considered significant. Bars represent means ± S.D.*

3   RESULTS AND DISCUSSION

Allysine and 2-aminoadipic acid were determined simultaneously by GC/MS in protein fractions from human lenses obtained at autopsy. An age-related linear increase in both oxidation products was observed, whereby mean allysine levels increased over the human life span five and three-fold from 4 to 30 µmol/mole lysine (p<0.002) and 12 to 50 µmol/mole lysine (p<0.0001) in the water soluble and insoluble fraction, respectively[16]. In this paper we show only the data for the water insoluble fraction A (Fig. 2A-2B). Overall, almost twice as much 2-AAA accumulated than allysine.

Most revealing were the effects of diabetes on these oxidation products. Diabetes increased allysine highly and significantly in both water soluble (P=0.002) and insoluble fractions (P<0.0001), but had no effect on 2-AAA (Fig. 2C). After age-adjustment, allysine was significantly increased in individuals with diabetes compared to controls, but not 2-AAA. The latter even tended to be lower in the water insoluble fraction.

The surprising lack of increase in 2-AAA in the diabetic lenses suggested that H_2O_2, a expected candidate oxidant for the conversion of allysine into 2-AAA, was not increased in the diabetic lens. In order to verify this, we determined the levels of methionine sulfoxide, which would also be oxidized by H_2O_2. The data showed no elevation of methionine oxidation in diabetic lens compared to non-diabetic individuals, confirming that neither H_2O_2 nor some other oxidant mechanism was increased in the diabetic lenses.

In order to understand the role of oxygen tension in the formation of allysine and 2-AAA, rabbits were exposed to hyperbaric oxygen for 48 hours, as previously described[18]. Lenses were processed with the same protocol as for human lenses and separated into cortical and nuclear water soluble and insoluble fractions. A two- to six-fold increase in allysine was noted in both protein fractions (Fig.3A). Surprisingly, however, none of these "protein carbonyls" were further oxidized into 2-AAA in the process (Fig.3B), i.e whatever 2-AAA was there, it was already present before the hyperbaric treatment.

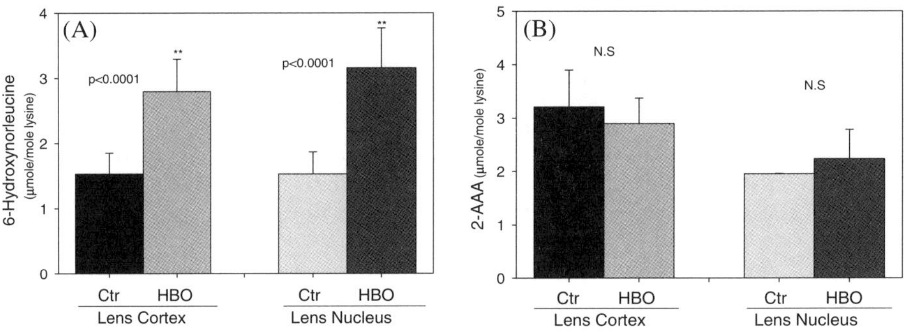

Figure 3   *Rabbit lens 6-hydroxynorleucine and 2-aminoadipic acid level before and after hyperbaric oxygen treatment (HBO) in water insoluble (WI) fractions of lens cortex and nucleus. (A) 6-hydroxynorleucine (p<0.0001). (B) 2-AAA (n.s). Student t test was used to compare the control (Ctr) and HBO. 2-AAA levels were not elevated by HBO (p = N.S.). Bars represent means ± S.D.*

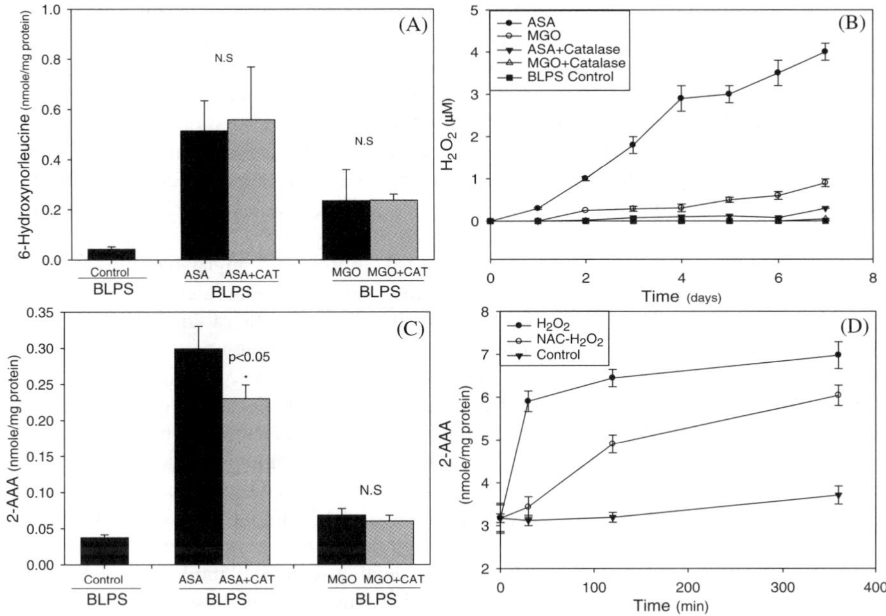

Figure 4 *Metal catalyzed oxidation (MCO) in vitro. Bovine lens water soluble protein (BLPS) was incubated with MGO or ASA with or without catalase (CAT) for 7 days. A and C: The 6-hydroxynorleucine and 2-AAA levels were determined by GC/MS; (B): the H_2O_2 was determined daily during incubation measured by Amplex-Red assay. D: Oxidized poly-L-lysine was incubated with H_2O_2 with or without N-acetyl-cysteine (NAC).*

The high resistance of protein carbonyls and methionine towards further oxidation in diabetes and hyperbaric O_2 treatment suggested to us the hypothesis that glutathione would prevent the oxidant from attacking the proteins. In order to test this hypothesis, we cultured for 72 h intact mouse lenses in presence of ascorbic acid without or with buthionine sulfoximine in order to inhibit GSH synthesis[19]. There was a massive increase in the methylglyoxal-derived AGEs CEL and MG-H1, including a tripling of 2-AAA levels, indicative of increased metal catalyzed damage to lysine residues. Evidence of active oxidation was reflected in the fact that methionine sulfoxide was 150%. Interestingly, the protein carbonyls (6-hydroxynorleucine) were very low, likely reflecting rapid and quantitative conversion of the semialdehyde into the more stable endproduct 2-AAA .

In order to evaluate the role of H_2O_2 amid several possible mechanisms for allysine and 2-AAA formation, we incubated bovine lens crystallins with either ascorbic acid (5 mM) or methylglyoxal (MGO, 5 mM) with or without catalase. The results in Fig.3A show that allysine, as expected, formed both from ASA and MGO, and that catalase only partially suppressed 2-AAA from ASA (Fig. 4C) even though H_2O_2 formation was nearly totally suppressed by catalase (Fig.4B). Moreover, there was minimal 2-AAA formation in presence of MGO, strongly suggesting that H_2O_2 is not necessary for allysine formation. However, addition of H_2O_2 to oxidized poly-L-lysine rapidly induced 2-AAA formation that slowed down but not suppressed by 10 mM *N*-acetyl-cysteine (Fig. 4D).

The above data unequivocally established the fact that protein carbonyls stemming from metal catalyzed oxidation are present in the aging human lens, and that these modifications are increased in presence of diabetes. The mechanistic studies with rabbit lenses exposed to hyperbaric oxygen confirm the critical role of oxygen in the formation of allysine.

The most intriguing finding is the uncoupling between the oxidation of lysine residues into allysine, and the total suppression of further oxidation to the stable endproduct 2-aminoadipic acid in diabetes. In skin, however, diabetes induces 2-AAA accumulation to a much larger extent than allysine [9], suggesting either a dramatic increase in some pro-oxidant species or an absence of defense mechanisms. Surprisingly, while capillary H_2O_2 production is increased in diabetes[20], levels are 100 times lower in serum (about 1 µM [21]) than the lens (about 100 µM[22]). However, vascular GSH is very low, i.e. 1 µM [23] while very high in the lens GSH (3-5 mM). This was a clue that GSH must be playing a crucial role in the inhibition of the further oxidation of allysine into 2-AAA, as demonstrated by the large increase in 2-AAA in lenses cultured in the presence of buthionine sulfoximine.

Finally, our results and hypothesis are tilted toward a role for H_2O_2 as the primary oxidant species. In the extravascular space, such a skin collagen, there is little doubt that other mechanisms besides H_2O_2 need to be considered, such as peroxinitrite, lipid peroxides, hypochlorous acid and free radicals. In the lens however, there is no evidence so far for 3-nitrotyrosine and chloramines suggesting ONOO- and hypochlorous acid should not play a role, although in vitro studies have been carried out [24]. Also, there is no evidence for the presence of myeloperoxidase in the lens [25]. Nevertheless, carbonyls and especially methionine residues, while most easily oxidized by H_2O_2 or possible other oxidants could conceivably be oxidized via different and less selective reactions which are in the competition with cysteine oxidation and cell's defense (antioxidants) and repair systems. Contrary to that, lysine oxidation into allysine goes through lysine-selective Suyama pathway and other less selective MCO systems.

4   SUMMARY

In summary, the data presented in this study suggest metal catalyzed oxidation of lysines occurs in the aging human lens, most likely by the Suyama mechanism involving Cu^{2+}, methylglyoxal whereby H_2O_2 most likely originates from ascorbic acid oxidation. The conversion of allysine to 2-aminoadipic acid does not readily occur in the lens due to the anaerobic environment. Thus, if high levels of 2-AAA are detected, this implicates a functional drop of glutathione and decreased ability to detoxify H_2O_2. Therefore 2-AAA is expected to accumulate during nuclear sclerosis together with the formation of protein-protein and protein-GSH disulfides[26]. The findings strongly implicate dicarbonyl/metal catalyzed oxidation of lysine to allysine, whereby low GSH combined with ascorbate-derived H_2O_2 likely contributes toward 2-AAA formation, since virtually no 2-AAA formed in presence of methylglyoxal instead of ascorbate. An important translational conclusion is that chelating agents might help delay nuclear sclerosis, and that glutathione plays a key role in preventing oxidation of protein carbonyls.

Acknowledgements

This research was supported by grants EY07099(VMM), EY 02027(FJG) and EY 014803 (FJG).

References

1.   D. Garland, *Exp Eye Res*, 1990, **50**, 677-682.
2.   A. K. Saxena, P. Saxena, X. Wu, M. Obrenovich, M. F. Weiss and V. M. Monnier, *Biochem Biophys Res Commun*, 1999, **260**, 332-338.
3.   P. Saxena, A. K. Saxena, X. L. Cui, M. Obrenovich, K. Gudipaty and V. M. Monnier, *Invest Ophthalmol Vis Sci*, 2000, **41**, 1473-1481.
4.   J. R. Requena and E. R. Stadtman, *Biochem Biophys Res Commun*, 1999, **264**, 207-211.
5.   M. Qian, M. Liu and J. W. Eaton, *Biochem Biophys Res Commun*, 1998, **250**, 385-389.
6.   S. Fu, R. Dean, M. Southan and R. Truscott, *J Biol Chem*, 1998, **273**, 28603-28609.
7.   J. R. Requena, R. L. Levine and E. R. Stadtman, *Amino Acids*, 2003, **25**, 221-226.
8.   R. L. Levine, N. Wehr, J. A. Williams, E. R. Stadtman and E. Shacter, *Methods Mol Biol*, 2000, **99**, 15-24.
9.   D. R. Sell, C. M. Strauch, W. Shen and V. M. Monnier, *Biochem J*, 2007, **404**, 269-277.
10.   E. R. Stadtman, *Am J Clin Nutr*, 1991, **54**, 1125S-1128S.
11.   M. Akagawa, T. Sasaki and K. Suyama, *Eur J Biochem*, 2002, **269**, 5451-5458.
12.   X. Fu, S. Y. Kassim, W. C. Parks and J. W. Heinecke, *J Biol Chem*, 2003, **278**, 28403-28409.
13.   S. Jalkanen and M. Salmi, *Embo J*, 2001, **20**, 3893-3901.
14.   R. Kurkijarvi, G. G. Yegutkin, B. K. Gunson, S. Jalkanen, M. Salmi and D. H. Adams, *Gastroenterology*, 2000, **119**, 1096-1103.
15.   P. H. Yu, L. X. Lu, H. Fan, M. Kazachkov, Z. J. Jiang, S. Jalkanen and C. Stolen, *Am J Pathol*, 2006, **168**, 718-726.
16.   X. Fan, J. Zhang, M. Theves, C. Strauch, I. Nemet, X. Liu, J. Qian, F. J. Giblin and V. M. Monnier, *J Biol Chem*, 2009.
17.   P. J. Thornalley, S. Battah, N. Ahmed, N. Karachalias, S. Agalou, R. Babaei-Jadidi and A. Dawnay, *Biochem J*, 2003, **375**, 581-592.

18. F. J. Giblin, L. Schrimscher, B. Chakrapani and V. N. Reddy, *Invest Ophthalmol Vis Sci*, 1988, **29**, 1312-1319.

19. J. Martensson, R. Steinherz, A. Jain and A. Meister, *Proc Natl Acad Sci U S A*, 1989, **86**, 8727-8731.

20. E. A. Ellis, D. L. Guberski, M. Somogyi-Mann and M. B. Grant, *Free Radic Biol Med*, 2000, **28**, 91-101.

21. F. Lacy, M. T. Kailasam, D. T. O'Connor, G. W. Schmid-Schonbein and R. J. Parmer, *Hypertension*, 2000, **36**, 878-884.

22. P. S. Devamanoharan and S. D. Varma, *Ophthalmic Res*, 1995, **27 Suppl 1**, 39-43.

23. G. Paolisso, G. Di Maro, G. Pizza, A. D'Amore, S. Sgambato, P. Tesauro, M. Varricchio and F. D'Onofrio, *Am J Physiol*, 1992, **263**, E435-440.

24. G. Thiagarajan, J. Lakshmanan, M. Chalasani and D. Balasubramanian, *Invest Ophthalmol Vis Sci*, 2004, **45**, 2115-2121.

25. L. J. Hazell, H. Fu, R. T. Dean, R. Stocker and R. J. Truscott, *Clin Exp Optom*, 2002, **85**, 97-100.

26. M. F. Lou, J. E. Dickerson, Jr., W. H. Tung, J. K. Wolfe and L. T. Chylack, Jr., *Exp Eye Res*, 1999, **68**, 547-552.

IMPAIRED OXYGEN METABOLISM IN DIABETIC NEPHROPATHY: ADVANCED GLYCATION, HYPOXIA, AND OXIDATIVE STRESS

T. Miyata[1] and M. Okamura[1]

[1] Center for Translational and Advanced Research, Tohoku University Graduate School of Medicine, Seiryo-Machi, Aoba-ku, Sendai, 980-8575, JAPAN

1 INTRODUCTION

The incidence of diabetic nephropathy remains worrisome despite major therapeutic advances. Classical factors contributing to its pathology, *e.g.,* hypertension, hyperglycemia, and hyperlipidemia are now amenable to treatment. Current therapies however do not fully prevent its renal complications. Novel agents able to interfere with several newer culprits should therefore provide additional benefits. More recently, experimental studies have incriminated newer culprits, such as advanced glycation, hypoxia, and oxidative stress, raising several questions from the clinical point of view. In this review, we utilize a unique rat model with hypertension and type 2 diabetes, and propose a central role in the genesis and progression of diabetic nephropathy of impaired oxygen metabolism, *i.e.,* hyopxia, and advanced glycation/carbonyl stress, with the hope to identify novel, more innovative therapeutic targets.

2 ANIMAL MODEL OF HYPERTENSIVE, TYPE 2 DIABETES

SHR/NDmcr-cp has the genetic background of the spontaneously hypertensive rat (SHR) and thus becomes hypertensive[1].   A mutation of the leptin receptor, in addition, generates obesity, hyperglycemia, hyperinsulinemia, and hyperlipidemia, all of which characterize human type 2 diabetes.   Eventually, this obese, diabetic rat develops significant renal damage, in agreement with our common clinical experience that hypertension plus metabolic derangements accelerate markedly diabetic renal damage. We tested several therapeutic approaches in this model, *i.e.,* obesity correction by caloric restriction, blood pressure normalization with antihypertensive drugs, glucose lowering by pharmacologic agents, and correction of hypoxia and of advanced glycation by novel approaches.

3 OBESITY CORRECTION BY CALORIC RESTRICTION

Restriction of the caloric intake of SHR/NDmcr-cp by 30% for 20 weeks corrected both obesity and hyperlipidemia without changes in blood pressure, hyperglycemia, and hyperinsulinemia[1]. Nevertheless, it prevented proteinuria and histological abnormalities of the kidney. Of note, renal damage correlated with body weight and with the renal content of advanced glycation end products (AGEs) and of oxidative stress[1]. Renoprotection in this model thus hinges upon the reduction of oxidative stress but remains independant of hypertension and hyperglycemia. Restriction of energy intake reduces oxidative stress in experimental animals[2,3]. Crujeiras *et al.*[4] demonstrated that energy restriction in obese subjects improves mitochondrial function through the reduction of oxidative stress.

4 BLOOD PRESSURE NORMALIZATION WITH ANTIHYPERTENSIVE DRUGS

Several clinical studies, mainly but not only in diabetic patients, have demonstrated that anti-hypertensive agents inhibiting the renin-angiotensin system (RAS), such as angiotensin converting enzyme inhibitors (ACEIs), angiotensin II type 1 receptor blockers (ARBs), or a direct renin inhibitor, achieve a better renoprotection than other anti-hypertensive drugs[5-7]. As a result, ARBs and ACEIs are now part of the standard treatment of patients with diabetic nephropathy, regardless of the presence of systemic hypertension. Interestingly, RAS inhibitors provide renoprotection independently of blood pressure (BP) lowering[5-7]. The effect on the kidney of several types of blood pressure-lowering agents was therefore tested[8,9]. An angiotensin receptor blocker (ARB), a calcium antagonist, or a beta-blocker was given for 20 weeks to SHR/NDmcr-cp. The agents normalized systolic blood pressure to the same extent, but only the ARB successfully decreased proteinuria. This finding fits with the clinical experience that renin-angiotensin system (RAS) inhibitors provide a better renoprotection than other types of antihypertensives. Renal benefits appear independent of systemic blood pressure lowering and of changes in metabolic abnormalities.

The optimal doses of ARB needed for renoprotection and for blood pressure lowering were further defined. Above 120 mg/kg/day, a dose exceeding the level necessary for angiotensin II receptor saturation, the ARB valsartan failed to further reduce blood pressure, a finding suggesting indeed complete blockade of angiotensin II receptor. Nevertheless, renoprotection witnessed by proteinuria progressed continuously in a dose-dependent manner, demonstrating further renal benefits of ARB independent of blood pressure lowering (our unpublished observation). Impressively, ARB, but not calcium antagonist or beta-blocker, markedly reduced the renal AGE content despite no modification of the concomitant metabolic syndrome including hyperglycemia[9]. Renal AGE content significantly correlated with proteinuria whatever the type of antihypertensive agent. ARB simultaneously corrected oxidative stress and hypoxia[8,9].

5   GLUCOSE LOWERING BY PHARMACOLOGIC AGENTS

The critical role of hyperglycemia in the genesis of diabetic renal injury is demonstrated by the fact that the strict glycemic control reduces microalbuminuria in diabetic patients with nephropathy. In addition, recent studies have implicated insulin resistance or hyperinsulinemia in its genesis[10]. In this context, it is of interest to ask which agent provides a better renoprotection, insulin or an insulin sensitizer such as pioglitazone. SHR/NDmcr-cp rats were thus given for 20 weeks either pioglitazone or insulin[11]. Neither treatment modified hypertension. Pioglitazone aggravated obesity and provided poorer glycemic control than insulin but, in contrast with insulin, it significantly decreased plasma insulin levels. As a result, renoprotection was markedly better with pioglitazone than with insulin treatment as shown by proteinuria reduction. Both pioglitazone and insulin reduced the renal accumulation of AGEs and markers of oxidative stress but only pioglitazone reduced renal expression of transforming growth factor (TGF)-beta[11]. Hyperinsulinemia and the attendant increase of TGF-beta expression might therefore prove useful therapeutic targets, independently of glycemic control, a conclusion supported by clinical evidence that pioglitazone enhances renoprotection in obese, diabetic patients with nephropathy.

6 HYPOXIA CORRECTION

Oxygen is essential to various bio-metabolic processes, including oxidative phosphorylation during mitochondrial respiration. All organs, including the kidney, thus depend on a sufficient and consistent supply of oxygen. The presence in diabetic nephropathy of chronic hypoxia has been confirmed[12,13]. Glomerular efferent arterioles enter the peritubular capillary plexus, which offers oxygen to tubular and interstitial cells. Diabetic glomerular and vascular lesions damage efferent arterioles and subsequently decrease the number of peritubular capillary, which in turn reduces oxygen diffusion to tubulointerstitial cells, eventually leading to tubular dysfunction and fibrosis. Anemia associated with chronic kidney disease also hinders oxygen supply. Together with significant decrease in oxygen supply, oxygen demand is increased in the diabetic kidney: remnant nephrons compensate for tubular loss of nephrons, with an attendant enhanced tubular transport and hence more energy consumption.

If hypoxia is the driving force for diabetic nephropathy, let us review the mechanisms of defence against hypoxia. Defence against hypoxia hinges upon the hypoxia-inducible factor (HIF), whose activation induces a broad range of genes that participate in erythrocytosis, angiogenesis, glucose metabolism, or cell proliferation/survival, with the eventual protection of hypoxic tissues[14,15]. HIF-a is constitutively transcribed and translated. Its level is primarily regulated by its rate of degradation. Oxygen determines its stability through its enzymatic hydroxylation by prolylhydroxylases (PHDs)[16]. The hydroxylated HIF-a is then recognized by Hippel-Lindau tumor suppressor protein (pVHL) and is rapidly degraded by the proteasome[17]. Non-hydroxylated HIF-a cannot interact with pVHL and is thus stabilized. It binds to its heterodimeric partner HIF-b and, in the nucleus, transactivates genes involved in the adaptation to hypoxic-ischemic stress.

Therapeutic approaches targeting HIF would be clinically of little benefit, if HIF is maximally activated under pathological conditions. This is not necessarily true. In the rat diabetic kidney, HIF activation is suboptimal: treatment with anti-oxidants augments it[18,19].

Thus, current studies focus on the inhibition of PHDs by small molecular compounds to modulate HIF activity.

Iron is essential for PHD activity, so that transition metal chelators potentially inhibit PHD activity. Cobalt chloride also inhibits the PHD activity through intracellular depletion of ascorbate necessary for iron (reduced) activity[20]. The renal benefit of cobalt chloride in the diabetic kidney has also been demonstrated in a hypertensive, type 2 diabetic rat model (SHR/NDmcr-cp)[21]. Cobalt chloride, given for 20 weeks, reduces proteinuria as well as histological kidney injury, despite sustained hypertension and metabolic abnormalities. Renal improvement is paralleled by a marked reduction in renal tissue expressions of HIF-regulated genes, including erythropoietin, VEGF, and HO-1, and also by a reduced renal production of TGF-beta and AGEs[21].

Unfortunately, cobalt chloride proved too toxic for further clinical use. Recently, less cumbersome, non-toxic small molecular inhibitors for oxygen sensor (PHDs) have been investigated[22]. We utilized a unique strategy, docking simulation based on the three dimensional protein structure of human PHD2, and synthetized two novel inhibitors for PHDs (TM6008 and TM6089)[23]. Both compounds bind to the active site within the PHD2 molecule where HIF binds. As anticipated, given orally, they stimulate HIF activity in various organs of transgenic rats expressing a hypoxia-responsive reporter vector and, given locally, they induce angiogenesis in a mouse sponge assay[23].

Non-specific inhibition of HIF degradation however augments vascular endothelial growth factor (VEGF) and erythropoietin production, both of which have proven detrimental in human diabetic retinopathy[24]. Dissociation of the benefits of HIF activation from its effects on VEGF and erythropoietin should prove helpful. Fortunately, three different PHD isoforms have been identified[25], *i.e.*, PHD1, PHD2, PHD3. The roles of the three PHD isoforms have recently been delineated by the specific disruption of each PHD gene. PHD2 primarily regulates angiogenesis and erythropoiesis[26,27]. By contrast, the specific disruption of PHD1 unexpectedly induces hypoxic tolerance in muscle cells without angiogenesis and erythrocytosis induction[28]. Basal oxygen metabolism is reprogrammed and generation of oxidative stress is decreased in hypoxic mitochondria. Inhibition of PHD1 likely arouses various protective mechanisms, including ATP production through enhanced glycolysis, and restriction of glycolytic intermediates entry into the oxidative phosphorylation of glucose through induction of pyruvate dehydrogenase kinase, which leads to attenuated entry of electrons into the electron transport chain. These reactions conserve energy, reduce oxidative damage, and protect the cell from hypoxic damage.

Table 1. *Summary of animal experiments in a hypertensive, type 2 diabetic rat with nephropathy (SHR/NDmcr-cp)*

		Treatments Anti-hypertensive agents							
		Caloric restriction	ARB	CCB	β-blocker	Pioglitazone	Insulin	Cobalt	AGE inhibitor(R147176)
Renoprotection		+	+	-	-				+
Oxygen metabolism	**AGE inhibition**	+	+	-	-	+	+	+	+
	Anti-oxidative stress	+	+	-	-	+	+	+	+
	Hypoxia correction	ND	+	-	-	ND	ND	+	ND
Haemo-dynamics	**BP lowering**	-	+	+	+	-	-	-	±
	RAS inhibition	-	+	-	-	-	-	-	±
Metabolic syndrome	**Obesity correction**	+	-	-	-	↓	-	-	-
	Glycemic control	+	-	-	-	+	+	-	-
	Lipid lowering	+	+	-	-	+	+	-	-
	Hyperinsulinemia correction	-	-	-	-	+	↓	-	-
Refs		1	8,9	9	9	11	11	21	30

ARB, angiotensin receptor blocker; CCB, calcium channel blocker: AGE, advanced glycation end product; BP, blood pressure; RAS renin angiotensin system; ND, not determined.

Specific inhibition of PHD1 may thus mediate tissue protection through a reduced oxidative stress, independent of HIF activation. Unfortunately, none of the current, thus far reported, PHD inhibitors is specific for a distinct PHD subtype[22].

7   INHIBITION OF ADVANCED GLYCATION END PRODUCTS

In addition to protective benefits by BP lowering and angiotensin II type 1 receptor (AT_1)-blockade, ARBs (or ACEIs) have unique abilities to correct oxidative stress, carbonyl stress, and advanced glycation[8,29].   In order to dissect the mechanisms of ARBs' protective benefits, we synthesized a novel, non-toxic ARB derivative R-147176, characterized by a weak affinity for the AT_1 (6,700 times less effective than olmesartan in AT_1 binding inhibition), but a striking inhibition of oxidative stress and advanced glycation[30].   Despite a minimal effect on blood pressure, it provides a significant renoprotection in two different experimental type 2 diabetic rat models, SHR/NDmcr-cp and Zucker diabetic fatty.   The renal benefit of ARB thus partly depends on its potent inhibition of oxidative stress and advanced glycation. R-147176, like ARBs, protects not only the kidney but also brain cells in an experimental rat stroke model[31].

8   CONCLUSION

The future prevention of diabetic nephropathy and of its dramatic consequences will undoubtedly rely on a multipronged approach.   The current therapies to correct classical factors, *i.e.*, obesity, hypertension, hyperglycemia, hyperinsulinemia, appear insufficient to fully prevent renal complications.   Our animal experiments demonstrate that renoprotection does not necessarily rely on blood pressure or glycemic control but appears consistently associated with a decreased AGE formation, hypoxia correction and/or a decreased oxidative stress, all of which are closely linked to oxygen metabolism (Table 1). Only time will tell us if renewed approaches can be extended to humans.

References

1.   Nangaku, M. *et al. Nephrol. Dial. Transplant.* 2005, **20**, 2661-2669
2.   Gredilla, R. & Barja, G. *Endocrinology* 2005, **146,** 3713-3717
3.   Cai, W. *et al. Am. J. Pathol.* 2008, **173**, 327-336

4.	Crujeiras, A. B. *et al. J. Physiol. Biochem* 2008,. **64**, 211-219.

5.	Jafar, T. H. *et al. Ann. Intern. Med.* 2001, **135**, 73-87.

6.	Brenner, B. M. *et al. N. Engl. J. Med.* 2001, **345,** 861-869.

7.	Parving, H. H. *et al. N. Engl. J. Med.* 2008,. **358**, 2433-2446.

8.	Nangaku, M. *et al. J. Am. Soc. Nephrol.* 2003, **14**, 1212-1222.

9.	Izuhara, Y. *et al. J. Am. Soc. Nephrol.* 2005, **16**, 3631-3641.

10.	Sarafidis, P. A. & Ruilope, L. M. *Am. J. Nephrol*2006,. **26**, 232-234.

11.	Ohtomo, S. *et al. Kidney Int.* 2007, **72**, 1512-1519.

12.	Nangaku, M. *J. Am. Soc. Nephrol.* 2006, **17**, 17-25.

13.	Singh, D. K., Winocour, P. & Farrington, K. *Nat. Clin. Pract. Nephrol.* 2008, **4**, 216-226.

14.	Semenza, G. L. *Nat. Rev. Cancer.* 2003,. **3**, 721-732.

15.	Marx, J. *Science* 2004, **303**, 1454-1456.

16.	Epstein, A. C. *et al.* 2001, *Cell* **107**, 43-54.

17.	Hon, W. C. *et al.* 2002, *Nature* **417**, 975-978.

18.	Katavetin, P. *et al. J. Am. Soc. Nephrol.* 2006, **17**, 1405-1413.

19.	Rosenberger, C. *et al. Kidney Int.* 2008,. **73**, 34-42.

20.	Salnikow, K. *et al. J. Biol. Chem.* 2004, **279**, 40337-4044.

21.	Ohtomo, S. *et al. Nephrol. Dial. Transplant* 2008,. **23**, 1166-1172.

22.	Fraisl, P., Aragonés, J. & Carmeliet, P. 2009, *Nat. Rev. Drug. Discov.* **8**, 139-152.

23.	Nangaku, M. *et al. Arterioscler. Thromb. Vasc. Biol.* 2007, **27**, 2548-2554.

24.	Watanabe, D. *et al. N. Engl. J. Med.* 2005, **353**, 782-792.

25.	Kaelin, W. G. Jr. & Ratcliffe, P. J. *Mol. Cell.* 2008, **30**, 393-402.

26.	Takeda, K., Cowan, A. & Fong, G. H. *Circulation* 2007, **116**, 774-781.

27.	Takeda, K. *et al. Blood* 2008, **113**, 3229-3235.

28.	Aragonés, J. *et al. Nat. Genet.* **40**2008,, 170-180.

29.	Miyata, T. *et al. J. Am. Soc. Nephrol.* 2002, **13**, 2478-2487.

30.	Izuhara, Y. *et al. Arterio. Thromb. Vas. Biol.* 2008, **28**, 1767-1773.

31.	Takizawa, S. *J. Cereb. Blood. Flow. Metab.* Epub ahead of print.

ROLE OF CARBONYL/OXIDATIVE STRESS ON THE PATHOGENESIS OF CHRONIC KIDNEY DISEASE AND THE METABOLIC SYNDROME

T. Mori[1,2], Q. Guo[1], T. Nakamichi[1], T. Miyata[3], M. Nakayama[4], S. Ogawa[1], K. Suyama[5], S. Ito[1]

[1]Division of Nephrology, Endocrinology and Vascular Medicine, Tohoku University Graduate School of Medicine, 1-1 Seiryocho, Aoba-ku, Sendai 980-8574, Japan
[2]Health Administration Center,
[3]Center for Translational and Advanced Research,
[4]Research Division of Chronic Kidney Disease, Tohoku University, Japan
[5]Department of Nutrition, Sendai University, Japan

1   INTRODUCTION

A large number of clinical studies have suggested that chronic kidney disease (CKD) is a risk factor of adverse outcomes, independent of classical risk factors, including hypertension, hyperlipidemia, and diabetes[1,2]. For example, micro-albuminuria is an independent risk factor of both cardiovascular and non-cardiovascular events and death[3,4], even subjects with no history of hypertension or diabetes. Consequently, renoprotective interventions would likely have far greater impact than merely the prevention of end stage renal disease. Both oxidative and carbonyl stress has been demonstrated to involve in the pathogenesis of CKD [5,6]. Enhanced oxidative/carbonyl stress induce many molecules and pathways that participate cellular injury and apoptosis[6]. In this paper, we review the some of the major effects of carbonyl/oxidative stress on kidney, and their role in the development and progression of CKD.

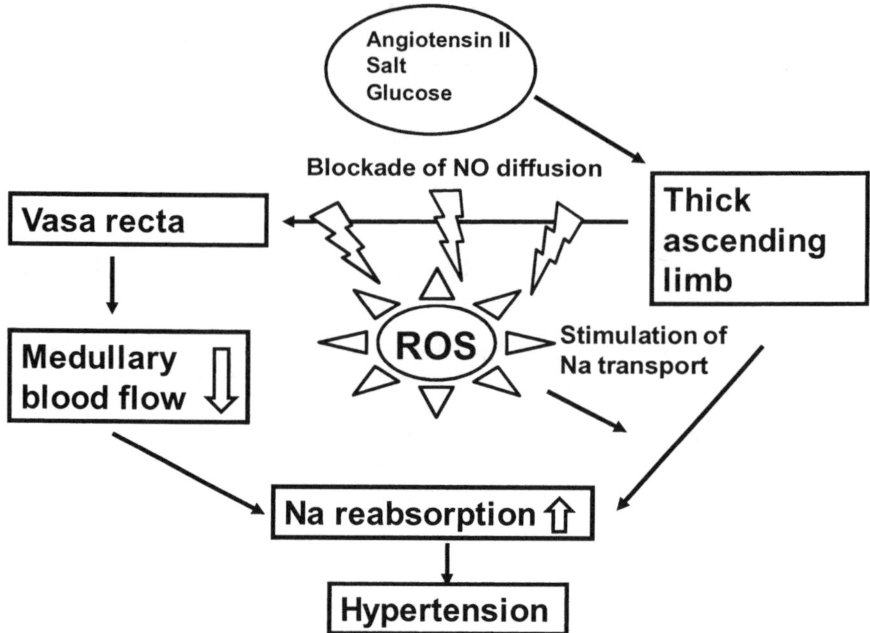

Figure 1   *Physiological role of oxidative stress in the renal medulla*

2   OXIDATIVE STRESS AND THE REGULATION OF SYSTEMIC BLOOD
       PRESSURE.

A number of animal studies have indicated that oxidative stress is involved in the pathogenesis of hypertension and renal injury. Enhanced oxidative stress is observed in many types of hypertensive rat models, such as angiotensin II infused rats, spontaneously hypertensive rats (SHR) and Dahl salt sensitive (Dahl S) rats[7,8,9]. When superoxide dismutase (SOD) mimetic 4-hydroxytetramethylpiperidine-1-oxyl (TEMPOL) is intravenously administered to angiotensin II infused rats or SHR, blood pressure levels are reduced, suggesting that ROS is partly responsible for development of hypertension[7,8]. The role of renal oxidative stress in the development of hypertension and renal injury has also been extensively studied in Dahl S rats[9,10,11]. For example, renal superoxide levels, as determined with lucigenin chemiluminescence, is increased in renal homogenate in high-salt fed DahlS rats compared to those of Dahl salt resistant rats[9]. Intravenous infusion of superoxide dismutase (SOD) mimetic 4-hydroxy-2,2,6,6-tetramethyl-piperidine-1-oxyl (TEMPOL) in high salt fed Dahl S rats for 3 weeks also results in the reduction of blood pressure, glomerulosclerosis and proteinuria[9]. Expression of nicotinamide adenine dinucleotide phosphate (NADPH) oxidase mRNA, a enzyme responsible for superoxide production was also found to be higher in DahlS rats as compared with SS13BN rats, a rat model in which chromosome 13 of normotensive Brown

Norway rats were introgressed into Dahl S background and salt sensitivity of blood pressure is reduced[12].

Direct evidence that specifically renal ROS can contribute to development of systemic hypertension was first demonstrated in studies in which TEMPOL and/or the SOD inhibitor diethyldithiocarbamic acid (DETC) were infused into renal medullary interstitial space from small implanted catheter in anesthetized Sprague Dawley (SD) rats. An infusion of DTEC reduced medullary blood flow and sodium excretion without altering cortical blood flow[13]. Conversely, an infusion of TEMPOL increased medullary blood flow and sodium excretion, indicating that superoxide was responsible for medullary circulation and sodium excretion[13]. Moreover, chronic r.i. infusion of DETC in SD rats reduced MBF and showed sustained hypertension[14]. How does ROS produced in hypertensive models? To answer this question, Mori etal. have previously determined using isolated thin tissue of outer medulla and fluorescent superoxide and NO indicators. As shown in **Figure 1**, superoxide can be produced in tubules such as medullary thick ascending limb in response to angiotensin II, salt and glucose [15,16]. This superoxide inhibits the diffusion of NO from tubules to the surrounded vasa recta [15,17]. With enhanced angiotensin II induced superoxide production in Dahl S rats, tubulo-vascular NO crosstalk is inhibited, thereby constricts vasa recta, reduces MBF and involve in development of hypertension[17].

The role of other ROS such as hydrogen peroxide (H_2O_2) was also determined. When H_2O_2 was infused r.i. directly to the renal medulla in anesthetized SD rats, reduction in MBF and urinary Na excretion was observed [18]. As is similar to DETC, chronic infusion of H_2O_2 r.i. in Sprague Dawley and SS13BN rats increased blood pressure[19,20]. Renal H_2O_2 is increased in Dahl S rats and infusion of catalase r.i. attenuated reduction of MBF and increase in high-salt induced hypertension, indicating that H_2O_2 was also responsible for regulation of medullary circulation [19,20].

Carbonyl stress is also enhanced in subjects with hypertension, diabetes and chronic kidney disease (CKD)[21,22]. Although methylglyoxal (MGO), an initiator of carbonyl stress, is a metabolite or glycolysis pathway, plasma level of MGO is increased in CKD patients regardless of diabetes [21]. Carbonyl stress is involved in the renal injury of non-diabetic CKD rats such as ischemia reperfusion model rats[23], and higher level of plasma methylglyoxal was observed in chronic thy-1 nephritis rats (unpublished observation). From these data, the renal clearance of the kidney may be responsible for the plasma methylglyoxal level [24]. Nakayama etal. have previously demonstrated in vitro using electron resonance spin-trapping method that large amount of ROS could be formed by the reaction of MGO and H_2O_2 [24]. Since high salt intake which does not induce hypertension enhances renal oxidative stress in SD rats, we hypothesized that administration of MGO in SD rats could induce salt sensitive hypertension by the interaction between carbonyl and oxidative stress. Neither 8% salt diet or 1% MGO in drinking water did not increase blood pressure in SD rats. We have chosen relatively high concentration of 1% MGO because this was the dose that SD rats could increase plasma MGO concentration equivalent to CKD stage 5 patients (approximately 10-15 times higher than healthy subjects). When this dose of MGO was administered in rats under 8% salt diet, significant increase in systolic blood pressure was observed, indicating that MGO induced salt sensitivity[25]. This was accompanied with increase in renal medullary oxidative/carbonyl stress as determined by 3-nitrotirosine as an indicator of oxidative stress and N^ϵ-carboxyethyl-lysine (CEL) as carbonyl stress (**Table 1**). Taken together with previous studies, carbonyl stress can induce salt sensitive hypertension by increase in renal oxidative stress.

Table 1   *Carbonyl stress induces salty sensitivity and insulin resistance*

	MGO	Salt	MGO+Salt
Renal CEL	⇧		⇧
Renal 3-Nitrotirosine		⇧	⇧
Urinary TBARS	⇧	⇧	⇧⇧
Salt sensitive hypertension			⇧
Insulin resistance			⇧

⇧ indicates increase in each parameters

Besides salt sensitivity, insulin resistance is also commonly observed in CKD. We tested whether MGO could increase insulin resistance in rats using glucose clamp technique. Increase in insulin resistance was observed in rats treated with MGO, which was completely reduced by administration of N-acetyl cysteine and carbonyl scavengers[25].

As described above, higher level of ROS including H_2O_2 is observed in the kidney of Dahl S rats. Similar with the *in vitro* study, we hypothesized that exogenous administration of MGO would react with renal H_2O_2 and increase blood pressure without salt loading in Dahl S rats. When 1% MGO in drinking water, a relatively high dose of which increases plasma MGO level equivalent to those of CKD stage 5, were administered in Dahl S rats, increase in blood pressure and renal injury was observed even with normal salt diet[26]. Increase in renal oxidative/carbonyl stress was observed as determined with 8-OHdG and CEL. Same dose of MGO did not increase blood pressure in SD rats, which has relatively lower level of renal H_2O_2. These results confirm that MGO could interact with oxidative stress and induce salt sensitive form of hypertension in rats with CKD model.

Not only the physiological action of carbonyl stress, have we recently focused on a novel mechanism that may participate in the renal injury. We paid attention to melamine induced nephropathy that is explained by the consumption of melamine and cyanulic acid. Since these molecules have amine and carbonyl residues, we hypothesized that melamine could also react with MGO and induce renal injury. Acute tubular injury was observed when melamine and MGO, that does not produce renal injury alone, was orally administered to Sprague-Dawley (SD) rats. This was inhibited with pyridoxamine treatment indicating that carbonyl substances were involved in this renal injury (Unpublished observation). In addition, MGO induced renal injury was enhanced when rats were treated with oxonic acid and raised uric acid, an amine produced in vivo. Thus, we propose this amine-carbonyl reaction as a novel mechanism that may also participate in the progression of CKD in which plasma level of both carbonyl and amine are high.

3   ROLE OF OXIDATIVE/CARBONYL STRESS IN CHRONIC KIDNEY DISEASE –FROM CLINICAL POINT OF VIEW

Hypertension and diabetes are the major causes of CKD and end-stage renal disease. It has been well known that inflammation and oxidative stress is involved in the pathogenesis of these diseases. Ogawa et al. have determined in diabetes nephropathy patients with hypertension that angiotensin II receptor blockers (ARB) could reduce albuminuria with reduction of inflammatory and oxidative stress markers [27]. In this study, patients with hypertensive diabetic nephropathy patients were treated with ARB (candesartan or valsartan) or thiazide diuretics trichloromethiazide. After 8 weeks of treatment, urinary albumin to creatinine ratio, oxidative stress and inflammatory markers were analyzed. There were no difference between reduction of blood pressure and HbA1c. Although urinary markers of 8epi-prostaglandin F2α and 8-OHdG were not reduced in diuretics group, these were significantly reduced in the ARB group. In addition, reduction of plasma level of MCP-1 and IL-6, both of which are the inflammatory markers, was observed only in the ARB group. Significant correlations were observed between reduction of oxidative stress markers and urinary albumin to creatinine ratio. Interestingly, subjects who had higher level of baseline oxidative stress, more the reduction of urinary albumin to creatinine ratio by ARB was observed. Moreover, those who have been treated with angiotensin converting enzyme inhibitors prior to the study, have reduced albuminuria, along with reduction of oxidative stress and inflammation even with the diuretics treatment. These results indicate that RAS induced oxidative stress could be responsible for protection from renal injury in diabetic nephropathy, and urinary marker of oxidative stress could be useful for predicting patients response well to ARB. These results were consistent to EUTOPIA study, which resulted in the reduction of inflammatory markers by the treatment of olmesartan in hypertensive patients[28]. These results strongly suggest the role of oxidative stress in patients with diabetes and hypertension.
Is oxidative/carbonyl stress playing a physiological role in human, and could carbonyl substances be a marker of CKD and metabolic syndrome? To answer this question we have recently determined urinary markers of oxidative and carbonyl stress in young adults (unpublished observation). In young male adults of BMI level of 30 and above, Na consumption is associated with systolic blood pressure, which was accompanied with higher urinary MGO and albumin excretion. In diabetes patients, plasma level of MGO in diabetes predicted 5 year progression of hypertension and intima media thickness of carotid artery (Unpublished observation).

4   ROLE OF RENAL PERFUSION PRESSURE AND OXIDATIVE STRESS
       ON THE PROGRESSION OF HYPERTENSIVE RENAL INJURY.

Clinical studies have indicated that hypertension is an independent risk factor of renal injury. As described above, ROS and carbonyl substances is playing a major role on pathogenesis of hypertensive renal injury. However, there is a difficulty separating the role of renal perfusion pressure to the renal injury. We have developed a system that could specifically determine the role of renal perfusion pressure to the renal injury in rats. The inflatable occluder cuff was chronically implanted on the aorta, in between two renal artery blanches. By servo-control of the cuff occlusion by the downstream arterial pressure, renal perfusion pressure of left kidney was controlled for two weeks [11,29]. Application of this system to angiotensin II infused and Dahl S rats, renal perfusion pressure of left kidney was controlled to a baseline normal level while right kidney was maintained to a higher pressure. Comparison of renal injury between right and left kidney attributed to renal perfusion pressure. With this system, more than 70% of renal medullary renal injury was attributed to renal perfusion pressure. Role of renal perfusion pressure was deferent between glomeruli. In this study, molecules differentially expressed between right and kidney was determined with cDNA microarray, real-time PCR and immunohistochemistry. As a result, molecules related to oxidative stress, inflammation and wound healing were highly expressed in the high pressure kidney compared to normal pressure left kidney [11]. The pressure induced induction of ROS was also confirmed in acute anesthetized rat experiments. Hydrogen peroxide level of renal interstitial fluid was determined using microdialysis technique. Renal perfusion pressure was step controlled by a occluder below and above renal artery, interstitial hydrogen peroxide concentration was increased along with increase in renal perfusion pressure[30]. These results indicate that oxidative stress is induced directly by increase in renal perfusion pressure. Although it has been demonstrated that renal aldehyde is high in SHR compared to control Wister Kyoto rats, there still remain unknown whether renal perfusion pressure could directly induce carbonyl stress[31]. However, as demonstrated in the rat experiments described above, we propose that pressure induced renal oxidative stress could also react with carbonyl compounds especially when in subjects with CKD.

5   PARADIGM SHIFT OF ANTI-AGING BETWEEN ANCIENT AND
       MODERN.

Why is oxidative/ carbonyl stress playing a role in CKD and metabolic syndrome? This can be explained when we consider life style and anti-aging in ancients. We speculate that animals were threatened with 1. starvation, 2. injury 3. infection in ancients. Those who had higher ability to protect from these had an advantage to live longer. As shown in **Figure 2**, enhanced Na reabsorption by renal oxidative stress may have participated in the protection from dehydration in starvation and infection. Those who had enhanced inflammatory mechanism may have an advantage for the infection. Wound healing was also involved in the injury. Thus, our ancestors who have survived enough to leave offspring had adapted to the tough life style in ancients, may have had higher level oxidative stress and inflammation compared to who did not survived. Despite of enhanced oxidative stress, inflammation and wound healing, they did not develop CKD or CVD. This is because the life style of our

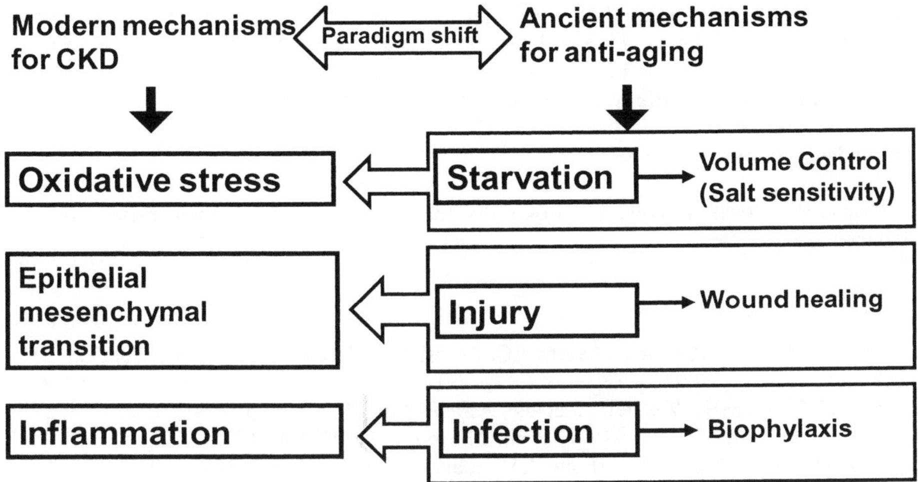

Figure 2   *Mechanism of CKD could be associated with the food and lifestyle of ancients*

ancestors were total deferent from us. There were no salt and pepper available for a steak. They ate low meat for many years. They had more excises since they had to hunt every day. Once upon a time, our ancestors have found salt as a spice, and fire to cook. These tools helped them to live longer on that time. Salted meat can preserve well, and taste better. Those who ate salty meet lived longer because of less chance of dehydration. Cooked meats are also tasted better by carbonyl compounds and preserved better. Thus, we speculate that our ancestors thought salt and fire as anti-aging tools. However, since after this finding of anti-aging tools, they did not have to go much hunting and thereby become less exercise. Invention of farming has also avoided starvation but resulted in lesser exercise with repletion. Many industries have influenced to our environments which may have also altered oxidative/carbonyl stress, inflammation and wound healing. Thus we propose that this paradigm shift of anti-aging could explain why these mechanisms are enhanced in CKD, CVD and metabolic syndrome. We do not seem to be adapted enough in modern environments since our genetical backgrounds and our body anatomy does not seem to be much different from our ancestors26. We might be able to say that CKD and CVD may be isolating subjects who had advantage in ancient life style. This reminded us of dinosaurs which was extinct by changes of environments, and of a fear for us human beings.

6   CONCLUSION

Oxidative/carbonyl stress, inflammation and wound healing is playing a key role in the pathogenesis of CKD. These mechanism have protected from life-threatening environments in ancients which was now become an evil to subjects with modern life style. Therefore, parameters related to these mechanisms such as markers of oxidative stress, inflammation and urinary albumin excretion could be a warning from modern life-threatening evil. Since there still is no medication that could completely protect CKD and CVD, this may be because we are not able to completely inhibit these ancient derived mechanisms. We may have to take into a consideration of ancient life style and counter act to stop the anti-aging paradigm shift.

Acknowledgements

We thank Ms Yoshimi Yoneki, Hiroko Ito and Kiyomi Kisu for their expert technical assistance.

References

1. Sarnak MJ, Levey AS, Schoolwerth AC, Coresh J et al., *Circulation.*2003, **108(17)**, 2154-69, 2003
2. Ibsen H, Olsen MH, Wachtell K, et al,. *Diabetes Care* 2006, **29**, 595-600.
3. Hillege HL, Fidler V, Diercks GF, et al,. *Circulation* 2002, **106**, 1777-1782.
4. Cubeddu LX, Hoffmann IS, Aponte LM, et al, *AJH* 2003, 16, 343-349.
5. Mori T, Cowley AW Jr, Ito S. *J Pharmacol Sci.* 2006, **100(1)**, 2-8.
6. Miyata T, Kurokawa K, van Ypersele de Strihou C. *Kidney Int Suppl.* 2000, **76**, S120-5.
7. Nishiyama A, Fukui T, Fujisawa Y, Rahman M, Tian RX, Kimura S, Abe Y. *Hypertension.* 2001, **37(1)**, 77-83.
8. Schnackenberg CG, Wilcox CS. *Hypertension.* 1999, **33(1 Pt 2)**, 424-8.
9. Meng S, Cason GW, Gannon AW, Racusen LC, *Hypertension.* 2003, **41(6)**, 1346-52.
10. Taylor NE, Glocka P, Liang M, Cowley AW Jr. *Hypertension.* 2006, **47(4)**, 692-8.
11. Mori, T, Polichnowski, A, Glocka, P, Kaldunski, M, Ohsaki, Y, Liang, M & Cowley, AW, Jr., High *J Am Soc Nephrol,* 2008, **19,** 1472-82.
12. Cowley AW Jr, Roman RJ, Kaldunski ML, Dumas P, Dickhout JG, Greene AS, Jacob HJ, *Hypertension.* 2001, **37(2 Part 2)**,456-61.
13. Zou A.P., Li N, Cowley A.W. Jr. *Hypertension* 2001, **37**, 547-553.
14. Makino A, Skelton MM, Zou AP, Roman RJ, Cowley AW Jr. *Hypertension.* 2002, **39(2 Pt 2)**, 667-72.
15. Mori T, Cowley AW Jr *Hypertension.* 2003, **42(4)**, 588-93.
16. Mori T, Cowley AW, Jr. *Hypertension.* 2004, **43(2)**, 341-346.
17. Mori T, O'Connor PM, Abe M, Cowley AW Jr. *Hypertension.* 2007, **49(6)**, 1336-41.
18. Chen YF, Cowley AW Jr, Zou AP. *Am J Physiol Regul Integr Comp Physiol.* 2003, **285(4)**, R827-33.
19. Makino A, Skelton MM, Zou AP, Cowley AW Jr. *Hypertension.* 2003, **42(1)**, 25-30.
20. Taylor NE, Cowley AW Jr. *Am J Physiol Regul Integr Comp Physiol.* 2005, **289(6)**, R1573-9.
21. Nakayama K, Nakayama M, Iwabuchi M, et al, *Am J Nephrol* 2008, **28**, 871-878.
22. Wang, X, Desai, K, Clausen, JT & Wu, L, *Kidney Int,* 2004, **66**, 2315-21.
23. Kumagai T, Nangaku M, Kojima I, Nagai R, Ingelfinger JR, Miyata T, Fujita T, Inagi R. *Am J Physiol Renal Physiol.* 2009 Apr , **296(4)**, F912-21.
24. Nakayama, M, Saito, K, Sato, E, Nakayama, K, Terawaki, H, Ito, S & Kohno, M, *Redox Rep,* 2007, **12,**125-33.
25. Guo, Q, Mori, T, Jiang, Y, Hu, C, Osaki, Y, Yoneki, Y, Sun, Y, Hosoya, T, Kawamata, A,

Ogawa, S, Nakayama, M, Miyata, T & Ito, S, *J Hypertens,* 2009, **27,** 1664-71.
26. Zhu WJ, Mori T, Yoneki Y, Nakayama K, Rahman M, Terawaki H, Nakayama M, Ito S. *Hypertension.* 2007, **50(4),** e135.
27. Ogawa S, Mori T, Nako K, Kato T, Takeuchi K, Ito S. *Hypertension.* 2006, **47(4),** 699-705.
28. Fliser D, Buchholz K, Haller H, *Circulation.* 2004, **110(9)**, 1103-7.
29. Mori T, Cowley AW Jr., *Hypertension* 2004, **43**, 752-759.
30. Jin C, Hu C, Polichnowski A, Mori T, Skelton M, Ito S, Cowley AW Jr. *Hypertension.* 2009, **53(6),** 1048-53.
31. Vasdev, S, Ford, CA, Longerich, L, Parai, S, Gadag, V & Wadhawan, S, *Artery,* 1998, **23,** 10-36.

ROLE OF THE GLYOXALASE SYSTEM IN RENAL SENESCENCE

Y.Ikeda[1], M. Nangaku[1], and R. Inagi[1]

[1]Division of Nephrology and Endocrinology, University of Tokyo School of Medicine, 7-3-1 Hongo, Bunkyo-ku, Tokyo 113-8655, Japan

1 INTRODUCTION

Because most diseases develop or progress with age, interventions that delay aging would greatly benefit both the prevention of age-associated diseases and the elongation of life span. Accumulating evidence from a variety of organisms has demonstrated that the causes and processes of senescence are multifactorial, involving both genetic and environmental factors, including telomere length, klotho gene, and caloric restriction (insulin resistance)[1-3]. Among them, the most important role in the aging process is played by oxidative stress. The oxidative (free radical) theory proposes that the exacerbation of reactive oxygen species (ROS) production in cells results in cumulative damage with subsequent senescence[4]. ROS cause oxidative or glycated modification (or both) of proteins associated with pathophysiological cellular dysfunction, emphasizing the linkage of protein glycation with longevity. This review summarizes the role of oxidative stress, subsequent oxidative protein modification (glycation), and the anti-glycation enzyme glyoxalase system in aging.

2   OXIDATIVE STRESS AND PROTEIN MODIFICATION BY GLYCATION IN LONGEVITY

The oxidative (free radical) theory is consistent with various studies of longevity. Hyperinsulinemia and metabolic diseases induce a reduction in life span associated with an increase in intracellular ROS production caused by insulin resistance[5, 6]. Conversely, caloric restriction extends life span in diverse species in association with a reduction in intracellular ROS production[7]. Further, longevity is also influenced by ROS production through the renin-angiotensin system. Mice deficient in angiotensin II type 1 (AT1) receptor [8] or rats administrated an AT1 receptor blocker (ARB)[9] show a prolongation of life span by amelioration of oxidative stress via suppression of the renin-angiotensin system. Oxidative theory also proposes a pathogenic role of posttranslational protein modification by glycation (carbonylation) with age and emphasizes the beneficial effect of anti-glycation enzymes that detoxify glycation precursors on longevity. Under oxidative stress condition, aldehydes and ketones, which are highly reactive glycation precursors, are generated by glycoxidation (or lipid oxidation) and bind to proteins to form advanced glycation endproducts (AGEs). AGEs enhance ROS production through AGE receptor (RAGE)–mediated signaling [10]. Glycation of mitochondrial protein significantly induces mitochondrial uncoupling, resulting in increased mitochondrial ROS production [10]. These findings are consistent with the idea that anti-glycation enzyme glyoxalase (GLO)1, which detoxifies the most potent glycation precursor, methylglyoxal (MGO), contributes to longevity. Overexpression of GLO1 extends life span in *C. elegans* by reducing glycation by MGO with a subsequent reduction in ROS production [11]. Taken together, pathogenic protein glycation decreases longevity by augmenting various ROS production pathways as well as via the loss of function of glycated proteins.

3   PATHOPHYSIOLOGY OF THE GLYOXALASE SYSTEM IN KIDNEY DISEASE

Glycated (carbonylated) modification of proteins by reactive aldehydes and ketones via the Maillard reaction is a frequent occurrence under hyperglycemic or oxidative stress conditions. Through a series of intermediates, this reaction finally leads to the formation of irreversible adducts, referred to as AGEs [12], which are associated with age-related diseases including diabetes, Alzheimer's disease, and atherosclerosis. MGO has received considerable attention as a highly reactive AGE precursor and its level is increased in various oxidative stress-related diseases. Oxidative stress significantly enhances MGO formation by glycoxidation or lipid oxidation and subsequent accumulation of MGO adducts, which lead to the loss of native biological function and induction of inappropriate cellular responses, including ROS production and apoptosis. Moreover, MGO is itself a toxic substance. MGO is mainly detoxified by the glyoxalase system, which is well conserved in various species.   The glyoxalase system is composed of two enzymes: GLOI, which metabolizes MGO to S-D-lactoylglutathione, and GLOII, which converts S-D-lactoylglutathione to D-lactate[13].   A series of studies on MGO and GLOI demonstrated their pathological contribution in diabetic complications (diabetic nephropathy and diabetic retinopathy), and neurological disorders, as well as the high risk of atherosclerosis by GLOI deficiency[14-18].   These findings suggest that the enhancement of GLO1 activity may show cellular- or organ-protective benefits against oxidative, hyperglycemic, or hypoxic damage.

The relationship between GLO1 and diabetic nephropathy has been the focus of intensive research.   MGO levels are elevated in diabetic subjects with more rapid progression of nephropathy [19].   On the other hand, the pathophysiological role of GLO1 in other kidney diseases remains to be elucidated. The most common cause of human acute kidney injury is ischemia due to hypotension or sepsis.   Moreover, ischemia/reperfusion injury is a significant problem in kidney transplantation. This background indicates the clinical importance of renal ischemia/reperfusion injury.   The mechanisms underlying renal ischemia/reperfusion injury are most likely multifactorial and interdependent, involving hypoxia, inflammatory responses and free radical damage.   Recently, several reports have demonstrated that MGO levels are increased in cardiac or cerebral ischemia/reperfusion: MGO levels are increased in heart homogenates [20], while methylglyoxal-arginine adducts (Argpyrimidine) are detected in cerebral arterial walls within the infarcted zone after 24 hr reoxygenation [21].   These findings spurred us to investigate the biological role of GLO1 in renal ischemia/reperfusion injury.

Investigations were conducted in rats by occlusion of the left renal vessels for 45 min, followed by 24 h reperfusion.   Of note, ischemia/reperfusion induced a reduction in GLOI activity in the kidney, which was associated with morphological damage and renal dysfunction.   Further, these manifestations were significantly ameliorated in rats overexpressing human GLOI (GLOI Tg rats) [22], which showed a reduction in tubulointerstitial injury and apoptosis of tubular cells, and improvement of renal function [23]. On immunohistochemistry, accumulation of the renal MGO adduct CEL induced by ischemia/reperfusion was also decreased in GLOI Tg rats as compared with wild-type rats. In *in vitro* studies, we exposed immortalized rat proximal tubular cells (IRPTC) or GLOI-knocked down IRPTC with siRNA to hypoxia (0.2% O_2) followed by reoxygenation (95% air). IRPTC knockdown of GLOI exacerbated the cellular injury estimated by LDH assay.   Taken together, these findings show that GLOI exerts renoprotective effects in ischemia/reperfusion injury by reducing the modification of protein by MGO in tubular cells [23].

4   ROLE OF GLOI IN RENAL SENESCENCE

Recent evidence has demonstrated the beneficial contribution of GLO1 to longevity. Senescence status appears to depend on the balance between the formation of intracellular aldehyde (e.g., MGO) and the activity of its detoxifying enzymes (e.g., GLO1) [24]. GLO1 activity is decreased with age in *C. elegans*[11], human lens[25], and human brain [26]. A reduction in life span is observed in high glucose-exposed *C. elegans* by glucose toxicity, an effect which includes an increase in ROS formation and glycation of mitochondrial proteins [27]. Importantly, GLO1 overexpression prevents this glucose toxicity-mediated life span reduction. Taken together with our studies demonstrating the renoprotective effect of GLO1 in kidney disease [23], we hypothesized that the decrease in AGE formation by GLO1 ameliorates renal senescence phenotypes.

To refine this hypothesis, we established a method to assess cellular senescence in human renal proximal tubular epithelial cells (RPTECs) at early or late passages by measuring changes in senescence markers such as cyclic dependent kinase inhibitors (p53, p21, or p16) or senescence associated beta-gal (SABG) staining. We then assessed if GLO1 reduces cellular senescence phenotypes by detoxification of the AGE precursor methylglyoxal. The results showed that cellular senescence, including expression levels of senescence markers, was significantly increased in RPTECs at late passage. Similar senescence phenotypes were induced by the chemical senescence inducer etoposide in cells at early passage. Most importantly, compared to control RPTECs, these changes were significantly less marked in cells overexpressing GLO1, but were more severe in GLO1-knock down RPTECs. In *in vivo* studies in rats overexpressing human GLO1 (GLO1-Tg), renal senescence phenotypic changes were determined by morphological changes, positive rate of SABG staining, and expression level of CDK inhibitors. Although wild-type and GLO1 Tg rats at the same age showed no differences in food intake, body weight gain, or blood pressure, aged (14-month-old) GLO1-Tg rats showed milder renal senescence phenotypes than the aged wild-type rats (p<0.05), as shown by the amelioration of age-related interstitial thickening with renal dysfunction, and a decrease in CDK inhibitor expression or SABG-positive rate. Importantly, renal GLO1 activity was significantly decreased with age in the wild-type rats but not GLO1-Tg rats, indicating a negative correlation between GLO1 and renal senescence. In conclusion, GLO1 plays a pivotal role in the amelioration of renal senescence, suggesting that AGE formation, which is enhanced by oxidative stress, contributes to the process of renal aging, in part at least.

5   CONCLUSION

Advanced glycation end-products (AGEs) accumulate in cells and organisms with aging, suggesting that AGE formation contributes to the aging process.   However, it remains unclear whether AGEs are a cause or a consequence of aging, particularly in the kidney. We previously demonstrated that GLOI exerts renoprotective effects in renal ischemia/reperfusion injury by reducing AGE formation in association with the suppression of oxidative stress in tubular cells.   Further, GLO1 also contributes to renal senescence by ameliorating senescence phenotypic changes, such as an increase in the expression of senescence markers (CDK inhibitors and SABG) and tubulointerstitial thickening with renal dysfunction.   We hypothesize that decreasing AGE formation by an enzyme which detoxifies AGE, precursor glyoxalase I (GLO1), ameliorates renal phenotypes with aging.   The mechanism of this amelioration of renal senescence by GLO1 might be via a reduction in MGO level and breaking of the vicious cycle of oxidative stress and AGE formation with age.   Further studies are required to elucidate the mechanism of the reduction in GLO1 activity by ischemia/reperfusion or age.   Our findings suggest that AGEs play an important role in senescence and GLO1 plays a pivotal role in the amelioration of renal senescence via, at least in part, the inhibition of AGE formation and subsequent breaking of the vicious cycle of oxidative stress and AGE formation. findings provide important insights into understanding the linkage of the development/progression of kidney disease with renal senescence, and suggest the possibility of renoprotective therapies targeting the glyoxalase system.

Acknowledgments

This work was supported by Grants-in-Aid for Scientific Research from Japan Society for the Promotion of Science (19590939 to RI and 21390260 to MN) and a grant from The Kidney Foundation, Japan (JKFB09-2 to RI).

References

1.   Shawi M, Autexier C. *Mech Ageing Dev.* 2008,**129**,3-10.

2.   Wang Y, Sun Z. *Ageing Res Rev.* 2009,8,43-51.

3.   Spindler SR. *Ageing Res Rev.* 2009 Oct 21.

4.   Harman, D. *J Gerontol.* 1956, **11** , 298-300.

5.   Bartke A. *Cell Cycle.* 2008, **7**, 3338-43.

6.   Sanz A, Pamplona R, Barja G. *Antioxid Redox Signal.* 2006, **8**, 582-99.

7.   Mair W, Dillin A. *Annu Rev Biochem.* 2008, **77**, 727-54.

8. Benigni A, Corna D, Zoja C, Sonzogni A, Latini R, Salio M, Conti S, Rottoli D, Longaretti L, Cassis P, Morigi M, Coffman TM, Remuzzi G. *J Clin Invest.* 2009, **119**, 524-30.

9. Basso N, Paglia N, Stella I, de Cavanagh EM, Ferder L, del Rosario Lores Arnaiz M, Inserra F. *Regul Pept.* 2005, **128**, 247-52.

10. Tan AL, Forbes JM, Cooper ME. Semin Nephrol. 2007,27,130-43.

11. Morcos M, Du X, Pfisterer F, Hutter H, Sayed AA, Thornalley P, Ahmed N, Baynes J, Thorpe S, Kukudov G, Schlotterer A, Bozorgmehr F, El Baki RA, Stern D, Moehrlen F, Ibrahim Y, Oikonomou D, Hamann A, Becker C, Zeier M, Schwenger V, Miftari N, Humpert P, Hammes HP, Buechler M, Bierhaus A, Brownlee M, Nawroth PP. *Aging Cell.* 2008, **7**, 260-9.

12. Brownlee M, Cerami A, Vlassara H. *N Engl J Med.* 1988, **318**, 1315-21.

13. Thornalley PJ. *Biochem. J.* 1990, **269**, 1-11.

14. Miller AG, Smith DG, Bhat M, Nagaraj RH. *J Biol Chem.* 2006, **281**,11864-71.

15. Kuhla B, Luth HJ, Haferburg D, Boeck K, Arendt T, Munch G. *Ann N Y Acad Sci.* 2005, **1043**,211-6.

16. Hovatta I, Tennant RS, Helton R, Marr RA, Singer O, Redwine JM, Ellison JA, Schadt EE, Verma IM, Lockhart DJ, Barlow C. *Nature.* 2005, **438**, 662-6.

17. Sakamoto H, Mashima T, Kizaki A, Dan S, Hashimoto Y, Naito M, Tsuruo T. *Blood.* 2000, **95**, 3214-8.

18. Miyata T, van Ypersele de Strihou C, Imasawa T, Yoshino A, Ueda Y, Ogura H, Kominami K, Onogi H, Inagi R, Nangaku M, Kurokawa K. *Kidney Int.* 2001, **60,** 2351-9.

19. Beisswenger PJ, Drummond KS, Nelson RG, Howell SK, Szwergold BS, Mauer M. *Diabetes.* 2005, **54**, 3274-81.

20. Bucciarelli LG, Kaneko M, Ananthakrishnan R, Harja E, Lee LK, Hwang YC, Lerner S, Bakr S, Li Q, Lu Y, Song F, Qu W, Gomez T, Zou YS, Yan SF, Schmidt AM, Ramasamy R. *Circulation.* 2006, **113**, 1226-34.

21. Oya T, Hattori N, Mizuno Y, Miyata S, Maeda S, Osawa T, Uchida K. *J Biol Chem.* 1999, **274**, 18492-502.

22. Inagi R, Miyata T, Ueda Y, Yoshino A, Nangaku M, van Ypersele de Strihou C, Kurokawa K. *Kidney Int.* 2002, **62**, 679-87.

23. Kumagai T, Nangaku M, Kojima I, Nagai R, Ingelfinger JR, Miyata T, Fujita T, Inagi R. *Am J Physiol Renal Physiol.* 2009, 296, F912-21.

24. Dmitriev LF, Titov VN. *Ageing Res Rev.* 2009 Oct 1. [Epub ahead of print]

25. Mailankot M, Padmanabha S, Pasupuleti N, Major D, Howell S, Nagaraj RH. *Biogerontology.* 2009 Feb 24. [Epub ahead of print]

26. Kuhla B, Boeck K, Lüth HJ, Schmidt A, Weigle B, Schmitz M, Ogunlade V, Münch G, Arendt T, *Neurobiol Aging.* 2006, 27, 815-22.

27. Schlotterer A, Kukudov G, Bozorgmehr F, Hutter H, Du X, Oikonomou D, Ibrahim Y, Pfisterer F, Rabbani N, Thornalley P, Sayed A, Fleming T, Humpert P, Schwenger V, Zeier M, Hamann A, Stern D, Brownlee M, Bierhaus A, Nawroth P, Morcos M. C *Diabetes.* 2009, **58**, 2450-6.

SERUM LOW MOLECULAR WEIGHT FLUORESCENT AGES ARE HIGHER IN NEONATES THAN IN ADULTS: ROLE OF KIDNEY METABOLISM

A. Gugliucci[1], S. Kimura[2], T.Menini[1], J. Taing[1] and M. Numaguchi[3]

[1]Glycation, Oxidation and Disease Laboratory, Department of Basic Sciences, Touro University-California, Vallejo, USA,
[2]Clinical Pathology Department Showa University School of Medicine, Yokohama, Japan
[3]Department of Obstetrics and Gynecology, Dokkyo University School of Medicine, Tokyo, Japan

1 INTRODUCTION

The Maillard reaction initiated by reactive dicarbonyls occurs much faster than when it is produced by glucose. A large body of evidence shows the importance of the Maillard reaction in animals and humans and has been correlated with the development of several disease entities, as recently reviewed [1]. Most of the information we have pertains to measurements performed in samples from adult individuals. Very little is known about the Maillard reaction *in utero*. In the recent 10th Maillard Reaction Symposium showcased in this volume, Thornalley's group presented data on the presence of AGEs molecules in rat embryo, using the state of the art technology for specific AGE adducts and free adducts (unpublished results, 10th Maillard Reaction Symposium meeting booklet, abstract 83, page 104). This seminal work shows that the Maillard reaction occurs early in life, at least in rodents. Data in humans are very scarce. Actually there is only one report in the literature demonstrating the presence of pentosidine concentrations in cord blood, using total serum hydrolisates [2].

A characteristic feature of chronic renal failure is a profound increase in plasma advanced glycation (AGE) free adducts and peptides [3-5]. AGE residues are formed on long and short-lived cellular and extracellular proteins. Cellular proteolysis forms AGE-peptide and AGE free adducts from these proteins, which are released into plasma for urinary excretion. Glomerular filtration rate and tubular functions are key determinants of their elimination and therefore of their circulating steady state concentration [6-7]. AGE free adducts are the chief molecular form by which AGEs are excreted in urine [7-11]. They display a high renal clearance, which distinctly declines in chronic renal failure patients, leading to accumulation of plasma AGE free adducts. This buildup is further augmented in end stage renal disease (ESRD) patients on hemodialysis

(HD) by increased AGE formation. The toxicity of AGEs resides not only on resistance of the extracellular matrix to proteolysis but mainly to AGE receptor mediated responses [12-13].
Kidney function develops in several stages in the embryo and fetus [14]. Essentially there is no glomerular filtration in the embryo and organogenesis takes a long time, in actual fact up to 4/5th of gestation is needed in sheep to achieve full nephrogenesis [15]. Filtration and urine excretion certainly occurs in the fetus, albeit it has only a fraction of the glomerular filtration rate (GFR) achieved in adult life. Renal blood flow, a key determinant of GFR is 2-3% in the fetus, whereas it reaches 20% in adults [15]. Filtration and urine excretion are important in fetal and amniotic fluid physiology. Nevertheless, at least from extensive data in sheep, the GFR at birth is less than 50% of adults [14-16]. In view of the importance of the kidney in AGEs metabolism and notably in low molecular weight and free adducts excretion, together with the evidence for the Maillard reaction occurring already at the embryo stage, we hypothesized that these compounds should be high in cord blood as compared to adult blood. To address this hypothesis we measured low molecular weight AGE fluorescence in cord blood from healthy neonates and healthy adults as controls.

2   MATERIAL AND METHODS

2.1   Patients

We performed a case-control study with 30 newborns (normal term deliveries, Apgar 8–9, male/female, 14/16). Gestational age was assessed from the date of last menstrual period and concurrent clinical assessment using the New Ballard Score. To classify infants as appropriate or small for gestation, reference of weight was made to Lubchenko's charts of intrauterine growth. The cord blood samples were obtained from the Obstetrics Unit Dokkyo University School of Medicine. The control adult subjects in the study (15 male and 15 female, aged 25–45 y) were selected from a healthy population of hospital workers at the Northern Yokohama Hospital, Showa University Tsuzuki-ku, Yokohama City. The nature and purpose of the study was explained to the parents and adult controls and informed consent was obtained from all subjects. The protocol was approved by the institutional review board of Showa and Dokkyo Universities, and investigations were performed in accordance with the principles of the Helsinki declaration. Immediately after delivery, a segment of the umbilical cord was clamped and four milliliters of venous blood sample from the umbilical cord was collected in evacuated dry tubes with separating gel just after delivery of the neonate. Blood was centrifuged at 800 g, at 4°C for 15 min and separated serum was immediately analyzed or frozen at 80° C until use. Chemicals: All chemicals are analytical grade and purchased from Sigma (St. Louis, MO, USA).

2.2   AGE fluorescence measurements

For total serum AGE fluorescence or pentosidine fluorescence, serum was diluted 1/100 in PBS, for low molecular weight AGEs, serum was diluted 1/2 with PBS, ultra-filtered through 10 kDa cut-off Amicon filters in a refrigerated centrifuge at 4°C and 4,000 g for 8 h and read without further dilution as previously described [17].

Table 1   *Clinical characteristics and serum AGEs in neonates and adult controls. Data are presented as mean ± SD and range. AGE fluorescence was measured as described in Methods. Low molecular weight serum fractions were obtained by ultrafiltration using 10 kDa Amicon filters as described in Methods. Significant differences between cord blood and adult control values are shown in the right column.*

Parameter	Neonate cord blood (n= 30, m/f 14/16)	Controls (n= 30, m/f 15/15)	p-value
Gestational age (weeks)	39 ± 2	---------	
Weight (g)	3010 ±310	---------	
Apgar	8-10	----------	
Total serum AGEs fluorescence (370 ex/440 em, AU)	4281 ± 456	4707 ± 652	0.03
LMW serum AGEs fluorescence (370 ex/440 em, AU)	128 ± 29	57 ± 18	0.0001
Ratio LMW/Total serum AGEs (370 ex/440 em, AU x 10^3)	30.5 ± 5.4	12.1±3.4	0.0001
Total serum pentosidine fluorescence (335 ex/385 em, AU)	869 ± 259	1115 ± 310	0.001
LMW serum pentosidine fluorescence (335 ex/385 em, AU)	66 ± 15	26 ± 7	0.0001
Ratio LMW/Total serum pentosidine (335 ex/385 em, AU x 10^3)	73 ± 24	25 ±8	0.0001

Determination of AGEs is based on the classic spectrofluorimetric detection 18-19. Fluorescence intensity was recorded at the emission maximum 440 nm upon excitation at 370 nm and at emission maximum 385 nm upon excitation at 335 nm for pentosidine fluorescence [18-19]. Fluorescence intensity is expressed in arbitrary units (AU) and corrected for dilutions. We employ a SPECTRAmax Gemini XPS spectrofluorometer with SOFTmax PRO software (Molecular Devices, Sunnyvale, CA).

2.3   Statistical analysis

One-sample Kolmogorov–Smirnov test was performed to confirm the assumption of normality in the 2 groups for 4 outcome variables. Results are expressed as the mean value ± SD. Differences were considered to be statistically significant if the null hypothesis could be rejected with 95% confidence. Correlations were assessed with the Pearson test. A p 0.05 was considered significant. The SPSS 12.0 statistical software package (SPSS Inc., Chicago, IL) was used for all data analyses.

3   RESULTS

Table 1 shows the characteristics of our neonate population and the AGE fluorescence for total serum and the low molecular weight fractions in both neonate cord blood and adult controls. Total serum AGEs fluorescence and total serum pentosidine-like fluorescence were significantly lower in neonates (p < 0.03 and < 0.001 respectively). The correlation between both parameters showed an r=0.45, p < 0.001. Conversely, low molecular weight (LMW) fluorescent AGEs were almost 3-fold higher in neonate cord blood than in adults (p < 0.00001) and a similar finding is apparent for low molecular weight pentosidine-like fluorescence. The correlation between both parameters showed an r=0.62, p < 0.0001. The ratio LMW/total serum fluorescent AGEs was also more than 2-fold higher in cord blood than in adults (p < 0.00001), as was the ratio LMW/total serum pentosidine-like fluorescence (p < 0.0001). No significant correlation was found between gestational age or weight and the above mentioned parameters, a natural consequence of the fact that all were normal-term deliveries.

4   DISCUSSION

AGEs and especially peptides and glycation free adducts, show a marked increase in uremia and have been linked to several pathophysiological mechanisms in this disorder [5-7]. Indeed, chronic renal failure results in much higher concentrations of circulating AGEs than diabetes. This retention is even more striking for glycation free adducts and peptides, since the kidney plays a major role in their metabolic and excretory disposal as our work, and others have shown in the past decade [4, 20-23].

If Maillard reaction products are formed in embryonic and, mainly, in fetal life (albeit at a slower pace as the natural biology of the reactions seems to indicate), their elimination by kidney filtration poses a problem, since the latter much lower in fetal life [14-16]. Indeed, proximal tubular function is also characteristically poor in the fetus, which is not surprising since the growth of the fetal kidney in the latter stages of gestation is characterized by intense proximal tubulogenesis [16-24-25]. Proximal reabsorption and further catabolism and re-excretion of low molecular weight AGEs are key elements in AGE elimination [4, 8-11].

Our results provide evidence supporting previous data on the formation of AGEs in fetal life and, for the first time, show that molecular weight AGEs are much higher in cord blood than in adults, which is in agreement with the key role the kidney plays in their catabolism. There is only one previous work in the literature, reporting the presence of pentosidine in cord blood, at lower concentrations than in adults [2]. In this study, total serum pentosidine was measured by HPLC after hydrolysis. In this sense our results are in agreement with this only previous report, while adding interesting and novel data shedding new light on the handling of the LMW and free adducts in the fetus. Both measurements performed in LMW serum fractions (pentosidine and 370/440 fluorescence) correlate with each other, which adds support to our hypothesis.

The classical 370/440 fluorescence is mainly associated to a collagen cross-link that was poorly characterized until recently, while pentosidine is a well-characterized AGE used as marker of the glyco-oxidation process [18-19]. To be sure, a limitation of our study is that we employed fluorescence, which has lower specificity and sensitivity than the state-of-the art techniques [21-22]. However, collagen-linked fluorescence at excitation/emission 370/440nm has widely been used as a marker for advanced glycation in studies of aging, diabetic complications, and end-stage renal disease. The nature of the major fluorophore has been recently partially elucidated; it has molecular weight of 623.2Da, a UV maximum at 348nm, and involves a lysine residue in an aromatic ring. It has a complex mechanism of formation, perhaps related to hypoxia since elevated levels were also found in non-diabetic individuals with chronic lung disease [26]. In this regard, it must be remembered that the fetal arterial pO_2 is about 20-30 mmHg, therefore a different Maillard chemistry than in adults may plausibly occur in these conditions [27]. In other words, although further studies are needed, our data showing a striking 3-fold increase in LMW fluorescent AGEs in cord blood provides a good starting point. The next phase is to confirm and extend these data by means of the application of liquid chromatography with tandem mass spectrometric (LC-MS/MS) in the detection of cord blood AGEs to provide a comprehensive and quantitative analysis of glycation adducts [20-21].

In Figure 1 we summarize our interpretation of the data presented. We depict the determinants of the steady state concentration of LMW AGEs and free adducts in adults (A) and in fetus (B). In both, production of AGEs by reactive dicarbonyls in cells seems to be the most favored pathway; alternatively some AGEs may come from food in adults [28-29]. AGE proteins may be catabolized by macrophages to yield LMW AGEs, although there is now doubt as to whether AGE-modified proteins really are metabolized by AGE receptors in vivo [21]. Rather, an important feature of the receptor for AGEs may be decrease in expression of glyoxalase 1 by its ligand S100A12 protein (which binds to the AGE receptor RAGE) leaving the tissues vulnerable to dicarbonyl stress and related AGE formation [21]. In any case, whichever the pathway, there is strong evidence that in adults LMW AGEs and free adducts are formed, circulate and are cleared by filtration and proximal tubule catabolism [6-13].

Our data showing a 3-fold increase of LMW AGEs and free adducts in cord blood (B), may be explained by the differences in the physiology of the fetal kidney, namely, null to low glomerular filtration coupled with low functionality of the proximal tubules [14-16,24-25]. In this regard, although predictably, formation of AGEs may be lower than in adults, clearance is much lower, which results in a clear-cut increase in these serum AGEs in the fetus. The role of the placental circulation and or the presence of maternal AGEs in the fetal circulation are points to consider and an open field for further research.

5   CONCLUSIONS

In conclusion, we show for the first time a 2-3 fold increase in low molecular weight AGEs in cord blood as compared to adults. Our finding is in agreement with the present paradigm giving the kidney a critical role in AGEs disposal, in this sense it provides a physiological model for low GFR and poor proximal tubule function in humans. Our results allow for two predictions: low molecular weight AGEs and free adducts should decrease after birth and they should be found in the amniotic fluid at increased concentrations during the latest phases of gestation.

A
AGE precursors in cells
AGE proteins
Macrophage and other cells catabolism to peptides
Renal filtration, proximal reabsorption and catabolism
LMW AGEs free adducts
Steady state concentration depends on the production and kidney function kinetics
AGE in food
Free adducts in urine
Gugliucci © 2010
B
Maternal AGEs?
AGE precursors in cells
AGE proteins
Macrophage and other cells catabolism to peptides
GFR reaches less than 50% of adult values at term. Proximal tubule nephrogenesis is completed last
LMW AGEs free adducts
Steady state concentration depends on the production, (which might be less in fetus) and kidney function, which is far less efficient in fetal life
No kidney clearance for half gestation, low clearance in 3rd trimester AGEs in amniotic fluid?
Gugliucci © 2010

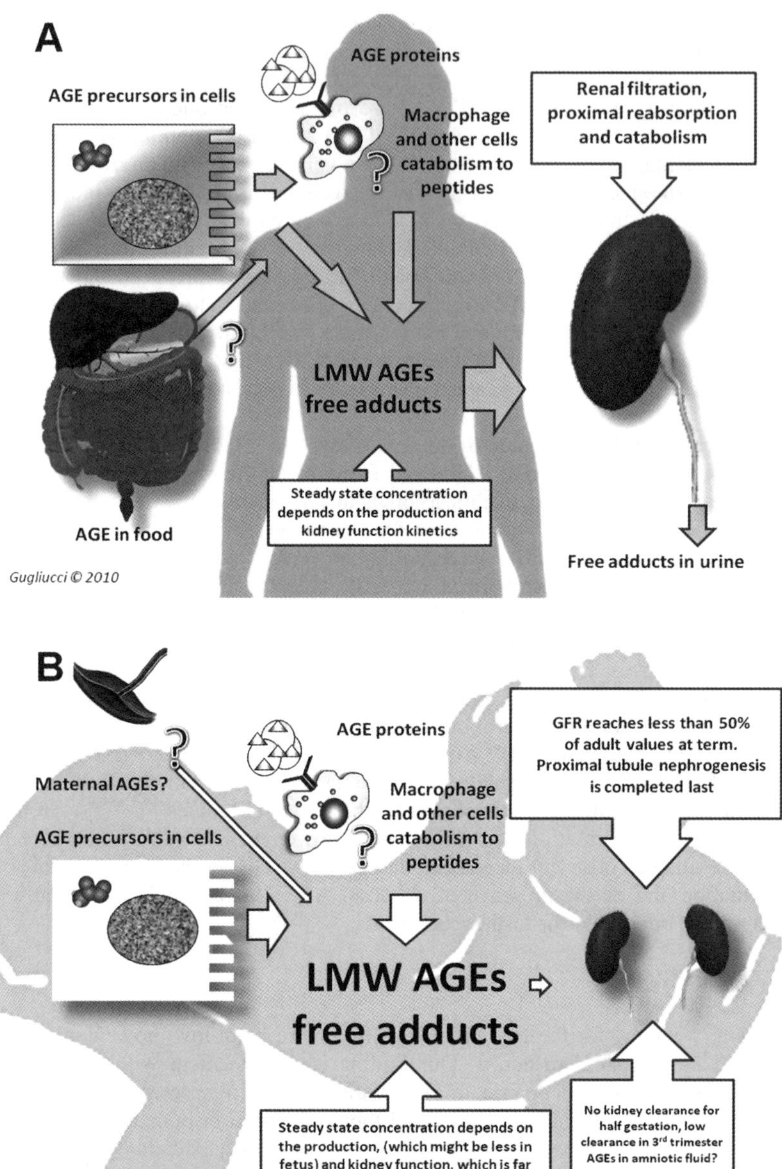

Figure 1

Figure 1   ***Proposed mechanism to explain higher steady-state level of low molecular AGEs in fetal blood as compared with adults.*** *In this simplified diagram we illustrate the determinants of the steady state concentration of LMW AGEs and free adducts in adults (A) and in fetus (B). In both, production of AGEs by reactive dicarbonyls in cells seems to be the most favored pathway and some AGEs may come from food in adults. AGE proteins may be catabolized by macrophages to yield LMW AGEs, although there is now doubt as to whether AGE-modified proteins are metabolized by AGE receptors occurs in vivo. Whichever the pathway, in adults LMW AGEs and free adducts are formed and cleared by filtration and proximal tubule catabolism. Our data showing a 3-fold increase of LMW AGEs and free adducts in cord blood (B), may be explained by the differences in the physiology of the fetal kidney, namely, low glomerular filtration coupled with low functionality of the proximal tubules. The role of the placental circulation and or the presence of maternal AGEs in the fetal circulation are points to consider and an open field for further research.*

Acknowledgements

This work was funded in part by a Grant-in aid in Scientific Research, (C) 14572186 from the Ministry of Education, Science, Culture and Sports of Japan to S.K. and by a Touro University grant to A.G. The authors are grateful to Dr P. P. Agagné for critical reading of this manuscript.

References

1.   Tessier FJ. *Pathol Biol (Paris).* 2009. Epub ahead of print
2.   Tsukahara H, Ohta N, Sato S, Hiraoka M, Shukunami K, Uchiyama M, Kawakami H, Sekine K, Mayumi M. *Free Radic Res* 2004, **38**, 691-5.
3.   Sourris KC, Forbes JM. *Curr Drug Targets* 2009, **10**, 42-50.
4.   Gugliucci A, Bendayan M. *Diabetologia* 1996; **39**:149–60
5.   Miyata T, Ueda Y, Yoshida A, Sugiyama S, Iida Y, Jadoul M, Maeda K, Kurokawa K, van Ypersele de Strihou C. *Kidney Int* 1997, **51**, 880-7.
6.   Rabbani N, Sebekova K, Sebekova K, Jr., Heidland A, Thornalley PJ. *Kidney Int* 2007, **72**, 1113-21.
7.   Forbes JM, Thorpe SR, Thallas-Bonke V, Pete J, Thomas MC, Deemer ER, Bassal S, El-Osta A, Long DM, Panagiotopoulos S, Jerums G, Osicka TM, Cooper ME. *J Am Soc Nephrol*, 2005, **16**, 2363-72.
8.   Thornalley PJ. J Ren Nutr 2006;**16**:178–84.
9.   Thornalley PJ. Pediatr Nephrol 2005;**20**:1515–22.38
10.   Ozdemir AM, Hopfer U, Rosca MV, Fan XJ, Monnier VM, Weiss MF. *Am J Nephrol* 2008, **28**, 14-24.
11.   Ozdemir AM, Hopfer U, Erhard P, Monnier VM, Weiss MF. *Ann N Y Acad Sci* 2005, **1043**, 625-36.
12.   Thomas MC, Tikellis C, Burns WM, Bialkowski K, Cao Z, Coughlan MT, Jandeleit-Dahm K, Cooper ME, Forbes JM. J Am Soc Nephrol 2005, **16**, 2976-84.
13.   Miyata T, Dan T. Diabetes Res Clin Pract 2008, **82 Suppl 1**, S25-9.
14.   Wintour EM, Riquelme R, Gaete C, Rabasa C, Sanhueza E, Silver M, Towstoless M, Lumbers ER. Reprod Fertil Dev 1995, 7, 415-26.

15.   Modi N. Br Med Bull 1988, **44**, 935-56.
16.   Smith FG, Lumbers ER. Biol Neonate1989, **55**, 309-16.
17.   Gugliucci A, Mehlhaff K, Kinugasa E, Ogata H, Hermo R, Schulze J, Kimura S. Clin Chim Acta 2007, **377**, 213-20.
18.   Monnier VM, Kohn RR, Cerami A. Proc Natl Acad Sci U S A 1984, **81**, 583-7.
19.   Sell DR, Nagaraj RH, Grandhee SK, Odetti P, Lapolla A, Fogarty J, Monnier VM. 1991. Pentosidine: a molecular marker for the cumulative damage to proteins in diabetes, aging, and uremia. Diabetes Metab Rev 7, 239-51.
20.   Thornalley PJ, Rabbani N. Semin Dial 2009, **22**, 400-4.
21.   Rabbani N, Thornalley PJ. Perit Dial Int 2009, **29 Suppl 2**, S51-6.
22.   Thornalley PJ. Perit Dial Int 2005;**25**:522–33.
23.   Gerdemann A, Lemke HD, Nothdurft A, Heidland A, Munch G, Bahner U, Schinzel R. Clin Nephrol 2000, **54**, 276-83.
24.   Elder JS, O'Grady JP, Ashmead G, Duckett JW, Philipson E. 1990. Evaluation of fetal renal function: unreliability of fetal urinary electrolytes. J Urol 144, 574-8; 93-4.
25.   Llanos A. Reprod Fertil Dev 1995, **7**, 1311-9.
26.   Sell DR, Nemet I, Monnier VM. Arch Biochem Biophys. 2009, Oct 30. [Epub ahead of print]
27.   Heisler D. Int Anesthesiol Clin 1993, **31**, 103-7.
28.   Cai W, He JC, Zhu L, Chen X, Zheng F, Striker GE, Vlassara H. Am J Pathol 2008, **173**, 327-36.
29.   Vlassara H, Striker G. Curr Diab Rep 2007, **7**, 235-41.

THE RENIN-ANGIOTENSIN SYSTEM AND ADVANCED GLYCATION END-PRODUCTS IN DIABETIC NEPHROPATHY HOW IMPORTANT ARE THESE PATHWAYS AS THERAPEUTIC TARGETS?

K.C. Sourris[1] and J.M. Forbes[1]

[1]JDRF Centre for Diabetic Complications, Baker IDI Heart and Diabetes Institute, Melbourne, Australia

1   BACKGROUND

The incidence of diabetes is increasing in all western societies irrespective of age, sex or ethnicity. Diabetes is a metabolic disorder characterised by chronic hyperglycaemia, hypertension, dyslipdaemia, microalbuminuria, and inflammation and is associated with numerous micro and macro vascular complications which include retinopathy, neuropathy, cardio/cerebrovascular disease and nephropathy [1,2]. Indeed, diabetic nephropathy, is the major cause of end-stage renal disease in Western societies [3,4]. Renal disease in diabetic patients is characterised haemodynamic (hyperfiltration and hyperfusion) as well as structural abnormalities (glomerulosclerosis, alterations in tubulointerstitium including interstitial fibrosis) and metabolic changes [5,6]. Within, glomeruli, there is thickening of basement membranes, mesangial expansion and hypertrophy and glomerular epithelial cell (podocyte) loss. In conjunction, disease progression is also seen in the tubulointerstitial compartment causing expansion of tubular basement membranes, tubular atrophy, interstitial fibrosis and arteriosclerosis. Not surprisingly, progression to end stage renal disease is enhanced by hyperglycaemia, hypertension and proteinuria, which are all common in diabetes [2,7,8]. This article will examine the role of the renin angiotensin system (RAS) in the development and progression of diabetic kidney disease, as well as its links with advanced glycation.

2   THE RENIN ANGIOTENSIN SYSTEM IN DIABETES

To date the most effective treatments for diabetic nephropathy target the RAS. It is clearly evident that the RAS play a fundamental role in the development of diabetic complications, specifically diabetic nephropathy. The introduction of ARBs and RAS antagonists into the treatment regimen of diabetic patients has markedly improved the rates of progression of diabetic nephropathy and subsequent end-stage renal disease. The renin-angiotensin system (RAS) is a co-ordinated hormonal cascade, which modulates vasoconstriction and facilitates renal-sodium absorption to maintain blood pressure control. The RAS cascade is initiated by the production of renin, which is released from renal juxtaglomerular cells, and converts angiotensinogen to angiotensin I. Angiotensin I, an inactive hormone, is subsequently cleaved into Angiotensin II by angiotensin converting enzyme (ACE). Angiotensin II drives the RAS and elicits its effects by binding to cellular receptors, the Angiotensin II type I (AT1) receptor and/or angiotensin type 2 (AT2) receptor. It was originally postulated that AT1 receptor was the primary receptor and modulator of the actions of Ang II, however in recent years it has been widely accepted that AT1 receptor and AT2 receptor elicit opposing actions upon ligand interaction with Ang II. The RAS is known to exist both systemically and locally in a number of different organs throughout the body including kidney, and vasculature [7, 9]. Activation of the local RAS appears to be independent of the systemic RAS. Intrarenal Ang II has a number of different roles including regulation of urine flow, maintenance of urine sodium excretion and regulation of glomerular filtration. The RAS in the kidney circulates on a positive feedback loop through the actions of Ang II on its precursor angiotensinogen. It is this positive feedback which is believed to drive tissue damage in the kidney leading to the development of diabetic nephropathy [1]. In diabetes, the local RAS has been found to be up-regulated, in particular within the kidney, whilst the systemic RAS appears to be down-regulated [8-11].

3   THE RENIN ANGIOTENSIN SYSTEM AND THE AGE/RAGE AXIS

AGEs are a heterogenous and complex group of compounds which play and important role in the development of diabetic nephropathy. The benefits of preventing AGE accumulation, in particular within the kidney, in experimental models of diabetic nephropathy has been previously demonstrated [12-15]. An important interaction between the RAS and AGEs has been the focus of investigation for some time. Our group, along with others, have demonstrated that ACE inhibitors [15, 16] and AT1 receptor antagonists [16-18] are potent inhibitors of AGE accumulation. Conversely, the administration of angiotensin II to rats, leads to the accumulation of AGEs [19]. We have also shown that administration of AGEs to rodents results in up-regulation of the renal RAS, similar to that seen in diabetes, suggesting that the interaction works both ways.

There is strong experimental evidence supporting a pathogenic role for the receptor for advanced glycation end products (RAGE) in diabetic renal disease. Indeed, RAGE knockout mice have less renal injury with diabetes [20, 21]. Conversely, diabetic mice genetically manipulated to over-express RAGE have significant glomerulosclerosis [22, 23]. AT1 receptor antagonists may also reduce the expression of both tissue RAGE [24 25] and soluble RAGE [17], possibly by reduceing Age ligand levels. By contrast we have shown that ACE inhibition may enhance soluble RAGE levels in both experimental models and in type 1 diabetic patients [15].

4   SYNERGYSM BETWEEN RAGE AND THE RENIN ANGIOTENSIN SYSTEM

Clinical studies have demonstrated that although effective, currently available drugs for the management of diabetic complications are able to completely abrogate renal disease in the context of diabetes. Given that more clinically desirable, earlier intervention with these compounds does not confer additional reno-protection [26, 27], the discovery of additional synergistic pathways as potential drug targets is of paramount importance. Ultimately, the most rational therapeutic approach to prevent diabetic complications is likely to be via combination therapies that draw on synergistic effects to optimise reno-protection. We have previously shown that there may be some important synergistic effects afforded by blockade of both the RAS and AGE accumulation [28-30]. More recently we have also shown in our diabetic mice, that RAGE deletion was more beneficial that AT receptor deletion in preventing of the decline in renal function seen with diabetic nephropathy [30]. Specifically, RAGE deletion significantly decreased the albumin excretion rate and renal superoxide generation. Furthermore, the increased renal membrane expression of RAGE seen in diabetes was not affected by AT1a deficiency in diabetes. This suggests that activation of RAGE may be upstream of the RAS in the development of diabetic nephropathy and likely other complications [30], which has been previously suggested by in vitro studies [31]. Indeed, further investigation of the interaction between these pathways is warranted to facilitate the development of more rational combination therapeutic approaches targeting diabetic complications. Taken together with previous research, these findings lead us to suggest that there is a significant interaction amongst the RAS and the AGE/RAGE axis. Such a combination may prove to be more potent and appropriate treatment strategy.

References

1. G. Giacchetti, L. A. Sechi, S. Rilli and R. M. Carey, *Trends Endocrinol Metab*, 2005, **16**, 120-126.
2. J. R. Rumble, R. E. Gilbert, A. Cox, L. Wu and M. E. Cooper, *J Hypertens*, 1998, **16**, 1603-1609.
3. E. J. Lewis, L. G. Hunsicker, R. P. Bain and R. D. Rohde, *N Engl J Med*, 1993, **329**, 1456-1462.
4. *Diabet Med*, 1988, **5**, 154-159.
5. M. E. Cooper, *Lancet*, 1998, **352**, 213-219.
6. M. E. Cooper, *Diabetologia*, 2001, **44**, 1957-1972.
7. G. Wolf, *Eur J Clin Invest*, 2004, **34**, 785-796.
8. J. M. Forbes, K. Fukami and M. E. Cooper, *Exp Clin Endocrinol Diabetes*, 2007, **115**, 69-84.
9. A. Wiecek, J. Chudek and F. Kokot, *Nephrol Dial Transplant*, 2003, **18 Suppl 5**, v16-20.
10. R. E. Gilbert, H. Krum, J. Wilkinson-Berka and D. J. Kelly, *Diabet Med*, 2003, **20**, 607-621.
11. B. F. Schrijvers, A. S. De Vriese and A. Flyvbjerg, *Endocr Rev*, 2004, **25**, 971-1010.
12. J. M. Forbes, M. E. Cooper, V. Thallas, W. C. Burns, M. C. Thomas, G. C. Brammar, F. Lee, S. L. Grant, L. A. Burrell, G. Jerums and T. M. Osicka, *Diabetes*, 2002, **51**, 3274-3282.
13. J. M. Forbes, T. Soulis, V. Thallas, S. Panagiotopoulos, D. M. Long, S. Vasan, D. Wagle, G. Jerums and M. E. Cooper, *Diabetologia*, 2001, **44**, 108-114.
14. J. M. Forbes, V. Thallas, M. C. Thomas, H. W. Founds, W. C. Burns, G. Jerums and M. E. Cooper, *Faseb J*, 2003, **17**, 1762-1764.
15. J. M. Forbes, S. R. Thorpe, V. Thallas-Bonke, J. Pete, M. C. Thomas, E. R. Deemer, S. Bassal, A. El-Osta, D. M. Long, S. Panagiotopoulos, G. Jerums, T. M. Osicka and M. E. Cooper, *J Am Soc Nephrol*, 2005, **16**, 2363-2372.

16.   T. Miyata, C. van Ypersele de Strihou, Y. Ueda, K. Ichimori, R. Inagi, H. Onogi, N. Ishikawa, M. Nangaku and K. Kurokawa, *J Am Soc Nephrol*, 2002, **13**, 2478-2487.

17.   K. Nakamura, S. Yamagishi, Y. Nakamura, K. Takenaka, T. Matsui, Y. Jinnouchi and T. Imaizumi, *Microvasc Res*, 2005, **70**, 137-141.

18.   J. M. Forbes, M. C. Thomas, S. R. Thorpe, N. L. Alderson and M. E. Cooper, *Kidney Int Suppl*, 2004, S105-107.

19.   M. C. Thomas, C. Tikellis, W. M. Burns, K. Bialkowski, Z. Cao, M. T. Coughlan, K. Jandeleit-Dahm, M. E. Cooper and J. M. Forbes, *J Am Soc Nephrol*, 2005, **16**, 2976-2984.

20.   M. T. Coughlan, D. R. Thorburn, S. A. Penfold, A. Laskowski, B. E. Harcourt, K. C. Sourris, A. L. Tan, K. Fukami, V. Thallas-Bonke, P. P. Nawroth, M. Brownlee, A. Bierhaus, M. E. Cooper and J. M. Forbes, *J Am Soc Nephrol*, 2009, **20**, 742-752.

21.   K. M. Myint, Y. Yamamoto, T. Doi, I. Kato, A. Harashima, H. Yonekura, T. Watanabe, H. Shinohara, M. Takeuchi, K. Tsuneyama, N. Hashimoto, M. Asano, S. Takasawa, H. Okamoto and H. Yamamoto, *Diabetes*, 2006, **55**, 2510-2522.

22.   Y. Yamamoto, I. Kato, T. Doi, H. Yonekura, S. Ohashi, M. Takeuchi, T. Watanabe, S. Yamagishi, S. Sakurai, S. Takasawa, H. Okamoto and H. Yamamoto, *J Clin Invest*, 2001, **108**, 261-268.

23.   R. Inagi, Y. Yamamoto, M. Nangaku, N. Usuda, H. Okamato, K. Kurokawa, C. van Ypersele de Strihou, H. Yamamoto and T. Miyata, *Diabetes*, 2006, **55**, 356-366.

24.   Q. Fan, J. Liao, M. Kobayashi, M. Yamashita, L. Gu, T. Gohda, Y. Suzuki, L. N. Wang, S. Horikoshi and Y. Tomino, *Nephrol Dial Transplant*, 2004, **19**, 3012-3020.

25.   M. Nangaku, T. Miyata, T. Sada, M. Mizuno, R. Inagi, Y. Ueda, N. Ishikawa, H. Yuzawa, H. Koike, C. van Ypersele de Strihou and K. Kurokawa, *J Am Soc Nephrol*, 2003, **14**, 1212-1222.

26.   M. Mauer, B. Zinman, R. Gardiner, S. Suissa, A. Sinaiko, T. Strand, K. Drummond, S. Donnelly, P. Goodyer, M. C. Gubler and R. Klein, *N Engl J Med*, 2009, **361**, 40-51.

27.   R. Bilous, N. Chaturvedi, A. K. Sjolie, J. Fuller, R. Klein, T. Orchard, M. Porta and H. H. Parving, *Ann Intern Med*, 2009, **151**, 11-20, W13-14.

28.   M. T. Coughlan, V. Thallas-Bonke, J. Pete, D. M. Long, A. Gasser, D. C. Tong, M. Arnstein, S. R. Thorpe, M. E. Cooper and J. M. Forbes, *Endocrinology*, 2007, **148**, 886-895.

29.   B. J. Davis, J. M. Forbes, M. C. Thomas, G. Jerums, W. C. Burns, H. Kawachi, T. J. Allen and M. E. Cooper, *Diabetologia*, 2004, **47**, 89-97.

30.   K. C. Sourris, Mibus, A.L, Koitka, A, Harcourt BE, Tan LY, Cooper , ME, Forbes, JM, *Diabetologia*, 2008, **51**, S73.

31.   K. Fukami, S. Ueda, S. Yamagishi, S. Kato, Y. Inagaki, M. Takeuchi, Y. Motomiya, R. Bucala, S. Iida, K. Tamaki, T. Imaizumi, M. E. Cooper and S. Okuda, *Kidney Int*, 2004, **66**, 2137-2147.

ATTENUATION OF DIABETES-ASSOCIATED ATHEROSCLEROSIS WITH LR-90, A NOVEL INHIBITOR OF AGE FORMATION

A Watson,[1] MC Thomas,[1] P Koh,[1] JL Figarola,[2] S Rahbar,[2] K Jandeliet Dahm[1]

[1]Danielle Alberti Memorial Centre for Diabetes Complications, Baker IDI Heart and Diabetes Institute, Melbourne , Victoria, AUSTRALIA
[2]Department of Diabetes and Endocrinology, Gonda Building, City of Hope National Medical Center, 1500 E. Duarte Road, Duarte, California 91010, USA

1   INTRODUCTION

Diabetes is associated with accelerated development and progression of atherosclerosis in large blood vessels that contributes to the increased burden of cardiovascular disease. Although a number of different factors contribute to accelerated atherosclerosis in diabetes, recent data show that the accumulation of advanced glycation end-products (AGEs) may have a significant role [1]. AGE-modifications have the potential to alter the structure and/or function of targets, as well as activate pro-atherogenic pathways including oxidative stress, inflammation and the renin- angiotensin system [2]. In this study, we explore the anti-atherosclerotic activity of a novel inhibitor AGE accumulation, LR-90 (figure 1). LR-90 was initially identified after a design-based screening a range of aromatic compounds for their potential to inhibit AGE formation and protein cross-linking[3]. In these studies, LR-90 was found to be a more effective in vitro inhibitor of AGE formation than the well known AGE inhibitor, aminoguanidine [3]. We have previously shown than aminoguanidine is an effective anti-atherosclerotic agent in the diabetic *apolipoprotein E* knockout (*apo E* KO) mouse[4]. Consequently, we hypothesized that LR-90 may show similar or enhanced efficacy in preventing atherogenesis in the same model.

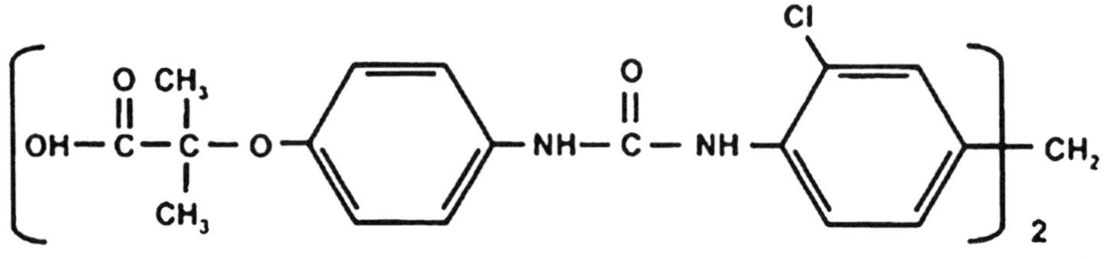

Figure 1.   The chemical structure of LR-90, methylene bis [4,4'-(2 chlorophenylureido phenoxyisobutyric acid)]

2   MATERIALS AND METHODS

2.1   Experimental model

This study utilised the well-characterized *apolipoprotein E* knockout (*apo E* KO) mouse model of accelerated atherosclerosis, whereby mice develop complex vascular lesions after 20 weeks of diabetes that resemble the morphology seen in human atherosclerosis [4-7]. Six-week-old male apoE KO mice (backcrossed 20 times to a C57BL/6 background; Animal Resource Centre, Canning Vale, Western Australia) were housed at the Precinct Animal Centre, Baker Heart Research Institute, and studied according to National Health and Medical Research Council (NHMRC) guidelines.

Mice (n=24/group) were rendered diabetic via 5 daily intraperitoneal injections of streptozotocin (Boehringer, Mannheim) 55 mg/kg per day, resulting in insulin deficiency[5]. All animals were followed for 20 weeks. Mice were allowed access to standard mouse chow and water *ad libitum*. Mice were culled by intraperitoneal injection of Euthal (100mg/kg) (Delvet Limited, Seven Hills, Australia), followed by exsanguination by cardiac puncture. Excised aortas (n=10-14) were placed in 10% neutral buffered formalin and quantitated for lesion area before being embedded in paraffin for immunohistochemical analysis. In the remaining mice (n=10-14), aortas were snapfrozen in liquid nitrogen and stored at minus 70°C for subsequent RNA extraction. At the conclusion of the study, glycated Hb (GHb) was measured by high-performance liquid chromatography[8]. Total plasma cholesterol, HDL, and triglyceride concentrations were measured in 9-10 mice /group by autoanalyzer. LDL cholesterol was calculated by the Friedewald formula.

2.2   Plaque Area Quantitation

Plaque area was quantitated as described previously [5, 6]. In brief, aortas (n=10 to 14 aortas/group) were cleaned of excess fat under a dissecting microscope and subsequently stained with Sudan IV-Herxheimer's solution (0.5% wt/vol) (Gurr; BDH Ltd, Poole, UK). Aortas were dissected longitudinally, divided into arch, thoracic, and abdominal segments, and pinned flat onto wax. Images were acquired with a dissecting microscope equipped with an Axiocam camera (Zeiss GmbH, Heidelberg, Germany). Total and segmental plaque area was quantitated as a percentage area of aorta stained (Adobe Photoshop version 7.0). Tissue was subsequently embedded in paraffin and sections cut for immunohistochemical analysis.

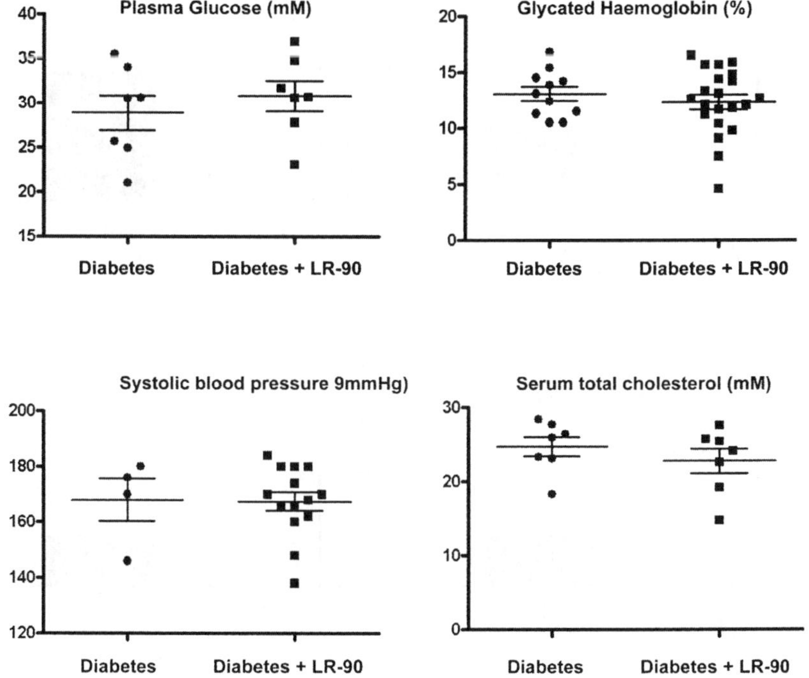

Figure 2   *The effect of the AGE-inhibitor, LR-90 on the metabolic profile and blood pressure of diabetic apoE KO mice*

3   RESULTS

3.1   General parameters

The induction of streptozotocin diabetes was associated with increased glucose and lipid levels. Treatment with the AGE-inhibitor, LR-90 had no effect on the glucose or lipid levels in diabetic *apoE* KO mice (**Figure 2**). No effects were also observed on systolic blood pressure as measured by tail cuff plethsmography (**Figure 2**). or weight gain during 20 weeks of study (data not shown).

3.2   Plaque area

The induction of streptozotocin diabetes was associated with increased accumulation of lipid rich plaque in the arch, thoracic and abdominal aorta, as detected by staining for fat with Sudan IV. Treatment with the AGE-inhibitor, LR-90 had a modest but significant effect on plaque accumulation, especially in the thoracic and abdominal segments (**Figure 3**). No significant effect was observed in the arch, consistent with the known haemodynamic dependence of arch lesions

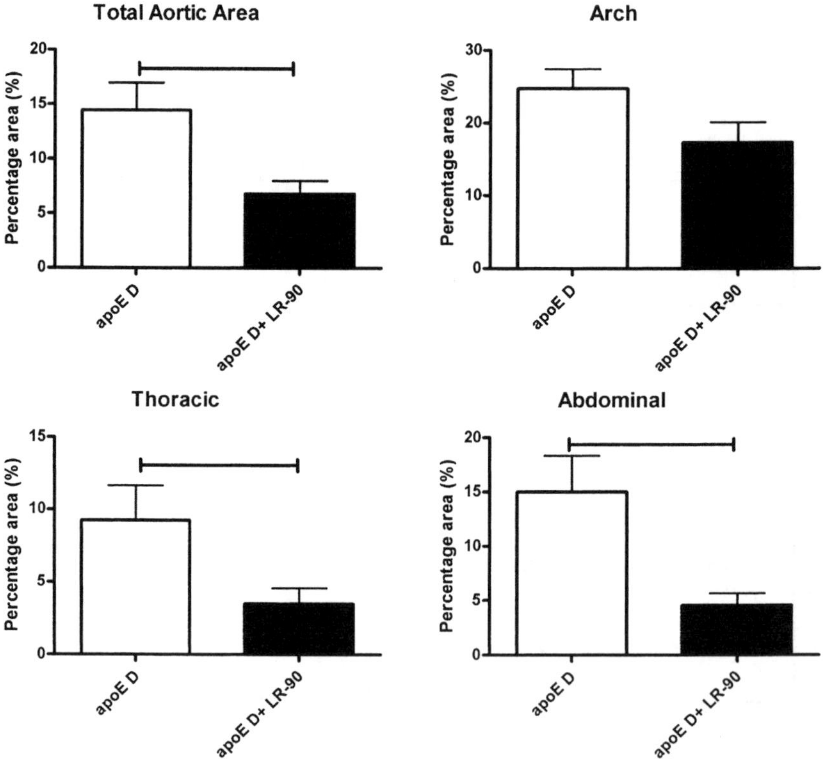

Figure 3   *The effect of the AGE-inhibitor, LR-90 on the plaque accumulation in diabetic apoE KO mice.*

4 DISCUSSION

Diabetes is associated with a significant increase in the extent and complexity of atherosclerotic plaque, contributing to increased cardiovascular events and premature mortality. Modification of traditional risk factors has failed to remove the residual risk associated with diabetes, and there remains an urgent need for new targets and interventions. Prolonged hyperglycaemia, dyslipidaemia and oxidative stress in diabetes result in the production and accumulation of AGEs. AGEs are thought to directly contribute to the development and progression of cardiovascular disease in diabetes by promoting vascular dysfunction and damage that leads to accelerated atherogenesis[1]. As a result, novel therapeutic agents to reduce the accumulation of AGEs in diabetes have gained interest as potential cardio-protective approaches[2]. In this study, we demonstrate that the AGE inhibitor, LR-90 pyridoxamine is able to prevent the development of atherosclerosis associated with diabetes. This is consistent with previous studies using a other AGE inhibitors (aminoguanidine and alegebrium chloride) [4], although the doses of LR-90 used here are 50-100 times less on a molar basis. The fact that each of these anti-AGEs agents are effective, despite their disparate mechanisms of action, strongly supports the keystone role of AGEs in diabetic vascular damage.

The anti-atherosclerotic effects of LR-90 appeared to be confined to the thoracic and abdominal aorta, with no significant effect observed in the aortic arch. This segment specific activity is similar to that observed for other AGE inhibitors including aminoguanidine and alegebrium chloride [4], which also demonstrated no reduction in plaque area or complexity within the aortic arch with treatment, in contrast to the thoracic and abdominal aortas, where significant attenuation was seen. It has been suggested that this potentially reflects the important role of hemodynamic factors over metabolic ones in the arch, and *vice versa* in the thoracic and abdominal segments. However, studies with soluble RAGE in diabetic apoE KO mice have clearly shown attenuation of arch lesions with this intervention[9]. Consequently, the full reasons behind this regional specifically for the effects of LR-90 remain to be established.

The mechanism(s) by which LR-90 exerts its their anti-atherosclerotic effects remains to be fully established. Certainly, LR-90 inhibits AGE formation and AGE-protein cross-linking *in vitro*, to a grater extent than aminoguanidine and pyridoxamine[3]. Moreover, LR-90 treatment is able to reduce diabetes associated increases in serum AGE and the in situ accumulation of immunoreactive AGE in collagen tissues and kidneys of diabetic rats[10, 11]. Based on its structure, it is possible to speculate that LR-90 has the ability to interact with reactive carbonyls, compounds that readily modify lysine, arginine and cysteine residues to form tissue AGEs in vivo. In addition, in vitro studies suggest that LR-90 is a potent chelator of Cu^{2+} and inhibits auto-oxidation of ascorbic acid, which represents a major source of AGEs *in vivo*.

It is now recognised that inflammation plays a pivotal role in early atherogenesis[12]. In particular, monocyte recruitment and adhesion to the nascent atherosclerotic lesion is generally regarded as one of the first steps toward plaque formation. Animal studies have shown that the majority of the cells in the lipid core of atherosclerotic plaques are derived from monocytes. Recent *in vitro* studies have shown that LR-90 can inhibit inflammatory responses in monocytes treated possibly by its ability to prevent oxidative stress and modulate NF-κB signalling[13].

LR-90 was derived from LR-16, a compound that is able to lowers cholesterol levels in rats. In our *apoE* KO model, no effects on lipid profile by LR-90 were detected, possibly as the effects of genetic deficiency of the lipoprotein profile overwhelms and drug associated actions. However, in the Zucker model of diabetes and dyslipidaemia lipid lowering effects have been propertied with LR-90 [11]

In summary, this study provides preliminary evidence suggesting that LR-90 could be an effective treatment for atherosclerosis associated with diabetes, where accumulation of AGEs and reactive intermediate compounds appear to be major contributors. In addition these finding support recent studies that have demonstrated that LR-90 may have beneficial effects on other diabetic complications, including retinopathy [14] and renal disease [10, 11]. Although the clinical utility of AGE inhibition remains to be firmly established, these findings provide evidence that targeting AGEs could be effective in combating established diabetic macrovascular disease and re-affirm the importance of ACE-inhibitor treatment in diabetes.

Acknowledgements

AMDW is currently supported by the National Health and Medical Research Council of Australia (NHMRC) Australian Biomedical Fellowship (472698). TJA and KAMJ-D are supported by NHMRC Senior Research Fellowships.

References

1.   K. Jandeleit-Dahm and M. E. Cooper, *Curr Pharm Des*, 2008, **14**, 979-986.

2.   M. C. Thomas, J. W. Baynes, S. R. Thorpe and M. E. Cooper, *Curr Drug Targets*, 2005, **6**, 453-474.

3.　　S. Rahbar and J. L. Figarola, *Arch Biochem Biophys*, 2003, **419**, 63-79.

4.　　J. M. Forbes, L. T. Yee, V. Thallas, M. Lassila, R. Candido, K. A. Jandeleit-Dahm, M. C. Thomas, W. C. Burns, E. K. Deemer, S. M. Thorpe, M. E. Cooper and T. J. Allen, *Diabetes*, 2004, **53**, 1813-1823.

5.　　R. Candido, K. A. Jandeleit-Dahm, Z. Cao, S. P. Nesteroff, W. C. Burns, S. M. Twigg, R. J. Dilley, M. E. Cooper and T. J. Allen, *Circulation.*, 2002, **106**, 246-253.

6.　　A. C. Calkin, J. M. Forbes, C. M. Smith, M. Lassila, M. E. Cooper, K. A. Jandeleit-Dahm and T. J. Allen, *Arterioscler Thromb Vasc Biol.*, 2005, **25**, 1903-1909.

7.　　L. G. Bucciarelli, T. Wendt, W. Qu, Y. Lu, E. Lalla, L. L. Rong, M. T. Goova, B. Moser, T. Kislinger, D. C. Lee, Y. Kashyap, D. M. Stern and A. M. Schmidt, *Circulation*, 2002, **106**, 2827-2835.

8.　　W. T. Cefalu, Z. Q. Wang, A. Bell-Farrow, F. D. Kiger and C. Izlar, *Clin Chem.*, 1994, **40**, 1317-1321.

9.　　L. Park, K. G. Raman, K. J. Lee, Y. Lu, L. J. Ferran, Jr., W. S. Chow, D. Stern and A. M. Schmidt, *Nat Med*, 1998, **4**, 1025-1031.

10.　　J. L. Figarola, S. Scott, S. Loera, C. Tessler, P. Chu, L. Weiss, J. Hardy and S. Rahbar, *Diabetologia*, 2003, **46**, 1140-1152.

11.　　J. L. Figarola, S. Loera, Y. Weng, N. Shanmugam, R. Natarajan and S. Rahbar, *Diabetologia*, 2008, **51**, 882-891.

12.　　P. Libby, P. M. Ridker and A. Maseri, *Circulation*, 2002, **105**, 1135-1143.

13.　　J. L. Figarola, N. Shanmugam, R. Natarajan and S. Rahbar, *Diabetes*, 2007, **56**, 647-655.

14.　　A. Bhatwadekar, J. V. Glenn, J. L. Figarola, S. Scott, T. A. Gardiner, S. Rahbar and A. W. Stitt, *Br J Ophthalmol*, 2008, **92**, 545-547.

N^ε–CARBOXYMETHYLLYSINE: IT'S ORIGIN IN SELECTED FOODS AND ITS URINARY AND FAECAL EXCRETION IN HEALTHY HUMANS'

F.J.Tessier[1], C. Niquet[1], L.Rhazi[1], K. Hedhili[1], P. Navarro[2], I. Seiquer[2] and C. Delgado-Andrade[2]

[1]Institut Polytechnique LaSalle Beauvais, Beauvais, France
[2]Instituto de Nutrición Animal, Estación Experimental del Zaidín, Granada, Spain

1   INTRODUCTION

Carboxymethyllysine (CML) was isolated and characterized almost at the same time *in vivo* [1] and in food products [2]. Its formation was originally described as an oxidative degradation of fructoselysine, the Amadori product found in glycated proteins. However, ten years later Fu et al. [3] found that CML was not only formed from glycoxidation but also from lipoxidation reactions. Due to the popular use of CML as a biomarker of the Maillard reaction in human tissues [4] most available studies regarding the chemical origin of CML are based on normal or pathological *in vivo* conditions. Bearing these limitations in mind, it has been speculated that CML would be formed primarily from lipid peroxidation reactions rather than glycation [5]. More recently a CML database subject to controversy shows that the highest CML levels have been found in food products with a high fat content (*e.g.* olive oil, butter) [6]. If some scientists consider that this may perhaps be a confirmation of what has been found *in vivo*[7], others, more skeptical, expressed doubts about the validity of the analytical method used [8,9,10]. The first objective of this paper was to take advantage of the great variability of the food composition of a single category of food (*i.e.* the chocolate drink mixes) to estimate which macronutrient is the main contributor to the formation of CML.
Despite the fact that CML is one of best-studied Advanced Glycation End-Products, there remain at least two open questions about what may be its potential toxicological effects and what is its *in vivo* origin. Food can provide a significant amount of CML, in addition to its endogenous synthesis. In 1987, Liardon et al. judged that serum and urinary CML could be mainly of exogenous origin and thus that CML excretion would be influenced by dietary CML levels [11]. The second objective of our research was to contribute to the validation of this old hypothesis with two human studies.

2   METHODS

2.1   Food analysis

Samples of chocolate drink mixes (n=24) were purchased from supermarkets in the US and Europe. Each sample was treated with sodium borohydride to stabilize the Amadori products and prevent their conversion to CML during the acid hydrolysis. A quantity of reduced sample, equivalent to 10 mg protein, was dissolved in 5 mL of 6 M HCl and incubated at 110°C for 20h. One hundred microliters of each acid hydrolysate was dried under vacuum and reconstituted in 250 µL of internal standard containing 0.25 µg of (D_2)-CML and 12.5 µg of $(^{15}N_2)$-lysine (solved in NFPA 20 mM) prior to analysis by LC-MS/MS. Liquid chromatography coupled to linear ion trap tandem mass spectrometry was used for the analysis of CML and lysine in the food samples. The following instrumentation and criteria were used: Surveyor HPLC system coupled to an LTQ mass spectrometer working in its tandem operation mode (ThermoFisher Scientific, Courtaboeuf, France); thermostat, 10°C; column, Hypercarb, 100 mm x 2.1 mm, 5 µm with a guard column Hypercarb, 10 mm x 2.1 mm, 5 µm; injection volume, 10 µL; flowrate, 0.2 mL/min; mobile phase, 20 mM NFPA in a water - acetonitrile gradient as follows: linear increase of acetonitrile from 0% to 50% over 20 min; ion source: electrospray ionization in positive mode; multiple reaction monitoring with the specific transitions *m/z* 205.0/130.0 and *m/z* 147.0/130.0 for CML and lysine, respectively, with a normalized collision energy of 37%.

2.2   Preliminary intervention study design

One healthy female volunteer aged 25 agreed to consume, over a period of 3 consecutive days, 1, 2 and 4 chocolate bars per day (containing mainly malt and cacao) in addition to her basic balanced diet consumed at the LaSalle Beauvais institute's cafeteria. Each bar contained 2.7 mg CML (1240 mg CML/kg protein). Thus the extra-exposure to CML coming from the bars was calculated to be from 2.7 to 10.9 mg/day. Twenty four-hour urine samples were collected for 6 days starting 2 days before the extra-exposure and ending one day after it. CML in urine was determined by LC-MS/MS according to the method described above for food analysis.

2.3   Adolescent intervention study design

Selection of the subjects, composition of diets, and study design have been described elsewhere [12]. Briefly, 20 male adolescents (12.4 ± 0.34 years of age, mean ± standard error) participated in a 2-wk randomized two-period crossover trial in which they consumed two different diets, with a 40-d washout period. Two 7-d menus of similar energy and nutrients content and containing the same servings per day of the different food groups were designed: a white diet (WD) free, as much as possible, of foods in which the Maillard reaction develops during cooking practices or those usually containing MRPs; and a brown diet (BD) rich in processed foods with an evident development of browning and thus rich in MRPs. Lunch and dinner were prepared by a local catering firm and distributed daily to the participants. The lunch and dinner 7-d menus, as well as the composition of breakfasts and afternoon snacks were widely described in Seiquer et al. [12].

The analysis of Maillard reaction markers in the diets [13] showed a greater development of the Maillard reaction in the BD than in the WD, according to significantly higher values of hydroxymethylfurfural and carboxymethyllysine, among others (hydroxymethylfurfural 0.94 ± 0.01and 3.87 ± 0.03 mg/kg, carboxymethyllysine 6.62 ± 0.25 and 15.72 ± 0.43 mg/100 g of protein, in the WD and BD, respectively). As an approximation to the CML absorption and excretion, a balance technique was followed. At the end of each 14-d dietary treatment, a 3-day collection of urine and faeces was performed from acidified containers. CML was analysed by the method previously described. This study was approved by the ethics committee of the San Cecilio University Hospital of Granada and was performed in accordance with the Helsinki Declaration of 2002, as revised in 2004. The informed consent was obtained from the parents of all the children participating in the study.

2.4 Statistical analysis

SPSS 13.0 for Windows (1999–2004, SPSS Inc., Chicago, IL, USA) was used for data entry and statistical analysis. The experimental data obtained after the crossover dietary treatments were analyzed by using the repeated-measures analysis of variance (ANOVA). When a significant effect between dietary treatments was found, post hoc comparison of means was made using Bonferroni's test. Differences were considered statistically significant at $P < 0.05$.

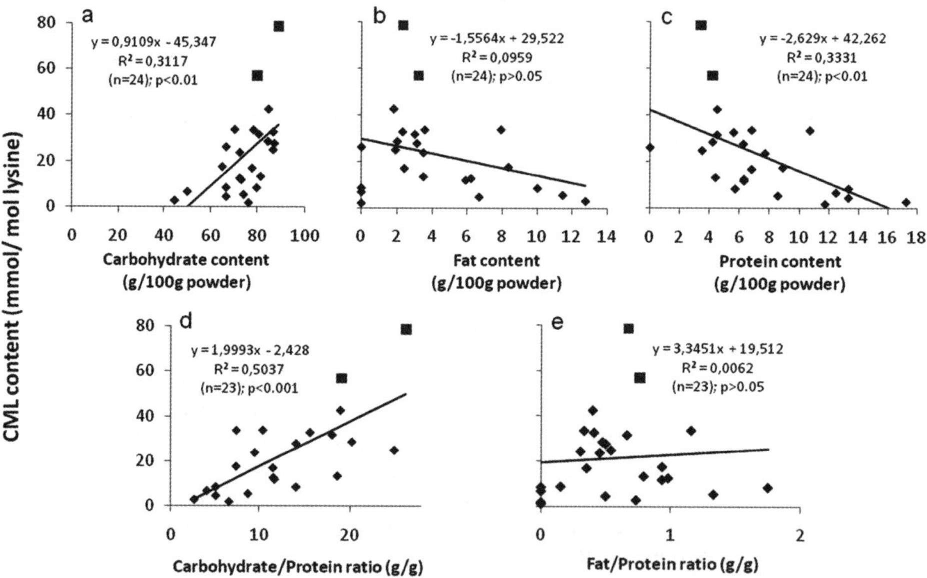

Figure 1. *Correlation plot of CML content versus carbohydrate content (a), fat content (b), protein content (c), carbohydrate/protein ratio (d) and fat/protein ratio (e) calculated from nutritional information provided on food labels. Symbol key:* ■ *chocolate drink mixes containing glucose;* ▲ *other chocolate drink mixes.*

3 RESULTS AND DISCUSSION

3.1 Origin of CML in chocolate-flavored drink mixes

The food samples analyzed in our study are all chocolate drink mixes. However, according to the food labels the proportion of the three major macronutrients varies greatly from sample to sample. This diversity of composition within a single food category was used to explore the origin of CML. The correlations between the macronutrient content indicated on the labels and the CML content expressed as mmol/ mol lysine suggest that CML could be formed essentially from carbohydrates but not from fat (Figure 1). Drink mixes high in fat do not seem to promote the formation of CML (Figure 1-b). Inversely, strong positive correlations were found between CML and the carbohydrate/protein ratio and the carbohydrate content (Figures 1-d and 1-a, respectively). In addition, of all drink mixes analyzed in this study, those enriched with glucose contained the highest levels of CML. Finally the result presented in Figure 1-c reveals that the rate of conversion of lysine to CML is inversely proportional to the food protein content.

To fully understand the origin of CML in drink mixes it would be necessary to quantify micronutrient levels and to compare them to the amount of CML. For instance ascorbic acid and iron have been suspected of promoting CML synthesis [14], and their importance in the formation of CML is under investigation in our laboratory. The level of Maillard reaction products in foods depends too on the conditions in which foods are treated (mainly the duration and the temperature of the heat treatment), information we do not have at our disposal. In spite of this the data presented in Figure 1 are sufficient to reject the relative role of fat versus carbohydrates in the formation of CML in drink mixes, and to reinforce the essential role of the "conventional" Maillard reaction in this process. A previous publication presented an excellent correlation (r = 0.96) between CML and furosine (indicator of fructose-lysine after acid hydrolysis of samples) with 27 UHT milk samples [15]. This result indicates that, in UHT milk, most CML comes from fructoselysine which is further evidence that carbohydrates are principally responsible for the formation of CML in food. Finally a more recent study on sugar- or lipid-casein model systems confirmed that after 8 hours of cooking at 95°C the presence of glucose leads to a formation of 15 times the quantity of CML as that produced in the presence of arachidonic acid [16].

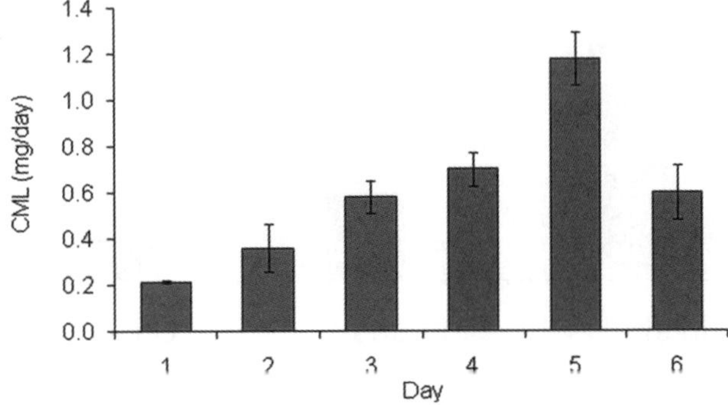

Figure 2. *Urinary elimination of CML after an increasing exposure to dietary CML. Apart from what is present in a regular diet, extra-quantities of CML of 2.7, 5.4 and 10.9 mg/day, via an intake of 1, 2 and 4 chocolate bars, were administered on day 3, 4 and 5 respectively. Values are mean ± standard error.*

3.2   Study of the urinary and faecal excretion of CML in human volunteers

The preliminary dietary study involving one healthy volunteer showed that the consumption of an increasing number of chocolate bars resulted in a proportional elevation of the urinary level of CML. Taking into account the average daily intake of CML of 6 mg at the LaSalle Beauvais institut's cafeteria (data not shown), and the extra amount of CML supplied with ingested chocolate bars, the total exposure was estimated to increase from 6 to 16.8 mg/ day from day 2 to day 5.This increasing intake of CML led to an increased excretion of urinary CML from 0.36 ± 0.11 mg/ day on day 2 to 1.18 ± 0.10 mg/ day on day 5 (Figure 2).

The second and more complete study was performed to confirm that the urinary and faecal excretion of CML are dependent on the diet. Since the BD had a greater CML content, the CML intake was also higher after this dietary treatment (11.29 ± 0.27 *vs.* 5.37 ± 0.14 mg/day, for the BD and WD, respectively) (Figure 3). Thus, CML faecal excretion was increased after consumption of the BD (3.52 ± 0.52 *vs.* 1.23 ± 0.30 mg/day, for the BD and WD, respectively), and the value supposed a 31.7% of ingested CML, while in the WD this percentage was 22.5%. Despite the increased CML faecal excretion after BD consumption, the possible absorption should be also quite higher. On the one hand the absorbed CML could enter into the organism in this form or having been metabolised previously by the intestinal cells, as has been evidenced for hydroxymethylfurfural, an intermediate product of the Maillard reaction which is quickly transformed into its sulfo-oxi-derivative within the enterocites catalyzed by sulfotransferases [17]. On the other hand the CML present in faeces could have been underestimated to some extent since it is well-established that Maillard reaction products can be degraded by intestinal microbial activity, and thus some of the derivative compounds are absorbed in the intestine and some others excreted in faeces [18].

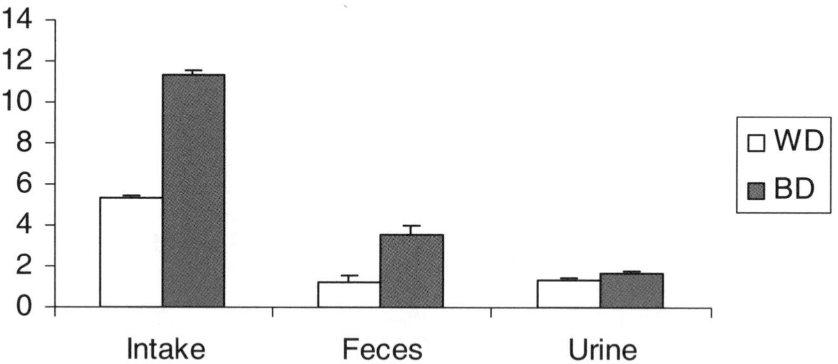

Figure 3.   *Intake, faecal and urinary elimination of CML after consumption of the WD and BD. Values are mean $\pm$ standard error. A symbol indicates significant differences between both dietary treatments.*

Based on relations established by different researchers between serum AGE (advanced glycated end products) levels and dietary CML intake [19,20], once in the organism, a fraction of the CML will become part of the circulating AGEs pool. Another portion could be deposited in target organs [21]. After its metabolic transit, still fairly unknown, a final part of the compound, or its metabolites are eliminated in urine. In this second study the excreted amount of CML in urine was slightly higher after the BD compared to the WD (1.63 ± 0.17 vs. 1.30 ± 0.14 mg/ day, respectively). However this difference was statistically not different and the urinary elimination rate was lower after BD consumption when calculated as a percentage of ingested CML (14.8 vs. 24.4%, for the BD and WD, respectively). These percentages suggest the compound suffers an intensive metabolism and hence its low recovery. This is in contrast to that observed for another AGE, pyrraline which is almost entirely released and absorbed from foods, and after its metabolic transit is eliminated rapidly and almost completely in the urine, indicating slight metabolization of this specific AGE within the body [22].

4 CONCLUSIONS

The first part of our study unambiguously demonstrates that the role of carbohydrates is dominant in the formation of CML in drink mixes and that the more protein there is in a food product the less will be the proportion of lysine converted to CML. However this work needs further studies on other types of foods to confirm the origin of CML formed during heat treatment of food. Among the data currently available in the literature the results of our two intervention studies permit to confirm that CML excretion is influenced by dietary CML levels. However the absorption mechanism and the metabolic transit of CML will necessitate further investigation for a full understanding of its utilization *in vivo*.

References

1. M.U. Ahmed, S.R. Thorpe, J.W. Baynes, *J. Biol. Chem.*, 1986, **261**, 4889-4894.
2   H.F. Erbersdobler, *Bibl. Nutr. Dieta.*, 1989, **43**, 140-155.
3   M.X. Fu, J.R. Requena, A.J. Jenkins, T.J. Lyons, J.W. Baynes, S.R. Thorpe, *J. Biol. Chem.*, **271**, 9982-9986.
4   S.R. Thorpe, J.W. Baynes JW, *Int. Congr. Ser.*, 2002, **1245**, 91-99.
5   A.S. Januszewski, N.L. Alderson, T.O. Metz, S.R. Thorpe, J.W. Baynes, *Biochem. Soc. Trans.*, 2003, **31**, 1413-1416.
6   T. Goldberg, W. Cai, M. Peppa, V. Dardaine, B. Baliga, J. Uribarri, H. Vlassara, *J. Am. Diet. Assoc.*, 2004, **104**, 1287-1291.
7   J. Baynes, *IMARS Highlights*, 2008, **3** (3), 19-20.
8   T. Buetler, *IMARS Highlights*, 2008, **3** (4), 8-9.
9   T.M. Buetler, E. Leclerc, A. Baumeyer, H. Latado, J. Newell, O. Adolfsson, V. Parisod, J. Richoz, S. Maurer, F. Foata, D. Piguet, S. Junod, C.W. Heizmann, T. Delatour, *Mol. Nutr. Food Res.*, 2008, **52**, 370-378.
10 M. Pischetsreider, *IMARS Highlights*, 2008, **3** (3), 17-18.
11 L. Liardon, D. De Weck-Gaudard, G. Philippossian, P.A. Finot, *J. Agric. Food Chem.*, 1987, **35**, 427-431.
12 I. Seiquer, J. Diaz-Alguacil, C. Delgado-Andrade, M. Lopez-Frias, M. Muñoz Hoyos, G. Galdo, M.P. Navarro, *Am. J. Clin. Nutr.*, 2006, **83**, 1082-1088.
13 C. Delgado-Andrade, I. Seiquer, M.P. Navarro, F.J. Morales, *Mol. Nutr. Food. Res.*, 2007, **51**, 341-351.

14 J. Leclère, I. Birlouez-Aragon, M. Meli, *Food Chem.*, 2002, **76**, 491-499.
15 W. Buser, H.F. Erbersdobler, R. Liardon., *J. Chromatogr. A*, 1987, **387**, 515-519.
16 M. Lima, P. Deo, J.M. Ames, presented in part at the 10th international symposium on the Maillard reaction. Palm Cove, Australia, 2009.
17 R.J Ulbricht, S.J. Northup, J.A. Thomas, *Fundamental and Applied Toxicology*, 1984, **4**, 843-853.
18 V. Somoza, *Mol. Nutr. Food Res.*, 2005, **49**, 663-672.
19 H. Vlassara, W. Cai, J. Crandall, T. Goldberg, R. Oberstein, V. Dardaine, M. Peppa, E. Rayfield, *Proc. Natl. Acad. Sci.*, 2002, **99**, 15596-15601.
20 R.J. Uribarri, M. Peppa, W. Cai, T. Goldberg, M. Lu, S. Baliga, J.A. Vassalotti, H. Vlassara, *Am. J. Kidney Dis.*, 2003, **42**, 532-538.
21 J. O'Brien, R. Walker, *Food Chem. Toxicol.*, 1988, **26**, 775-783.
22 A. Foerster, T. Henle, *Biochemical Society Transactions*, 2003, **31**, 1383-1385.

OXIDATIVE STRESS AND THE MAILLARD REACTION IN FOOD

M. Pischetsrieder[1]

[1] Department of Chemistry and Pharmacy, Food Chemistry, Emil Fischer Center, University of Erlangen-Nuremberg, Erlangen, Germany.

1   BACKGROUND

Oxidation and glycation are two of the most important chemical reactions taking place during food processing, which lead to protein and lipid damage, flavor generation, browning and the formation of potentially harmful contaminants. Oxidation is promoted mainly by reactive oxygen species (ROS) or other radicals. Early glycation products, such as the Amadori product or Schiff bases, are formed from reactive carbonyl compound and an amine group of a protein or an amino acid. However, there is growing evidence that both reactions are not independent processes. Oxidation and glycation reactions influence each other strongly as summarized in figure 4. Thus, it is difficult to differentiate between both processes, when properties and composition of processed foods are regarded. Consequently, it is an intricate task to define marker compounds that monitor both reactions separately during food manufacturing. For example, AGEs comprise a wide range of compounds, producted not only via glycation but also including glycoxidation. Glycoxidation is a process that requires at least two reaction steps: the reaction of the amine component with a reactive carbonyl and an oxidation step. The most prominent example for a glycoxidation product is N^ε-carboxymethyllysine (CML), which is formed by oxidative cleavage of the C2-C3 bond of an Amadori product [1]. Another important glycoxidation product, which has been identified in processed food, is pentosidine [2]. The present review summarizes several different pathways how oxidation and glycation reactions can interact during food processing.

2   OXIDATION PROMOTES GLYCATION BY THE FORMATION OF DICARBONYL INTERMEDIATES

Apart from the formation of glycoxidation products, oxidation can also enhance protein glycation indirectly (figure 1): Oxidation of sugars leads to the formation of sugar degradation products, which often show a much higher reactivity towards proteins than the sugar educts. Oxidation products of sugars with a reactive α-dicarbonyl structure are, for example, glyoxal and glucosone [3, 4]. The higher tendency of sugar degradation products to modify proteins has been shown, for instance, for peritoneal dialysis fluids. Basically, these medical products are concentrated glucose solutions, which have been heat sterilized. During the thermal treatment, glucose undergoes considerable degradation leading to the formation of reactive carbonyls. Thus, about 500 µM glucose degradation products are formed in a heat sterilized peritoneal dialysis fluid with a glucose concentration of about 4% [5]. When the solutions were tested in the presence of proteins to determine their glycation potential, however, up to 80% of the AGEs were formed from the reactive carbonyls in the fluids, whereas only 20% of the AGEs were formed from glucose itself, although the latter is present in a 500 fold molar excess over reactive carbonyls [6]. In another study, up to 45% inhibition of protein glycation was observed when the incubation was performed in the presence of a metal chelating agent preventing sugar oxidation [7]. Thus, the oxidation of sugars yielding reactive carbonyls, such as glyoxal or glucosone, can considerably increase protein glycation.

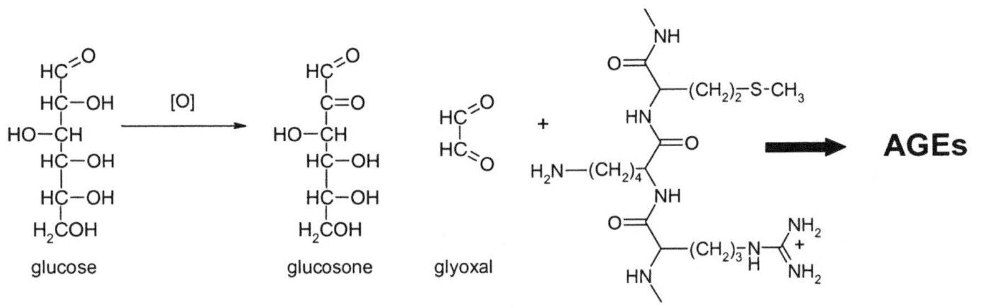

Figure 1   *Oxidative conditions promote glycation via the formation of reactive sugar oxidation products*

3   MAILLARD REACTION PROMOTES PROTEIN OXIDATION

On the other side, glycation can also promote oxidation. When protein damage was studied in a model of thermal milk processing, protein oxidation products, such as methionine sulfoxide or aminoadipic semialdehyde, were identified as major protein modifications apart from the well studied Maillard products [8]. Surprisingly, hardly any protein oxidation products were detected, when the samples were heated in the absence of the sugar. These results indicate that the presence of the sugar enhances protein oxidation. Two reaction mechanisms can be suggested to explain this phenomenon. It can be hypothesized that aminoadipic semialdehyde, in which the ε-amino group is formally oxidized to an aldehyde group, is the product of an apparent oxidation (figure 2). In the first step, the α-dicarbonyl compound generated during the heating of the milk sugar forms a Schiff base with the ε-amino group of lysine. Then, tautomerization takes place in a reaction analogous to the Strecker degradation

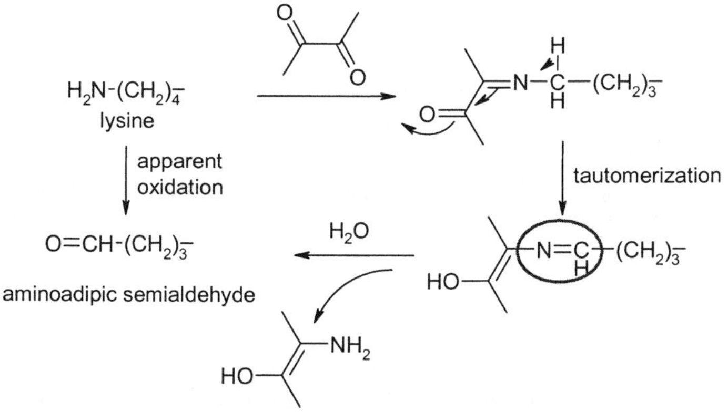

Figure 2 *Apparent oxidation of lysine by a Strecker degradation-type reaction*

of amino acids. The resulting Schiff base is hydrolyzed releasing aminoadipic semialdehyde and an enamineol. Thus, without any electron transfer, a formal change of the oxidation numbers is caused merely by tautomerization reactions. On the other hand, the formation of methionine sulfoxide cannot be explained by an apparent oxidation. An alternative reaction mechanism will be suggested in the following section.

4 MAILLARD REACTION PRODUCTS GENERATE ROS

A cytotoxic activity and the ability to induce nuclear translocation of transcription factor NF-κB were identified as important cellular reactions when cells are stimulated by Maillard reaction mixtures or food items such as coffee, which are rich in Maillard products [9,10]. Both cellular effects, however, were fully or at least partially abolished, when coffee extracts or Maillard reaction mixtures were administered to the cells together with catalase. These results indicate that Maillard reaction products are able to generate H_2O_2, which then induces cellular reactions, such as cell death or immunomodulation.
Subsequently, the presence of H_2O_2 could be determined directly in the Maillard reaction mixtures. Thus, in a mixture of ribose and lysine, which had been heated for 30 min at 120 °C, 286.9 μM H_2O_2 was measured [9]. In the meantime, it was further shown that H_2O_2 production by Maillard reaction products or coffee is very fast and depends on the concentration, the pH value, and the temperature. Furthermore, H_2O_2 generation can be clearly attributed to reaction products between sugars and amines, because the heated or unheated sugar or amino acid alone did not show any effects. Formation of ROS has also been observed for other Maillard systems [11-13]. Thus, an alternative mechanism can be proposed how glycation enhances protein

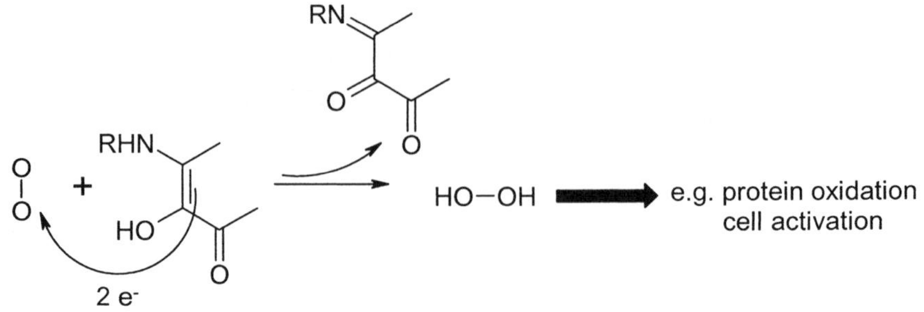

Figure 3 *Possible mechanism for the generation of hydrogen peroxide by Maillard products with amino reductone structure*

oxidation, such as the formation of methinone sufoxide: During the thermal processing of food, sugars glycate proteins or amino acids, leading to glycation products. The latter are able to generate ROS and, as a consequence, induce oxidative protein damage. Furthermore, it was suggested that CML-adducts increase the metal binding capacity of proteins favoring metal-catalyzed protein oxidation [14].

5 AMINOREDUCTONES AS POTENT H_2O_2-GENERATORS IN MAILLARD REACTION MIXTURES

In order to identify components of Maillard reaction mixtures, which cause ROS-dependent cellular effects, a substructure library of Maillard products was prepared [15]. The substructure library was designed in a way that the components contained major structural elements, which, on the one hand, are often found in Maillard products and, on the other hand, have been previously linked to the generation of ROS. Thus, fructosyllysine -the Amadori product of glucose and lysine-, 3-deoxyglucosone -an α-dicarbonyl intertermediate-, CML, the reductone 2,3-dihydro-3,5-dihydroxy-6-methyl-4H-pyran-4-one, and the C_4-amino reductone were synthesized and tested for their ability to induce nuclear translocation of NF-κB in macrophages. Among the tested compounds, only the C_4-amino reductone showed a biological effect in the test system. It was also demonstrated that the effect was not specific to the C_4-amino reductone, but could also be induced by several other reaction products containing an amino reductone substructure. The cellular reaction was further mediated by the ability of the amino reductones to produce H_2O_2. Thus, it can be hypothesized that the enamineol group, which has a high electron density, is able to reduce oxygen directly to give hydrogen peroxide (figure 3).

6   MAILLARD REACTION PRODUCTS INHIBIT OXIDATION

The last pathway for Maillard reaction and oxidation to interact has probably been studied most extensively. Whereas the prooxidative activity of Maillard reaction products is rarely described, their antioxidative activity is well established, both in model mixtures and in processed food [16-18]. Maillard products cannot only inhibit lipid oxidation in food systems, but are also able to adverse oxidative cell damage [19, 20]. Furthermore, a diet rich in Maillard products significantly reduced consumers' susceptibility to lipid oxidation [21, 22]. The antioxidative activity of Maillard products seems to be contrary to the prooxidative effect discussed above. However, both phenomena are based on the reducing properties of certain Maillard products, such as the aminoreductones. Thus, an anti-oxidative Maillard product can directly reduce oxidizing agents, such as peroxides or radicals, to give redox-inactive products. As a consequence, oxidative damage is inhibited. But in a similar way, Maillard products can also reduce oxygen to give ROS, such as superoxide anion, hydrogen peroxide, or hydroxyl radical, which are stronger oxidants then oxygen itself. The decision how a Maillard system will react, either as antioxidant or pro-oxidant, seems to be largely dependent on the reaction conditions, mainly on the concentration of the reducing compounds [23]. A similar behavior has been discussed for other antioxidants before.

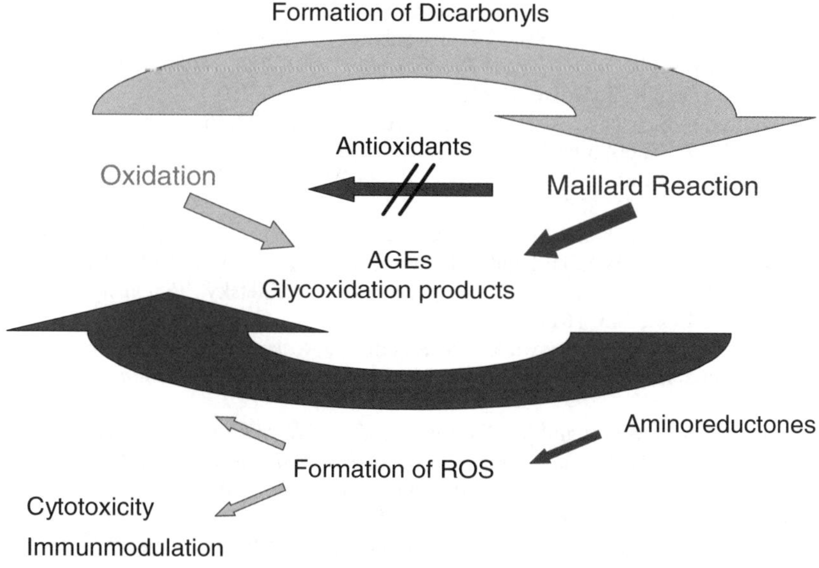

Figure 4   *Interactions between oxidation and Maillard reaction during food processing*

7   CONCLUSIONS

Aminoreductones have been identified as potent components of Maillard reaction mixtures that induce cellular effects through the formation of hydrogen peroxide. On the other hand, the antioxidative effect of Maillard products is well established. The extent of antioxidative and pro-oxidative activity of Maillard products seems to depend on the reaction conditions. Thus, it is difficult to differentiate clearly between the physiological and technological effects of oxidation and glycation reactions in processed food.

References

1.   M. U. Ahmed, S. R. Thorpe and J. W. Baynes, *J Biol Chem*, 1986, **261**, 4889-4894.
2.   T. Henle, U. Schwarzenbolz and H. Klostermeyer, *Z. Lebensm. Unters. Forsch. A*, 1997, **204**, 95-98.
3.   S. P. Wolff, Z. Y. Jiang and J. V. Hunt, *Free Radic Biol Med*, 1991, **10**, 339-352.
4.   T. Usui, M. Yoshino, H. Watanabe and F. Hayase, *Biosci. Biotechnol. Biochem.*, 2007, **71**, 2162-2168.
5.   M. Frischmann, J. Spitzer, M. Funfrocken, S. Mittelmaier, M. Deckert, T. Fichert and M. Pischetsrieder, *Biomed Chromatogr*, 2009, **23**, 843-851.
6.   A. Tauer, T. Knerr, T. Niwa, T. P. Schaub, C. Lage, J. Passlick-Deetjen and M. Pischetsrieder, *Biochem Biophys Res Commun*, 2001, **280**, 1408-1414.
7.   S. P. Wolff and R. T. Dean, *Biochem J*, 1987, **245**, 243-250.
8.   J. Meltretter, S. Seeber, A. Humeny, C. M. Becker and M. Pischetsrieder, *J Agric Food Chem*, 2007, **55**, 6096-6103.
9.   S. Muscat, J. Pelka, J. Hegele, B. Weigle, G. Munch and M. Pischetsrieder, *Mol Nutr Food Res*, 2007, **51**, 525-535.
10.   J. Hegele, G. Munch and M. Pischetsrieder, *Mol Nutr Food Res*, 2009, **53**, 760-769.
11.   Z. Y. Jiang, A. C. Woollard and S. P. Wolff, *FEBS Lett.*, 1990, **268**, 69-71.
12.   B. J. Ortwerth, H. James, G. Simpson and M. Linetsky, *Biochem. Biophys. Res. Commun.*, 1998, **245**, 161-165.
13.   J. Zu, J. Morita, S. Nishikawa and N. Kashimura, *Carbohydr. Lett.*, 1996, **1**, 457-464.
14.   J. R. Requena and E. R. Stadtman, *Biochem Biophys Res Commun*, 1999, **264**, 207-211.
15.   A. Wuehr, M. Deckert and M. Pischetsrieder, *Mol Nutr Food Res*, accepted.
16.   K. Kawashima, H. Itoh and I. Chibata, *J. Agric. Food Chem.*, 1977, **25**, 202-204.
17.   S. M. Monti, A. Ritieni, G. Graziani, G. Randazzo, L. Mannina, A. L. Segre and V. Fogliano, *J Agric Food Chem*, 1999, **47**, 1506-1513.
18.   A. Papetti, M. Daglia, C. Aceti, M. Quaglia, C. Gregotti and G. Gazzani, *J Agric Food Chem*, 2006, **54**, 1209-1216.
19.   J. S. Smith and M. Alfawaz, *J. Food Sci.*, 1995, **60**, 234-240.
20.   L. Goya, C. Delgado-Andrade, J. A. Rufian-Henares, L. Bravo and F. J. Morales, *Mol Nutr Food Res*, 2007, **51**, 536-545.

21.   R. Dittrich, C. Dragonas, D. Kannenkeril, I. Hoffmann, A. Mueller, M. W. Beckmann and M. Pischetsrieder, *Food Res. Int.*, 2009, **42**, 1315-1322.
22.   I. Seiquer, B. Ruiz-Roca, M. Mesias, A. Munoz-Hoyos, G. Galdo, J. J. Ochoa and M. P. Navarro, *J. Sci. Food Agric.*, 2008, **88**, 1245-1252.
23.   M. Pischetsrieder, F. Rinaldi, U. Groß and T. Severin, *J. Agric. Food Chem.*, 1998, **46**, 2945-2950.

DICARBONYLS IN COLA DRINKS SWEETENED WITH SUCROSE OR HIGH FRUCTOSE CORN SYRUP

P. J Thornalley[1] and N Rabbani[1]

[1]Clinical Sciences Research Institute, Warwick Medical School, University of Warwick, University Hospital, Coventry CV2 2DX, U.K.

1   INTRODUCTION

Sweetened soft drinks are a popular beverage. Globally, Coca-cola™ products alone are consumed at a rate > 1.5 billion drinks per day. In the USA, per capita consumption of sweetened soft drinks is > 60 litres per year [1]. Most sweetened soft drinks worldwide are sweetened with sucrose derived from cane or beet sugar. High fructose corn syrup ("isoglucose" or glucose-fructose; HFCS) is a corn syrup-derived product that has been treated with invertase to convert part of the endogenous glucose to fructose. High fructose corn syrup contains 42% or 55% fructose - HFCS-42 and HFCS-55 products - depending on manufacturing process, with the remaining sugar mainly glucose and some higher sugars. HFCS production is only *ca.* 8% of sucrose production. In the USA, however, the use of HFCS has grown to approximately equal that of sucrose; HFCS represents 40% of caloric sweeteners added to foods and beverages. Beverages sweetened with HFCS account for 80% of added sugars in the US diet and account for 80% of the recent increase in calorific intake of the global diet. HFCS-55 has sweetness equivalent to sucrose and is used in carbonated soft drinks in the USA – such as colas. HFCS-42 is less sweet and is used in many fruit-flavoured noncarbonated beverages, bakery and other products. In contrast, little HFCS is used in the Europe. Hence, sweetened cola drinks produced in the USA contain mainly HFCS whereas cola drinks produced in Europe contain sucrose [2].
Recent concern has arisen from the concurrent increase of use of HFCS in sweetened cola drinks in the USA with increase in obesity, renal disease and other disorders[3,4]. In the USA, the consumption of HFCS increased 10-fold between 1970 and 1990 - exceeding changes in intake of any other food in the diet. For all Americans aged $\geq$ 2 years old, an average energy intake of 132 kcal per day in the diet comes from HFCS with *ca.* 316 kcal per day in the diet for the top quintile[3]. There has been added concern of a claimed high content of reactive dicarbonyl glycating agents – glyoxal, methylglyoxal (MGO) and 3-deoxyglucosone 4 (3-DG) - in sweetened cola and other soft drinks[5]. Dicarbonyl compounds may lead to increased protein and nucleotide damage associated with diabetes, renal failure and ageing-related disorders[6]. Degradation of glucose, fructose and sucrose during thermal processing of soft drinks leads to the formation of these dicarbonyls. The greater susceptibility of glucose and fructose to thermolysis relative to sucrose may lead to

higher dicarbonyl content in HFCS sweetened soft drinks relative to sucrose-sweetened soft drinks. As samples were heated in pre-analytic processing of dicarbonyl measurement in the study by Lo *et al.* [5], the apparent concentrations of dicarbonyls in soft drinks were likely overestimates. The aim of this study was to re-evaluate the level of dicarbonyl glycating agents, glyoxal, MGO and 3-DG in cola drinks and compare the levels in cola drinks sweetened with sucrose or HFCS.

2   MATERIALS AND METHODS

2.1   Materials

Glyoxal (40% w/v) and 1,2-diaminobenzene (DAB; purified by sublimation and zone refined; ≥99%) were purchased from Sigma. [$^{13}C_2$]. Glyoxal was prepared by enzymatic oxidation of [13C2]ethylene glycol[7]. MGO was prepared by acid hydrolysis of freshly distilled methylglyoxal dimethylacetal and purified by distillation under reduced pressure as described[8]. [13C2]MGO was prepared by the oxidation of [$^{13}C_3$]acetone with selenium dioxide and purified by distillation under reduced pressure[9]. 3-DG was prepared by the 3-DG bis(benzoylhydrazone) method and purified by elution through mixed-bed anion/cation exchange resin and preparative reversed phase HPLC[10]. [13C6]3-DG was prepared by a similar method starting with [13C6]glucose. Concentrations of stock solutions of dicarbonyls was determined by derivatisation with aminoguanidine and spectrophotometric assay of the 3-amino-1,2,4-triazine adducts formed[11].

2.2   Cola drinks

Retail cola drinks, 330 ml cans, from 3 different batches produced in the UK sweetened with sucrose (Cola-S) and 3 different batches produced in the USA (Cola-GF) were analysed.

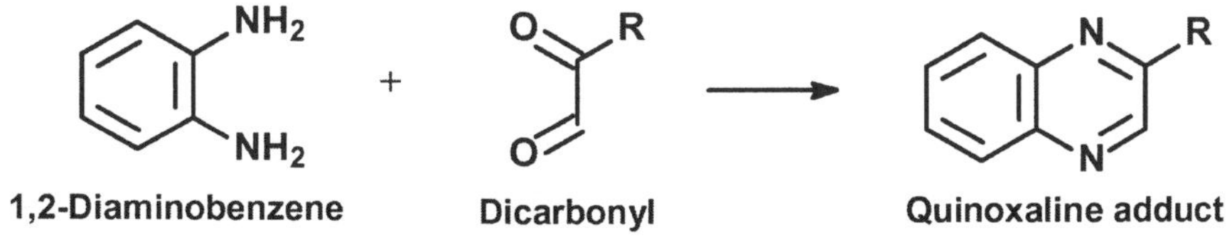

Figure 1.   *Derivatization of dicarbonyl compounds with 1,2-diameinobenzene to quinoxaline derivatives. Key: R = H, glyoxal to quinoxaline; R = Me, MGO to 2-methylquinoxaline; and R= CH2(CHOH)2CH2OH, 3-DG to 2-(erythro-2,3,4-trihydroxybutyl)quinoxaline.*

2.3   Assay of dicarbonyls

Glyoxal, MGO and 3-DG were determined in cola by derivatisation with 1,2-diaminobenzene (DAB - **Figure 1**) and quantitation by stable isotopic dilution analysis liquid chromatographytandem mass spectrometry (LC-MS/MS). Test sample: cola (2 µl),

20% w/v trichloroacetic acid with 0.9% w/v NaCl (10 µl), isotopic standards – 2 pmol [13C2]glyoxal, [13C2]MGO and [13C6]3-DG; water (23 µl) and 0.5 mM DAB with 0.5 mM diethylenetriaminepenta-acetic acid in 0.2 M HCl (10 µl). Samples were incubated for 4 h in the dark. The samples were then analysed by LC-MS/MS using positive ion electrospray ionisation and multiple reaction monitoring on an Acquity™ UPLC-Quattro Premier tandem mass spectrometer (Waters, Elstree, UK). The column was BEH C18 1.7 µm particle size, 2.1 x 100 mm column and guard column (5 x 2.1 mm). The mobile phase was 0.1% trifluoroacetic acid in water with a linear gradient of 0 - 50% methanol over 0 - 20 min and isocratic 50% methanol thereafter; the flow rate was 0.2 ml/min. The capillary voltage was 3.5 kV, the cone voltage 50 V, the interscan delay time 100 ms, the source and desolvation gas temperatures 120°C and 350°C and the cone gas and desolvation gas flows were 150 and 550 l/h respectively. Analyte/internal standard peak response ratios were calibration by assay of authentic standards: 1 – 10 pmol glyoxal, MGO and 3-DG. For determination of 3-DG in Cola-GF, cola was diluted 500-fold with water prior to analysis.

3   RESULTS

3.1   Assay of dicarbonyls by stable isotopic dilution analysis LC-MS/MS

Derivatisation of dicarbonyls under non-oxidising, acidic conditions with DAB provides for efficient derivatisation, high recoveries and low coefficients of variation. LC-MS/MS provides for optimum detection of high specificity and high sensitivity – **Table 1**.

Table 1. *Analysis characteristics for the assay of dicarbonyl compounds by derivatisation with 1,2-diaminobenzene and stable isotopic dilution analysis liquid chromatography – tandem mass spectrometry.*

Analyte	Retention time (min)	Molecular ion to fragment ion (Da)	Neutral fragment	Collision energy (eV)	Cone voltage (V)	Limit of detection (fmol)	Recovery (%)
Glyoxal	18.7	130.9 to 77.1	NC-CH=NH	28	24	32	99
[13C2]Glyoxal	18.7	132.9 to 77.1	NC-CH=NH	28	24	32	99
MGO	20.8	144.8 to 77.1	NC-CMe=NH	28	24	8	94
[13C3]MGO	20.8	147.8 to 77.1	NC-CMe=NH	28	24	8	94
3-DG	14.6	235.0 to 199.2	$2H_2O$	16	21	34	76
[13C6]3-DG	14.6	241.0 to 205.2	$2H_2O$	16	21	34	76

3.2 Dicarbonyls content of cola drinks sweetened by sucrose and by high fructose corn syrup

The dicarbonyl content of Cola-GF was markedly higher than of Cola-S. The increase in dicarbonyl concentration in Cola-GF was: glyoxal 5-fold, MGO, 8-fold and 3-DG 114-fold (P<0.001). Inter-batch coefficients of variation for estimations were: glyoxal 5%, methylglyoxal 6 - 8% and 3-DG 5 – 17 % - **Table 2**.

Table 2. *Dicarbonyl content of cola drinks sweetened with sucrose or high fructose corn syrup. Data are mean ± SD (n = 3).*

	Glyoxal (µM)	MGO(µM)	3-DG (µM)
Cola-S (sucrose sweetened)			
Batch #1	0.51±0.01	0.29±0.01	2.47±0.04
Batch #2	0.56±0.04	0.35±0.01	2.25±0.07
Batch #3	0.56±0.05	0.31±0.01	2.31±0.06
Cola-GF (HFCS sweetened)			
Batch #1	2.51±0.12	2.41±0.06	226±16
Batch #2	2.71±0.08	2.54±0.14	316±26
Batch #3	2.79±0.14	2.70±0.12	257±30

4 DISCUSSION

4.1 Importance of appropriate pre-analytic processing in dicarbonyl estimation

Previous estimates of glyoxal, MGO and 3-DG in similar drinks in the USA by Lo et al[5] were 10 – 100 fold, 10 – 30 fold and 3 – 10 fold higher, respectively, than estimates herein. Heating and non-acidic conditions in sample processing may have contributed to these apparent marked overestimates. Repeating the protocol of Lo *et al.* of sample heating in preanalytic processing led to marked overestimation of dicarbonyl content of the cola samples analysed herein. Although it may seem counterintuitive to employ acidic conditions for the derivatisation of dicarbonyls by DAB with protonation of the amino groups (pK$_{a1}$ = 4.6, pK$_{a2}$ = 0.8), the rate-limiting step of quinoxaline formation is acid-catalysed dehydration and hence derivatisation proceeds faster under acidic conditions[12]. Acidic conditions also stabilize DAB and related derivatising agents from oxidative degradation[8].

4.2 Physiological relevance of dicarbonyl content of sweetened cola drinks

The concentration of glyoxal and MGO in cola sweetened with sucrose are similar to plasma concentrations of diabetic patients, and the concentration of 3-DG is *ca.* 10-fold higher than the plasma concentration of diabetic patients[13,14]. The concentration of glyoxal,

MGO and 3- DG in cola sweetened with HFCS are similar to concentrations found in a glucose-containing, heat sterilised dialysis fluid[15]. The reactivity of 3-DG for glycation of proteins and nucleotides tends to be 100 – 200 fold less than the glycation reactivity of glyoxal and MGO[16,17]. Therefore, glyoxal and MGO in sweetened cola drinks probably pose the most significant threat of macromolecular damage by dicarbonyl compounds consumed.

The amount of dicarbonyl in a 330 ml serving may be markedly lower than the endogenous flux of dicarbonyl formation in the body. For example, most formation of MGO comes from degradation of triosephosphates; *ca.* 0.089 % glucotriose is converted to MGO per day[18]. This equates to 3 - 4 mmol MGO produced endogenously in adult human subjects leading an active lifestyle per day; *cf.* a 330 ml can of sweetened cola that contains *ca.* 0.2 – 1 μmol MGO which is less than 0.1% of the endogenous rate of formation. These and similar arguments have been used in attempts to assess the significance of dicarbonyls in HFCS sweetened colas in the USA[2].

Further studies have been reported recently on the attempts to assess the effect of methylglyoxal in cola drinks in plasma MGO in human subjects[19]. A liquid chromatography method for MGO assay was used although it is not clear that the method was reliable since – by the authors' own admission - their method overestimated the concentration of MGO in human blood plasma in a previous study[19]. The concentration of MGO in regular, sweetened cola was 7.2 μM – which may have been an overestimate of MGO in Cola-GF (- the sugar sweetener in the cola studies was not stated). Six healthy subjects consuming 300 ml sweetened cola and two subjects had a *ca.* 20% increase of plasma MGO concentration at 30 min after consumption of the cola drink; in one of these subjects, there was *ca.* 40% increase in plasma glucose concentration at this time that may have led to the increased plasma MGO concentration indirectly. Further research is required to assess definitively if the content of MGO and other dicarbonyl glycating agents in sweetened cola drinks increases the dicarbonyl burden in human subjects significantly. Local deleterious effects on upper gastrointestinal tract epithelia may be the most likely possibility of cola-derived dicarbonyls causing harm[20].

We conclude that dicarbonyl compounds are present in retail cola drinks. Cola drinks have markedly higher dicarbonyl contents when sweetened with high fructose corn syrup than with sucrose. Previous estimates of glyoxal, methylglyoxal and 3-DG in similar drinks in the USA were 10 – 100 fold, 10 – 30 fold and 3 – 10 fold higher, respectively, than estimates herein. Heating and non-acidic conditions in sample processing may have contributed to these apparent marked overestimates.

Acknowledgements

The authors received funding to cover expenses for analysis of dicarbonyl compounds in soft drinks from the American Corn Refiners Association, USA.

References

1. A. Wolf, G. A. Bray, and B. M. Popkin, *Obesity Reviews*, 2008, **9**, 151-164.
2. John S. White, *Journal of Nutrition*, 2009, **139**, 1219S-1227.
3. G. A. Bray, S. J. Nielsen, and B. M. Popkin, *American Journal of Clinical Nutrition*, 2004, **79**, 537-543.
4. D. A. Shoham, R. Durazo-Arvizu, H. Kramer, A. Luke, S. Vupputuri, A. Kshirsagar, and R. S. Cooper, *PLoS ONE*, 2008, **3**, e3431.

5. C. Y. Lo, S. Li, Y. Wang, D. Tan, M. H. Pan, S. Sang, and C. T. Ho, *Food Chemistry*, 2008, **107**, 1099-1105.

6. P. J. Thornalley, *Drug Metab & Drug Interact*, 2008, **23**, 125-150.

7. K. Isobe and H. Nishise, *Biosci.Biotech.Biochem.*, 1994, **58**, 170-173.

8. A. C. McLellan and P. J. Thornalley, *Anal.Chim.Acta*, 1992, **263**, 137-142. 10

9. J. D. Clelland and P. J. Thornalley, *J.Label.Comp.Radiopharm.*, 1990, **28**, 1455-1464.

10. M. A. Madson and M. S. Feather, *Carbohydr.Res.*, 1981, **94**, 183-191.

11. P. J. Thornalley, A. Yurek-George, and O. K. Argirov, *Biochem.Pharmacol.*, 2000, **60**, 55-65.

12. S. Ohmori, M. Mori, M. Kawase, and S. Tsuboi, *Journal of Chromatography-Biomedical Applications*, 1987, **414**, 149-155.

13. A. C. McLellan, P. J. Thornalley, J. Benn, and P. H. Sonksen, *Clin.Sci.*, 1994, **87**, 21-29.

14. P. J. Beisswenger, S. Howell, A. Touchette, S. Lal, and B. S. Szwergold, *Diabetes*, 1999, **48**, 198-202.

15. Dawnay, A., Argirova, M., Millar, D. J., Holmes, C., and Thornalley, P. J., *J.Am.Soc.Nephrol.*, 1999, **10**, 312A.

16. N. Ahmed and P. J. Thornalley, *Diabetes Obesity & Metabolism*, 2007, **9**, 233-245. 11 17.

17. P. J. Thornalley, *Biochem.Soc.Trans.*, 2003, **31**, 1372-1377.

18. P. J. Thornalley, *Biochem.J.*, 1988, **254**, 751-755.

19. K. Nakayama, M. Nakayama, M. Iwabuchi, H. Terawaki, T. Sato, M. Kohno, and S. Ito, *American Journal of Nephrology*, 2008, **28**, 871-878.

20. S. Kapicioglu, A. Baki, A. Reis, and Y. Tekelioglu, *Diseases of the Esophagus*, 1999, **12**, 306-308.

FORMATION OF MUTAGENS/CARCINOGENS UNDER PHYSIOLOGICAL CONDITIONS AND THE INHIBITORY EFFECTS OF DAILY FOODS ON THEIR FORMATION AND THE INDUCTION OF GENOTOXICITY

N. Kinae[1], M. Hirano[1], T.Urahira[1]M. Iio[1], S. Masumori[2]and S.Masuda[1]

[1]Department of Food and Nutritional Sciences, Graduate School of Nutritional and Environmental Sciences and Global COE program, University of Shizuoka, 52-1 Yada Suruga-ku Shizuoka-shi, Shizuoka 422-8526, Japan
[2]Biosafety Research Center, Foods, Drugs and Pesticides, 582-2, Shioshinden, Iwata-shi, Shizuoka, 437-1213, Japan

1   INTRODUCTION

It is well-known that the Maillard reaction, proceeds not only in daily foods with and without heating, but also in our body such as lens crystallins and skin collagen.   As the results, melanoidins and advanced glycation end products (AGEs) are formed *in vitro* and *in vivo*.   These AGEs often show biological properties such as antioxidative, antibacterial, anticarcinogenic and carcinogenic activities.   Through the Maillard reaction, 2-amino-3,8-dimethylimidazo[4,5-f] quinoxaline (MeIQx) and other heterocyclic amines have been identified as mutagens/carcinogens in cooked beef, chicken and mutton samples[1].   We have previously reported that several mutagens/carcinogens containing heterocyclic amines are formed in the model reaction mixtures of carbohydrates and amino acids under physiological conditions, without heating[2].   For example, acrylamide and glycidamide are potent mutagen and proximate carcinogens that form DNA-adducts with guanine and adenine residues of DNA[3-5].   In the present work, we report on the formation of the mutagen/carcinogen, acrylamide, under physiological conditions and on the inhibitory effects of Japanese daily foods on the formation by using several genotoxicity tests.   In addition, we explore that inhibitory effect of green tea on the formation of these genotoxic Maillard reaction products and their antimutagenic activities.

2   MATERIALS AND METHODS

2.1   Chemicals, reagents and analysis

All chemicals were of the best grade and purchased from Wako Pure Chemical Inc. (Osaka, Japan).   A mixture of D-glucose, asparagine and creatine was dissolved in phosphate buffered saline (pH7.4) and incubated at 37℃ for 100-1000 days.   The brown reaction solution was treated with Oasis HLB and Max. The content of acrylamide was determined by LC/MS/MS analysis.

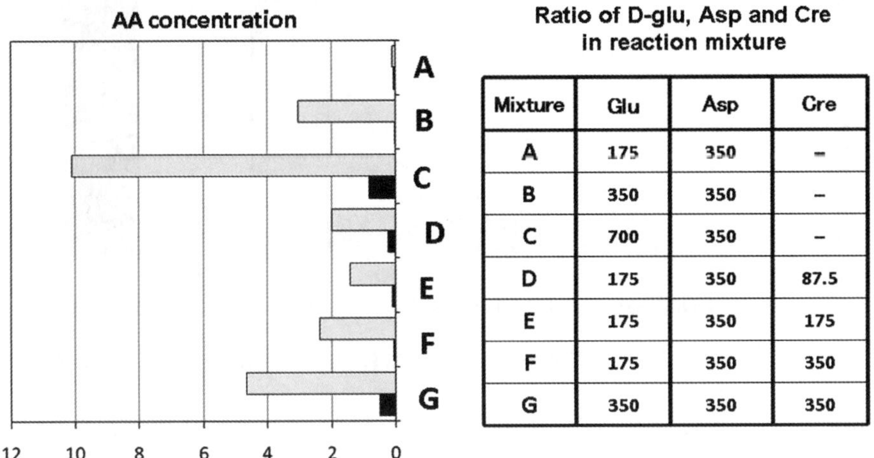

Ratio of D-glu, Asp and Cre in reaction mixture

Mixture	Glu	Asp	Cre
A	175	350	–
B	350	350	–
C	700	350	–
D	175	350	87.5
E	175	350	175
F	175	350	350
G	350	350	350

Figure 1   *Isolation of AA from the reaction mixtures under physiological conditions*

2.2   Animal models

Wistar rats (5weeks, male) were administered acrylamide (50mg/kg body weight) after the induction of diabetes with streptozotocin.   To examine the inhibitory effect of green tea on the induction of the genotoxicity of acrylamide, green tea extract (20-100mg/kg body weight) was administered to mice following the induction of diabetes. Several organs were submitted to the comet assay and micronucleus test to examine the DNA damaging potency and chromosomal aberration, respectively.

3   RESULTS

3.1   In vitro studies

As previously described[6], significant amounts of acrylamide were generated in the reaction mixture, incubated under physiological conditions at the mole ratio of D-glucose : asparagine (175~700:350) with and without creatine (Figure 1).   As the concentration of D-glucose increased, the yield of acrylamide also increased.

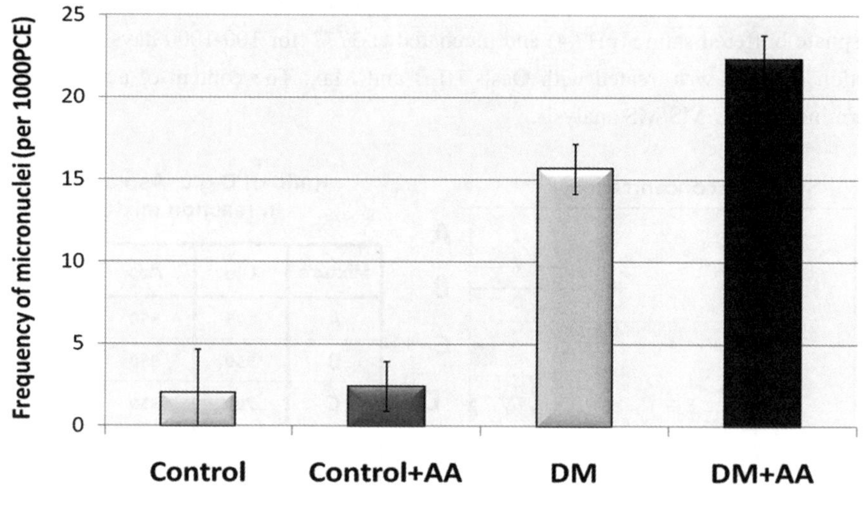

Figure 2   *Frequency of micronuclei in bone marrow of normal and diabetic rats treated with or without acrylamide*

3.2   In vivo studies

The induction diabetes was associated with an increased frequency of micronuclei in bone marrow of rats.   Co-treatment with acrylamide significantly increased the frequency of micronuclei (**Figure 2**).   However, this effect was no observed in control mice treated with acrylamide.   Treatment with acrylamide also induced markers of cellular DNA damage in liver, kidney and peripheral blood in both normal and diabetic rats (**Figure 3**).

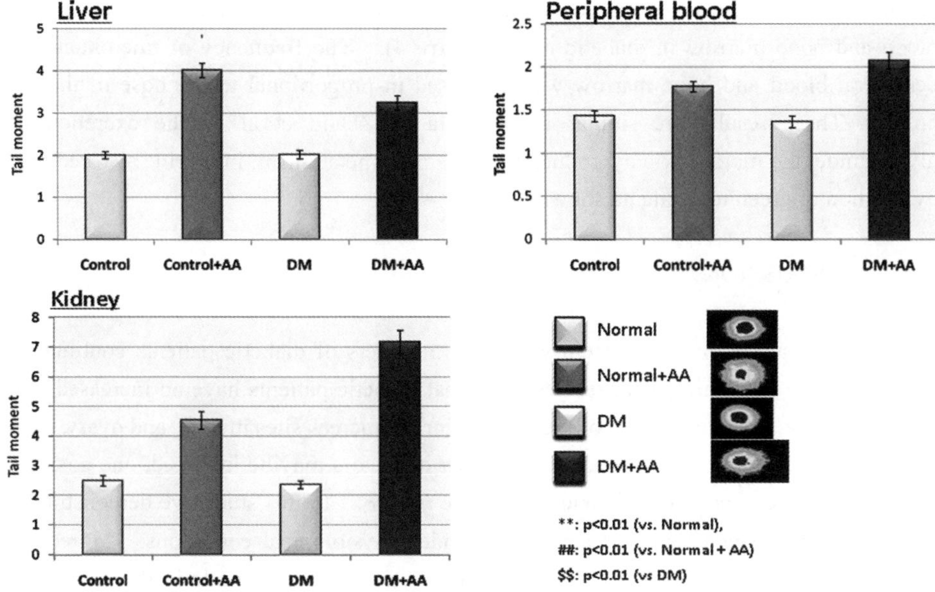

Figure 3 *Cellular DNA damage of normal and diabetic rats treated with or without acrylamide (AA)*

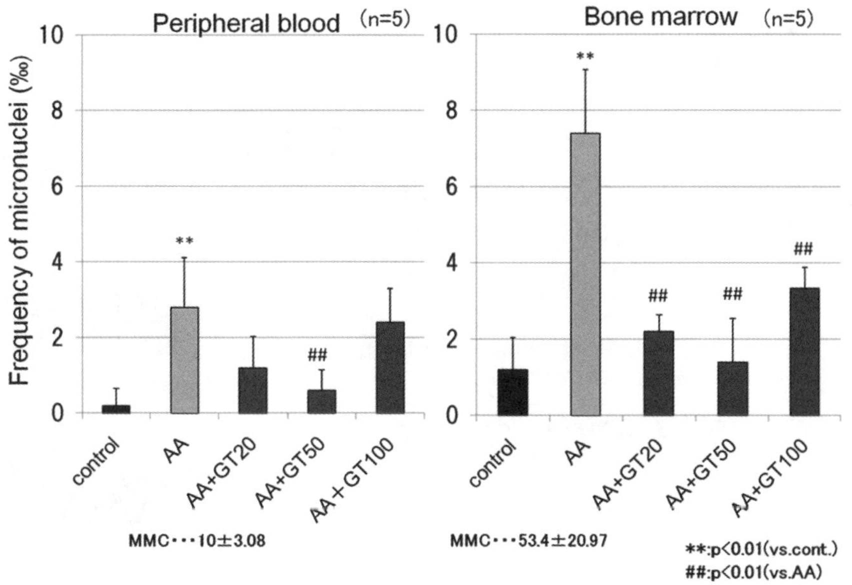

Figure 4 *Inhibitory effect of green tea on the frequency of micronuclei induced by acrylamide (AA)*

Administration of green tea extracts reduced the frequency of micronuclei in peripheral blood and bone marrow in diabetic mice (**Figure 4**).   The frequency of micronuclei in peripheral blood and bone marrow was decreased in proportional to the dose of the tea powder. These results are similar to the data of Zhang et al[7].   The excretion of glycidamide, the major toxic byproduct of acrylamide metabolism, into urine also reduced by addition of green tea (data no shown).

4   DISCUSSION

In Japan, as in the Western world, the numbers of diabetic patients continue to increase.   The epidemiological studies show that diabetic patients have an increased risk of some cancers, including those of the liver, kidney, pancreas sites in men, and ovary, liver, stomach cancer in women.   Although a number of factors may be involved, one possible contributor is the generation of toxic acrylamide *in vivo*.   In this study, we demonstrate a glucose dependent-generation of acrylamide under physiological conditions.   Moreover, we show increased susceptibility to the genotoxic effects of acrylamide in the context of diabetes. Taken together these data point to a potentially important role for acrylamide and other products of the Maillard reaction in diabetes-associated cancers.

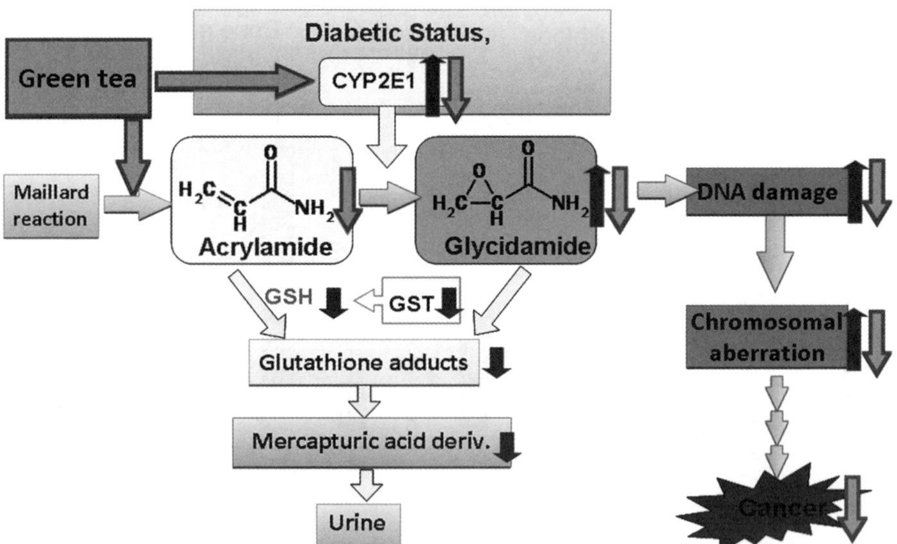

Figure 5   *Possible pathways for enhancement and reduction of genotoxicity of acrylamide (AA) in diabetic rats*

The means by which the toxicity of acrylamide is enhanced in diabetes remains to be established. We hypothesize that accumulation of acrylamide in diabetes, leads to increased formation of glycidamide (figure 5), produced by CYP2E1. This toxic byproduct then induces DNA damage and chromosomal aberration. We have previously demonstrated that the expression of CYP2E1 is increased in diabetic rats. Notably, green tea also inhibits the activity of CYP2E1 *in vivo*, as well as inhibiting the formation of acrylamide *in vitro*. These results suggest that several daily foods including green tea and wasabi may act as inhibitors for the formation of acrylamide and the induction of genotoxicity, and therefore may be a useful tool to prevent the cancer in diabetic patients.

References

1. Sugimura T, Wakabayashi K, Nakagama H, Nagao M, *Cancer Sci*, 2004, **95(4)**, 290-299
2. Kinae N, Mori C, Kujirai K, Mausmori S. and Masuda S. *Ann NY Acad Sci,* 2005, 1043, 80-84.
3. IARC International Agency for Research on Cancer, 1994, **60**, 389-433,
4. Tareke E, Rydberg P, Karlsson P, Eriksson S, Tornqvisa M, *J Agric Food Chem*, 2002, **50**, 4998-5006.
5. Koyama N, Sakamoto H, Sakuraba M, Koizumi T, Takashima Y, Hayashi M, Matsufuji H, Yamagata K, Masuda S, Kinae N. and Honma M, *Mutat Res,* 2005, **603**, 151-158.
6. Mottram DS, Wedzicha BL, Dodson AT, *Nature*, 2002, **419**, 448-449.
7. Zhang Y, Ying T, Zhang Y, J. *Food Sci,* 2008, **73(2)**, 60-66.

ANTITUMOR EFFECTS OF THE EARLY MAILLARD REACTION PRODUCTS

V.V. Mossine, V.V. Glinsky[1,2] and T.P. Mawhinney[3]

[1]Departments of Biochemistry University of Missouri-Columbia, Columbia, MO 65211, USA
[2]Departments of Pathology and Anatomical Sciences, University of Missouri-Columbia, Columbia, MO 65211, USA
[3]Harry S. Truman Memorial Veterans Hospital, Columbia, mo 65201, USA

1   INTRODUCTION

Multiple studies suggest that many cancers are induced by environmental factors and that more than two thirds of human cancers could be prevented by lifestyle changes including dietary modification.[1] For example, increased intake of vegetables and fruits is consistently associated with lower cancer risk as compared to other diet modifications.[2] Sifting through the publications of epidemiological studies linking dietary fruit and vegetable intake and prostate cancer risk, we stumbled upon an intriguing fact that processed or dehydrated fruit consumption is associated with lower urological cancer risk, as compared with raw fruit consumption (data compiled in Table 1). Fruits and vegetables are rich in carbohydrates, both simple monosaccharides such as glucose or fructose, and polysaccharides such as cellulose, pectins, xylans etc. Processing (peeling, cutting, thermal treatment) brings about major changes in the physical state of plant food, largely due to enzymatic fragmentation of the rigid cell wall polysaccharides into smaller and more bioavailable oligosaccharides, thus increasing nutritional value of food.

Dehydration and thermal treatment also cause formation of the Maillard reaction products, primarily D-fructosamines, also referred to as Amadori rearrangement products, in relatively large quantities (Table 2), as well as products of their degradation, such as brown polymeric and volatile aroma molecules. Although the Maillard reaction products were implied as responsible for the enhancement of antioxidant properties of processed foods,[3] it is not known, whether this class of compounds does contribute to the protective potential of processed foods against cancer. However, recent laboratory data suggest that Amadori rearrangement products, although not cytotoxic themselves, may target specific molecular mechanisms associated with tumorigenesis and interact cooperatively with other phytochemicals and therapeutics against tumor growth and dissemination. Here we present a compilation of such data and posit that early Maillard reaction products may have a significance for the prevention of tumorigenesis.

Table 1. *Epidemiological data on comparative prostate cancer risk for processed and raw fruits.*

Type of Study	No. Subjects	Fruit	RR, HR, or OR @ 95% CI	Risk lowering effect
Case-control[4]	320 cases	raw tomato	0.65 (0.40-1.0) *P*=.12	Suggestive positive
	246 controls	processed tomato	0.52 (0.33-0.83) *P*=.005	Significantly positive
Case-control[5]	628 cases	raw tomato	0.93 (0.67-1.30) *P*=.76	No effect
	602 controls	processed tomato	0.73 (0.48-1.10) *P*=.13	Suggestive positive
Prospective cohort[6]	392 cases*	raw tomato	1.58 (1.10-2.25)	Negative
	1202 total*	tomato sauce	0.56 (0.38-0.82)	Positive
Prospective cohort[7]	97 cases	raw tomato	1.04 (0.60-1.80) *P*=.89	No effect
	1975 subcohort	processed tomato	0.67 (0.38-1.16) *P*=.13	Suggestive positive
Prospective cohort[8]	180 cases	fresh fruits	0.78 (0.50-1.20) *P*=.46	Suggestive positive
	14000 total	dried fruits	0.51 (0.31-0.85) *P*=.01	Significantly positive
Prospective cohort[9]	642 cases	total fruits	1.01 (0.99-1.04)	No effect
	1525 subcohort	dried fruits	0.49 (0.18-1.32)	Positive
Prospective cohort[10]	569 cases**	total fruits	0.99 (0.96-1.01)	No effect
	3123 subcohort	dried fruits	**0.40 (0.11 -1.50)	Positive

*Cases of cancer progression among diagnosed with prostate cancer

**Urothelial cancers

Table 2.   *Estimates for D-fructosamine and D-lactulosamine content in some dehydrated foods.*

Food product	Analytes	Total content	Reference
Tomato powder	14 D-fructose-amino acids	9.2 g / 100 g	[22]
Bell pepper	---- // -----	4.6 g / 100 g	[22]
Asparagus	---- // -----	2.1 g / 100 g	[22]
Carrot	---- // -----	0.93 g / 100 g	[22]
Raisins	5 D-fructose-amino acids as 2FMAA*	141 mg / 100 g	[23]
Dark malt	10 D-fructose-amino acids	130 mg / 100 g	[22]
Milk powder	D-lactulose-lysine as 2FMAA	70 mg / 100 g protein	[24]
Milk-crumb	D-fructose- and D-lactulose-lysine as 2FMAA	4 g / 100 g protein	[24]
Cheese	---- // -----	0.2-0.4 g / 100 g protein	[24]

*2FMAA – 2-furoylmethyl-amino acid or furosine

2   FRUCTOSE-AMINO ACIDS

Dehydrated fruits, vegetables and other foods may contain significant amounts of D-fructose-amino acids (Table 2). Experimental evidence shows that these compounds are partially absorbed intestinally into circulation but are not readily metabolized.[11] Biological activities of D-fructose-amino acids were a subject of several studies. Some, such as D-fructose-L-arginine[12] and D-fructose-L-histidine,[13] were recognized as potent antioxidants, while others, such as D-fructose-L-proline,[14] displayed an immunostimulating potential. The ability of D-fructosamine to inhibit human tumor cell proliferation in vitro has been noted as early as in 1956.[15] There were a number of attempts to improve pharmacokinetic parameters of several anti-tumor drugs and reporter molecules by conjugation with carbohydrate unit via the Amadori rearrangement.[16,17] However, the fructosamine part in these molecules was not considered a pharmacophore but rather a means of improving solubility and stability. More recently, we tested anti-tumor activity of several D-fructose-amino acids in experimental *in vitro* and *in vivo* models of melanoma,[18,19] rhabdomyosarcoma,[18] breast,[19,20] and prostate[13] cancer (Tables 3 and 4).

Table 3. *Induction of apoptosis and inhibition of cancer cell aggregation, proliferation, or clonogenic growth by D-fructosamines and D-lactulosamines.*

Cell line	Inhibitor, at 2mM	% of proliferating or aggregating cells*	References
MDA-MB-435	D-fructose-glycine	67 (C);** 68 (NA)	[19,20]
human breast carcinoma	D-fructose-D-leucine	22 (C); 63 (NA)	
	D-fructose-L-phenylalanine	17 (C)	
	D-lactulose-L-leucine	48 (A); 19 (C); 61 (NA)	
B16 murine	D-fructose-γ-aminobutyric acid	50 (A)	[18,19
Melanoma	D-fructose-L-lysine	37 (A)	
	D-fructose-L-phenylalanine	52 (A)	
	D-lactulose-L-phenylalanine	61 (A)	
MAT-LyLu rat	D-fructose-L-aspartic acid	97 (P)	[13]
prostate	D-fructose-L-histidine	72 (P)	
adenocarcinoma	D-fructose-L-histidine/ lycopene	1 (P)	

*Controls – 100%

**A – aggregates number; C – colony number; NA – non-apoptotic; P – proliferating

Although the exact mechanism(s) of the *in vivo* antitumor activity of FAs is not known, a number of *in vitro* experiments showed that D-fructose-amino acids may target cancer cell adhesion by competing for specific cell surface glycoprotein-lectin interactions, particularly those involving β-galactoside-specific lectins expressed on metastatic cells.[19] Thus, D-fructose-L-leucine and D-fructose-D-leucine displayed significant inhibition of metastatic human breast cancer cell aggregation and related clonogenic survival *in vitro*.

Table 4. *Inhibition of primary or secondary tumorigenesis in vivo by D-fructosamines and D-lactulosamine.*

Model, treatment	Inhibitor	% of tumors or incidence	References
MDA-MB-435 in nude mice, i.p.	D-fructose-L-leucine	lung metastases: 105 (T);[**] 86 (I)	[20]
	D-fructose-D-leucine	LM: 0.5 (T); 22 (I)	
	D-lactulose-L-leucine	LM: 2.5 (T); 37 (I)	
	D-lactulose-L-leucine	LM: 2 (T); 74 (I)	[37]
	D-lactulose-L-leucine/ Taxol	LM: 0 (T); 35 (I)	
B16 in BALB/c mice, pre-incubation	D-fructose-γ-aminobutyric acid	LM:14 (T)	[18]
	D-fructose-L-leucine	LM:29 (T)	
	D-fructose-L-isoleucine	LM:13 (T)	
	D-lactulose-L-leucine	LM:19 (T)	
MAT-LyLu in syngeneic rat, pre-incubation	D-fructose-L-histidine	subcutaneous: 71 (T); 86 (I)	[13]
	D-fructose-L-histidine/ lycopene	SC: 6 (T); 21 (I)	
rat prostate carcinogenesis, diet	D-fructose-L-histidine/ lycopene	prostate: 17 (I)	[13]

[*]Controls – 100%;   [**]T – average/median number or volume; I – incidence

In vivo, these compounds demonstrated the ability to markedly reduce metastasis of MDA-MB-435 xenografts in nude mice.[20] A different mechanism of antitumor activity has been proposed for D-fructose-L-histidine.[13] FruHis stands out of other naturally occurring D-fructose-amino acids in respect of its exceptional metal-binding and antioxidant activities.[21] Remarkably, FruHis completely protected DNA degradative oxidation by copper-promoted hydroxyl radicals *in vitro*, while none of other tested major antioxidants from tomato (ascorbic acid, quercetin, caffeic acid etc.) could match the protective potential of FruHis. Even more intriguing was a discovery of an exceptionally strong synergism between FruHis and lycopene against advanced prostate cancer cell proliferation at physiologically achievable concentrations, both *in vitro* and *in vivo*,[13] given that this type of antitumor interaction represents the first one ever observed for rather extensively studied lycopene. A possible practical significance of FruHis, and perhaps other food-related D-fructose-amino acids, for cancer prevention has been established in a rat prostate carcinogenesis prevention experiment,[13] when androgen/carcinogen-induced rats were given diets including tomato paste (no D-fructose-amino acids), tomato powder (contains naturally occurring FAs), and tomato paste/FruHis. The tomato paste/FruHis combination and, to a lesser extent, tomato powder significantly improved survival of the experimental animals from cancer, by specifically modulating carcinogenesis in the prostate site. These results provide a justification for the hypothesis about a contribution of the D-fructose-amino acid fraction to the cancer risk-lowering effect of dehydrated fruits and vegetables in humans.

3   LACTULOSE-AMINO ACIDS

Dehydrated milk-based foods is the major source of D-lactulose-amino acids in human nutrition. The dairy industry widely uses milk dehydration to manufacture powdered milk as a base for numerous dairy products, such as infant formulas, confectionaries, reconstituted milk etc. During the process of heating, drying and storage, lactose in milk can readily interact with amino compounds that are naturally present, primarily lysine residues in milk proteins.[25,26] Estimated contents of D-lactulose-amino acids in selected products are compiled in Table 2. Commercial dairy products may contain, therefore, up to 40% of protein lysine in form of lactulose-lysine. Dietary availability of D-lactulose-amino acids is similar to that of the fructosamine derivatives. An ample evidence exists that the Amadori-type lysine glycoconjugates are not available to mammals as a nutrient and that lactulose-lysine is partially absorbed into the bloodstream and excreted unchanged.[26,27]
Our early studies on the anti-tumorigenic potential of D-lactulose-amino acids had a specific focus on testing the hypothesis that carbohydrate-mediated homo- and heterotypic cancer cell aggregation and adhesion is one of the key events in the metastatic process.[28] As a proof that β-galactoside participated in mediating such interactions, D-lactulose-L-leucine (β-D-Gal-(1→4)-D-FruNH-L-Leu, LctLeu) inhibited asialofetuin-induced aggregation of B16-F1 murine melanoma cells, as well as adhesion of peanut agglutinin (β-D-Gal-specific lectin) to human breast carcinoma MDA-MB-435 cells.[20] These initial results turned our attention to galectins, as potential targets for D-lactulose-amino acids in tumor models. Galectins are a group of about 15 mammalian β-galactoside-binding lectins that are abnormally expressed in breast, prostate and other cancers. Galectins have been implicated in the tumor proliferation and dissemination, and suspected in cancer immune protection.[29] Aberrant galectin-3 expression has been correlated with metastatic potential of advanced tumors in a number of cancers. Galectin-3 demonstrated antiapoptotic and invasive activities in cancer cells,[30] and dissemination of advanced carcinoma cells requires their docking in blood capillaries *via* galectin-3 – βGal interaction.[31,32] Galectins-1 and -3 have been recently identified as major immunosuppressive factors released by several epithelial malignancies.[29,30] Consequently, galectins thus have been recognized as important novel targets for cancer therapies. Physiological galectin ligands, that are thought to participate in the metastatic spread of advanced prostate and other cancers, include matrix glycoproteins displaying *N*-acetyl-

lactosamine (such as laminin, fibronectin) or Thomsen-Friedenreich antigen[32] (mucins). TFA is a cancer-associated antigen which is commonly expressed in many adenocarcinomas and for years has been considered as an attractive target for prostate cancer immunotherapy.[33] Incidentally, D-lactulose-amino acids are structurally mimicking TFA (β-D-Gal-($1{\rightarrow}3$)-α-D-GalNAc-O-Thr/Ser) molecules. Galectins-1, -3, and -4 readily interact with lactulosamines, as estimated by both indirect and direct methods.[34,35] For example, thermodynamic dissociation constants K_D measured for the interactions between galectin-1 and D-lactulose-amino acids by the surface plasmon resonance varied in the range of 70 μM (Lct-L-Pro) through 380 μM (Lct-L-Thr), which makes them potentially better galectin blockers than traditionally used lactose or *N*-acetyl-lactosamine. Interestingly, dehydration and oxidative degradation of lactulosamines leads to an array of advanced glycation endproducts (AGEs), which have also been identified as ligands for galectin-3,[36] although the exact galectin-3-binding AGE species is yet to be established.

In vitro testing showed that D-lactulose-amino acids inhibited, in a dose-dependent fashion, binding of galectins-1 and -3 to laminin, 90K glycoprotein, and immunoglobulin E.[35] The ability of lactulosamines to bind to galectins and inhibit interactions of galectins with their physiological glycoprotein ligands suggested that these carbohydrates might modulate some galectin-mediated events in mammalian cells. Most of the work with cell cultures was performed using D-lactulose-L-leucine, which was the first LA that demonstrated potentially beneficial effects in various *in vitro* and *in vivo* cancer models. It was determined that spontaneous aggregation of human MDA-MB-435 breast carcinoma cells was reduced by half upon addition of LctLeu to the medium. In addition, the reduction was accompanied by a 5-fold increase in the level of apoptosis of the MDA-MB-435 cells[20] after 2 days of cultivation on plastic. In 0.9% agarose, LctLeu caused more that 5-fold decrease in colony formation of this cell line.[20,31] In another experiment, LctLeu demonstrated ability to induce apoptosis and dose-dependent inhibition of clonogenic growth of an endothelial cancer-derived murine hemangiosarcoma SVR cell line, with IC_{50} about 350 μM.[38] Notably, LctLeu synergized, in a dose-dependent manner, with chemotherapeutic drug doxorubicin. This interaction between a D-lactulose-amino acid and a chemotherapeutic drug suggested that LA may modulate drug resistance of cancer cells by blocking a galectin-3-dependent anti-apoptotic signaling pathway. This was confirmed in our most recent study, which examined interaction between LctLeu and Taxol against proliferation of MDA-MB-435.[37] Specifically, LctLeu inhibited the galectin-3-induced phosphorylation of proapoptotic Bad protein in cells treated with Taxol. At the same time, LctLeu did not effect Taxol-induced phospho-rylation of antiapoptotic Bcl-2. The sensitization of MDA-MB-435 cells to Taxol by LctLeu led to their pronounced synergistic interaction both in cell culture and in a xenograft model *in vivo* (Table 4).

Disabling the cell adhesion activity of galectins by D-lactulose-amino acids is another prospective approach to exploit their antimetastatic potential. Dissemination of metastatic cancer requires entrapment of circulating blood-borne tumor cells in blood vessels in order for them to initiate successful formation of secondary tumors. Our studies have demonstrated importance of TFA interactions with galectin-3 translocated to the surface of vascular endothelial cells in mediating the formation of the initial endothelial cell-cancer cell contacts in flow.[31,32] Such glycoprotein-lectin interaction might be targeted then by blocking galectin-3 or TFA with the appropriate antibodies or other galectin ligands, such as lactulosamines. Indeed, LctLeu significantly inhibited rolling on, and abrogated adhesion to HUVECs of, suspended MDA-MB-435 cells under conditions of laminar flow.[32]

Yet another mechanism, by which lactulosamines can enact the antitumor potential through galectin blocking is related to involvement of galectins in tumor immune tolerance. Galectins has been identified as inducers of apoptosis in activated cytotoxic T-lymphocytes (CTLs),[29,30] and solid tumors may use the galectins to inactivate infiltrating CTLs and escape the immune surveillance. Blocking galectins, therefore, could potentially aid cancer immunotherapies. For example, a divalent lactulosamine protected TFA-activated CTLs from galectin-1-induced apoptosis and restored activity of the cells in regard with interferon-γ production.[39] *In vivo*, this Amadori compound has significantly improved efficiency of dendritic cell immunotherapy in a transgenic murine model of breast cancer.

4 CONCLUSIONS

Epidemiological data suggest a more pronounced protective effect of dehydrated foods against cancer risk, as compared to raw products. The main difference in chemical composition of raw and dehydrated foods is a significant amount of the early Maillard reaction products, such as fructose- and lactulose-amino acids, that emerge in the latter. These Amadori compounds possess antioxidant, metal-binding, and lectin-blocking properties, which enable them to modulate cancer cell adhesion, resistance to apoptosis and immune surveillance. In particular, fructose-amino acids showed high antioxidant potential, modulated cancer cell aggregation and clonogenic survival, interacted cooperatively with lycopene against prostate carcinogenesis. Lactulose-amino acids, on the other hand, have demonstrated the ability to competitively inhibit interactions of galectins-1 and -3 with their natural glycoprotein ligands, including pancarcinoma Thomsen-Friedenreich antigen. In spite of systemic toxicity lack, Amadori compounds may act as a therapeutic antimetastatic agents and tumorigenesis inhibitors, especially in combination with other anti-cancer modalities. These results open a new perspective towards the development of novel adjuvant anticancer therapies based on dietary glycoaminoconjugates that are non-toxic and target dissemination of malignant tumors - the primary cause of death in the advanced cancer disease. Finally, our data hint that a novel paradigm may emerge at the intersection of the Maillard reaction and the cancer chemoprevention fields, which defines optimized food processing technologies as an additional resource for improving the nutritional value of foods in regard with lowering cancer risk in populations.

References

1. V.L.W. Go, D.A. Wong and R. Butrum, *J.Nutr.*, 2001, **131**, 3121S-3126S.
2. E. Riboli and T. Norat. *Am.J.Clin.Nutr.*, 2003, **78**, 559S-69S.
3. M. Anese, M.C. Nicoli, C.R. Lerici and R. Massini, *Spec.Pub.Roy.Soc.Chem.*, 2000, **255**(Dietary Anticarcinogens and Antimutagens), 260-265.
4. A. Tzonou, L.B. Signorello, P. Lagiou, J. Wuu, D. Trichopoulos and A. Trichopoulou, *Int. J. Cancer*, 1999, **80**, 704-708.
5. J.H. Cohen, A.R. Kristal and J.L. Stanford, *J.Natl.Cancer Inst.*, 2000, **92**, 61-68.
6. J.M. Chan, C.N. Holick, M.F. Leitzmann et al., *Cancer Causes & Control*, 2006, **17**, 199-208.
7. G.L. Ambrosini, N.H. de Klerk, L. Fritschi, D. MacKerras and B. Musk, *Prostate Cancer Prostatic Dis.*, 2008, **11**, 61-66.

8. P.K. Mills, W.L. Beeson, R.L. Phillips and G.E. Fraser, *Cancer*, 1989, **64**, 598-604.
9. A.G. Schuurman, R.A. Goldbohm, E. Dorant and P.A. van den Brandt. *Cancer Epidemiol. Biomark. Prev.*, 1998, **7**, 673-680.
10. M.P. Zeegers, R.A. Goldbohm, and P.A. van den Brandt *Cancer Epidemiol. Biomarkers Prev.*, 2001, **10**, 1121-1128.
11. V. Faist and H.F. Erbersdobler, *Ann.Nutr.Metab.*, 2001, **45**, 1-12.
12. K. Ryu, N. Ide, H. Matsuura, and Y. Itakura, *J.Nutr.*, 2001, **131**, 972S-6S.
13. V.V. Mossine, P. Chopra and T.P. Mawhinney, *Cancer Res.*, 2008, **68**, 4384-4391.
14. M. Tarnawski, R. Kulis-Orzechowska and B. Szelepin, *Int.Immunopharmacol.*, 2007, **7**, 1577-1581.
15. E. Sorkin and A. Fjelde, *G.Ital.Chemioter.*, 1956, **3**, 355-361.
16. Z. Mazerska, B. Woynarowska, B. Stefanska, E. Borowski and S. Martelli, *Drugs Exp.Clin.Res.*, 1987, **13**, 345-351.
17. C. Schweinsberg, V. Maes, L. Brans, P. Blauenstein, D.A. Tourwe, P.A. Schubiger, R. Schibli and E.G. Garayoa, *Bioconjugate Chem.*, 2008, **19**, 2432-2439.
18. T.V. Denisevitch, R.A. Semyonova-Kobzar, V.V. Glinsky and V.V. Mosin, *Exp.Oncol.*, 1995, **17**, 111-117.
19. G.V. Glinsky, V.V. Mossine, J.E. Price, D. Bielenberg, V.V. Glinsky, H.N. Ananthaswamy and M.S. Feather, *Clin.Exp.Metastasis*, 1996, **14**, 253-267.
20. G.V. Glinsky, J.E. Price, V.V. Glinsky, V.V. Mossine, G. Kiriakova and J.B. Metcalf, *Cancer Res.*, 1996, **56**, 5319-5324.
21. V.V. Mossine and T.P. Mawhinney, *J.Agric.Food Chem.* 2007, **55**, 10373-10381.
22. K. Eichner, M. Reutter and R. Wittmann in: P.A. Finot, H.U. Aeschbacher, R.F. Hurrell and R. Liardon, (Eds.), *The Maillard Reaction in Food Processing, Human Nutrition and Physiology*, [Proc.4th Int.Symp.Maillard React.], Birkhauser Verlag, Basel, 1990, pp. 63-77.
23. M.L. Sanz, M.D. del Castillo, N. Corzo and A. Olano, *J.Agric.Food Chem.* 2001, **49**, 5228-5231.
24. E. Dworschák, V. Tarján and S. Turos, *ACS Symp.Ser.*, 1983, **215**, 159-168.
25. P.A. Finot, R. Deutsch and E. Bujard, *Progr.Food Nutr.Sci.*, 1981, **5**, 345-355.
26. A. Rérat, R. Calmes, P. Vaissade and P.-A. Finot, *Eur.J.Nutr.* 2002, **41**, 1-11.
27. V. Schwenger, C. Morath, K. Schönfelder, W. Klein, K. Weigel, R. Deppisch, T. Henle, E. Ritz, M. Zeier, *Nephrol.Dial.Transplant.*, 2006, **21**, 383-388.
28. G.V. Glinsky, *Crit.Rev.Oncol.Hematol.*, 1993, **14**, 1-13.
29. F.-T. Liu and G.A. *Nat.Rev.Cancer*, 2005, **5**, 29-41.
30. S. Nakahara, N. Oka and A. Raz, *Apoptosis*, 2005, **10**, 267-275.
31. V.V. Glinsky, G.V. Glinsky, O.V. Glinskii, V.H. Huxley, J.R. Turk, V.V. Mossine, S.L. Deutscher, K.J. Pienta, and T.P. Quinn *Cancer Res.* 2003, **63**, 3805-3811.
32. S.K. Khaldoyanidi, V.V. Glinsky, L. Sikora, A.B. Glinskii, V.V. Mossine, T.P. Quinn, G.V. Glinsky and P. Sriramarao, *.Biol.Chem.*, 2003, 278, 4127-4134.
33. G.F. Springer, *J.Mol.Med.*, 1997, **75**, 594-602.
34. G.A. Rabinovich, A. Cumashi, G.A. Bianco, D. Ciavardelli, I. Iurisci, M. D'Egidio, E. Piccolo, N. Tinari, N. Nifantiev and S. Iacobelli, *Glycobiology*, 2006, **16**, 210-220.
35. V.V. Mossine, V.V. Glinsky and T.P. Mawhinney, In: A. Klyosov, D. Platt and Z.J. Witczak, (Eds.), *Galectins*, Wiley, 2008, pp. 235-270.

36. F. Pricci, G. Leto, L. Amadio, C. Iacobini, G. Romeo, S. Cordone, R. Gradini, P. Barsotti, F.-T. Liu, U. Di Mario and G. Pugliese, *Kidney Int.*, Suppl. 2000, **77**, S31-S39.
37. V.V. Glinsky, G. Kiriakova, O.V. Glinskii, V.V. Mossine, T.P. Mawhinney, J.R. Turk, A.B. Glinskii, V.H. Huxley, J.E. Price and G.V. Glinsky, *Neoplasia*, 2009, **11**, 901-909.
38. K.D. Johnson, O.V. Glinskii, V.V. Mossine, J.R. Turk, T.P. Mawhinney, D.C. Anthony, C.J. Henry, V.H. Huxley, G.V. Glinsky, K.J. Pienta, A. Raz and V.V. Glinsky, *Neoplasia*, 2007, **9**, 662-670.
39. M.E. Huflejt, V.V. Mossine and M. Croft, *PCT Int. Appl.*, 2003, WO 0326494 A2 20030403, 41 pp.

AGES FLUORESCENCE OF PLASMA, URINE AND SKIN REFLECTS DIETARY EXPOSURE TO MAILLARD PRODUCTS IN FORMULA-FED INFANTS

K. Šebeková[1], G. Saavedra[2], K.Klenovicsová[1,3], P. Boor[1] and I. Birlouez-Aragon[2]

[1] Department of Clinical and Experimental Pharmacotherapy, Slovak Medical University, Bratislava, Slovakia
[2] SPECTRALYS Innovation, BIOCITECH, Romainville, France
[3] 2nd Department of Paediatrics, Medical Faculty, Comenius Univestity, Bratislava, Slovakia.

1   INTRODUCTION

Maillard reaction products (MRPs) arise during thermal processing of foods, via non-enzymatic glycation, i.e. spontaneous reaction between reducing sugars and free amino groups of proteins, peptides or amino acids. In comparison with human breast milk, infant formulas contain, among others, higher concentrations of proteins, lactose, iron and vitamin C. This creates a suitable milieu for formation of MRPs during industrial heat sterilization and spray atomization, essential for antimicrobial safety of infant formulas [1-3]. Infant formulas contain higher amounts of MRPs than breast milk[4,5]. In powdered formulas MRP levels further increase during their storage[1,2,4].

Advanced glycation end products (AGEs) are *in vivo* analogues of MRPs. Hyperglycemia, and enhanced oxidative- and carbonyl-stress favor AGEs formation[6]. In vivo, AGEs may exert their toxic effects via modification of the structure and function of proteins (particularly cross-linking), or after the interaction with their specific cell surface receptors, mainly RAGE (receptor for advanced glycation end products),[7]. AGE/RAGE interaction results in oxidative and inflammatory reactions, as well as proatherogenic actions[7]. Dietary MRPs are at least partially absorbed into circulation[4,5,8], thus they contribute to biological pool of AGEs *in vivo*. Several studies suggest that exaggerated intake of highly thermally processed diets (rich in MRPs) exert negative health effects, including diabetogenic, nephrotoxic and pro-oxidative effects, and induction of low-grade inflammation (reviewed in [9]).

Feeding of infants with formulas represents a considerable MRPs load. N^ε-(carboxymethyl)lysine (CML) represents a chemically-defined MRP, and widely used indicator of the Maillard reaction in foods[3]. Theoretical daily burden of dietary CML in 6-month-old breast-fed infants represents 0.01-0.03 mg/d, while in the formula-fed children 4.6±4.2 mg/d[4]. In comparison with the breast-fed children, high dietary intake of CML in formula-fed infants results in significantly higher plasma CML concentration (+46%), and

70-fold higher urinary excretion of CML[4]. These data suggest that after oral exposure, infant formulas-derived CML is partially absorbed into circulation, and rapidly excreted via urine. However, the mentioned data do not give any insight into the question of the potential bioavailability of the absorbed MRPs. Their rapid excretion might ensure only limited impact on glycation of tissue/biological fluids proteins, or AGE/RAGE interaction. Autofluorescence of skin is specific, validated, and accepted non-invasive method to assess *in vivo* accumulation of AGEs in skin[10-13]. Many AGEs, such as pentosidine, argpyrimidine and pyrraline, are fluorescent, thus can be globally measured at excitation wavelengths around 350 nm, and emission around 430 nm. Skin autofluorescence correlates with skin-collagen-linked fluorescence from skin biopsies, and with their content of specific AGEs[10,11]. Thus, it represents a clinically applicable marker of AGE-modification of tissues.

In the former study, we determined the CML content of infant formulas, and CML levels in plasma and urine of breast- and formula-fed babies[4]. CML has no intrinsic fluorescence, thus its contribution to protein glycation is not directly assessed by measuring skin autofluorescence. In the present study, a fluorimetric technique was used to determine the MRPs content of infant formulas and human milk. AGEs/MRPs-related fluorescence was also measured in plasma and urine in a large cohort of healthy 3-12 month-old infants, either breast- or infant formula-fed at the time of investigation, or formerly. To reveal whether the intake of a MRPs-rich diet in early childhood affects the glycation of long-lived proteins, skin autofluorescence was assessed in a different cohort of healthy 1-29 month-old infants.

2   MATERIALS AND METHODS

2.1 Dietary study

Healthy 3-12 month-old infants (n=224) were examined. Exclusion criteria for infants were: pathological findings during physical examination, elevated inflammatory markers, anamnestic data of acute/recurrent inflammatory or chronic diseases, and positivity of antibodies against HCV/HIV. Based on medical records and interviews with mothers/legal representatives, detailed feeding pattern of the infant was recorded, focusing on the length of breast feeding, infant formula feeding, or their combination. After exclusion of 17 children (either meeting the exclusion criteria or they could not be unequivocally assigned according to feeding regimen), plasma samples obtained from 207 healthy infants were taken into analyses. Spot urine samples were collected from 137 out of them.

The study was carried out in accordance with the Declaration of Helsinki and after the approval by the Institutional Ethics Board (SMU Bratislava), in cooperation with pediatricians of the primary care in Bratislava and surroundings, and the Pediatric Departments of Children's Faculty Hospital in Bratislava, after obtaining the signed informed consent form the parents/legal representatives of infants.

Breast milk was obtained from 56 mothers of the infants. According to feeding regimen, infants were classified as breast-fed (n=91) and formula/mixed-fed (n=116). Each group was further subdivided according to age into: 3-6 month-old babies, 7-9 months-olds weaning infants, and 10-12 month-old toddlers on diversified diet. Thirty-eight babies were exclusively breast-fed (MM, gender: 16 females (F)/22 males (M)), and 43 babies (20 F/23 M, IF) were infant formula/mixed fed. Mixed-fed babies represented 3 babies receiving concurrently IF and mother milk, and 6 babies that were initially breast-fed for maximally 3 weeks. Fourteen of formula-fed babies were given hydrolyzed- and 29 non-hydrolyzed formulas. Thirty eight weaning infants (15F/23M) received mother milk and 44 (27F/17M) were administered formulas (15 received hydrolyzed and 29 non-hydrolyzed formulas). From among toddlers on diversified diet, 15 (6F/9M) received breast milk as a supplement to diversified diet, and 29 received formula (7 of them hydrolyzed and 22 non-hydrolyzed formulas; 11F/18M). Twelve infants aged 7-12 months had been fully or partially breast-fed for no longer than 10 weeks. Group characteristics are detailed in Table 1.

Table 1 *Characteristics of the infants included in the dietary study*

	Feeding regimen	3-6 month-olds	7-9 month-olds	10-12 month-olds
Age (months)	BF	5.2±0.9	7.9±0.9	11.0±1.0
	IF	4.8±0.9	7.6±0.9	11.0±0.9
Gestational age (weeks)	BF	38.8±2.6	38.5±2.8	38.4±1.8
	IF	38.2±3.2	37.9±2.7	38.0±3.3
Birth weight (g)	BF	3481±632[**]	3188±741	3215±511
	IF	2958±662	2969±726	3040±771
Body weight (g)	BF	7427±991[**]	8276±1086	9051±1441
	IF	6596±1096	7692±1153	8951±1181
Mean daily weight gain (g)	BF	24.5±4.6	20.6±3.0	17.4±3.3
	IF	24.4±5.8	20.5±4.8	17.5±3.4
Albumin (g/l)	BF	40.7±3.6	42.6±2.8	42.5±3.9
	IF	40.5±4.5	41.9±2.6	42.0±2.6
MRP-F/Albumin (AU/g)	BF	2.2±0.4[***]	2.7±0.8	2.9±1.1
	IF	2.7±0.5	2.8±0.8	2.9±1.1
MRP-F/Creatinine (AU/mmol)	BF	51.1±27.5[**]	58.3±37.2	62.5±45.2
	IF	64.2±40.2	65.7±43.5	77.2±40.5

BF: breast-fed; IF: infant formula-fed; MRP-F/Albumin: fluorescence of advanced Maillard products/fluorescence of soluble tryptophan in plasma expressed in arbitrary units (AU) per gram albumin ; MRP-F/Creatinine: fluorescence of advanced Maillard products per mmol creatinine in spot urines; **: $p < 0.01$, ***: $p < 0.001$ (vs. IF)

2.2   AGEs/MRPs fluorescence in infant formulas and biological fluids

Biological fluids MRPs-specific fluorescence was measured in collected biological material. The fluorescence intensity of diluted urine and plasma samples was measured as follows: 500µl of urine sample were centrifuged during 10 minutes at 3000 rpm and filtered on 0,25µm Nylon filters. An aliquot of 100µl was resuspended in 10 ml water (dilution 1:100) and analyzed in acryle fluorescence cuvettes (Sarstedt, Nümbrecht, Germany). Plasma samples were centrifuged, diluted (1:50) in water, filtered and analyzed at excitation 350 nm and emission 440 nm. Autoanalyzer (Vitros 250, J&J, USA) was used to determine plasma albumin and urinary creatinine concentrations.

All fluorescence analysis were performed on a Spex fluorimeter (Jobin-Yvon, Edison, NJ, USA). Infants included in the dietary study consumed 9 different non-hydrolyzed and 7 hydrolyzed formulas (11 starting and 5 follow on formulas) of 4 brands. The content of fluorescent MRPs was analyzed in the non hydrolyzed samples (5 starting, 2 follow-on) only, using the FAST method (14). This method gives the global intensity of the MRPs fluorescence at 330/420 nm in the soluble whey fraction, indicative of the degree of advancement of the reaction in the total milk sample. This fluorescence intensity is divided by that measured at 290/350 nm refering to tryptophan, in order to correct the MRPs fluorescence by the protein concentration in the pH 4,6 supernatant of the milk, providing the FAST index. [(FAMP/FTrp)×100] . The method cannot be applied to hydrolyzed formulas because peptides are not denatured and can therefore not precipitate at pH 4.6, impeding to get a clear supernatant. The formula samples were analyzed immediately after their purchase, and at the end of storage period (2-4 weeks) recommended by each manufacturer. Correlations between FAST index and formula content of total protein, fat and saccharides, whey protein, casein, vegetable fat, milk fat, lactose, linoleic and linolenic acid, vitamin C, iron, copper and zinc (as indicated by the manufacturers) were performed.

2.3   Skin autofluorescence

Tissue AGEs accumulation was assessed non-invasively using the AGE Reader (DiagnOptics Technologies B.V., Groningen, The Netherlands) in 93 healthy term infants (1 to 29 month-olds). Exclusion criteria were identical with those in the dietary study. Single measurement of skin autofluorescence was performed directly on intact healthy skin, on a lower forearm of each child. The autofluorescence was calculated as a ratio of mean fluorescence intensities detected from the skin at 420-600 nm and 300-420 nm. According to the instructions of the AGE Reader manufacturer, in subjects with skin reflectance < 12% (representing skin phototype V and VI), and in those with recently applied skin oils/ointments, the measurements are not sufficiently reliable. Thus 8 children displaying skin reflectance < 12%, and 4 in whom skin oil was applied before visiting the pediatrician were excluded. Finally, 79 infants were taken into evaluation (38 females/41 males; mean age: 11.9±8.4 months, gestational age: 39.0±1.6 months).
Detailed feeding pattern of the infants was recorded, as described above. Infants were divided into breast-fed (n=44) and formula/mixed-fed (n=35) group. Breast-fed group consisted of exclusively breast-fed babies (infants younger than 6 months of age); weaning toddlers receiving breast milk (7-12 moths of age), and older infants receiving mother milk as a supplement to diversified diet. Infants assigned to formula/mixed-fed group were either never breast-fed, or were mixed-fed (formula combined with mother milk, n=3), the elder infants received infant formulas as a supplement to diversified diet.

2.4   Statistics

Data are presented as mean ±SD. Data were tested for normal distribution. Multiple group comparison was carried out using ANOVA or Kruskal-Wallis test, with post-hoc Scheffe's test, or Mann-Whitney U test, where appropriate. Pearson or Spearman correlation coefficients were calculated, as appropriate. General linear model was employed to study the impact of independent variables on skin autofluorescence levels. P<0.05 was considered significant. Statistical program SPSS 16 was used.

3   RESULTS

The FASTindex is a rapid and simple method to assess the extent of Maillard products formed in milk products as a result of the heat process. Analysis of the FAST index confirmed that MRPs content of human breast milk is much lower in comparison with infant formulas (5.0±0.9 vs. 115.3±89.8, respectively), the latter exhibiting a great variability depending on the brand (Fig 1). No significant difference in the FAST index was observed between the starting and follow-on formulas (156±116 vs. 138±37, respectively). In the starting formulas FAST index rose during their shelf-life (by 10 % to 916 %), while in the follow on formulas a decrease by 2% and 21% was detected (Fig 1). In the evaluated formulas, FAST index (determined in the samples from just opened packages) correlated directly with the casein content (r=0.691, p<0.05), and inversely with that of vegetable fat (r=-0.963, p<0.001). No significant relationship between content of total proteins, whey proteins, fats, sugar, linoleic acid, linolenic acid, vitamin C, or iron, was revealed.

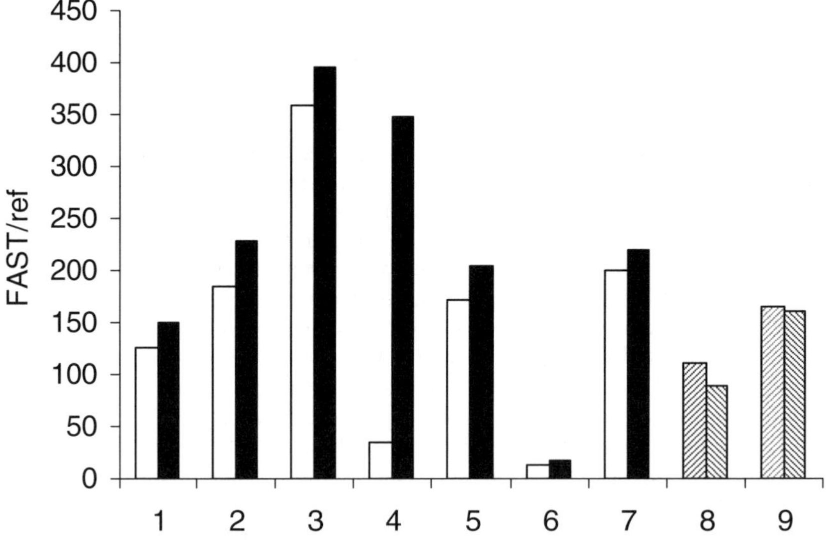

Figure 1   *FAST index in 9 non-hydrolyzed infant formulas and changes during storage. Solid bars: starting formulas (n=7); hatched bars: follow-on formulas (n=2); first bar: sample taken immediately after purchase; second bar: sample taken at the end of recommended storage period; FAST/ref: fluorescence of advanced Maillard products/fluorescence of soluble tryptophan*

Breast- and infant formula-fed children did not differ significantly by age, gestational age and albuminemia. Birth weight and the weight at the investigation tended to be lower in the formula-fed groups, but significance was reached only in the 3-6-months-old infants. Mean daily weight gain was comparable in all 3 age-groups (Table 1). Breast-fed babies displayed significantly lower plasma MRPs fluorescence levels if compared with the formula-fed infants. Plasma MRPs fluorescence remained almost unchanged with ageing in the infant formula-fed children, while it gradually increased in the breast-fed group (ANOVA: F=6.19, p=0.003). Compared to breast-feeding, consumption of infant formulas was associated with higher plasma AGEs-specific fluorescence in 3-6 months old infants, but not in older children (Tab. 1). Urinary excretion of fluorescent AGEs was higher in the formula- than in the breast-fed babies, and the difference in older children diminished, correspondingly with findings in plasma (Tab. 1). No significant difference between children on hydrolyzed versus non-hydrolyzed formulas was observed (data not shown).

In the infants, skin autofluorescence rose in an age-dependent manner (Fig 2). Till 6th month of age the breast-fed infants had significantly lower skin autofluorescence compared with the formula-fed ones. With increasing age the difference in skin autofluorescence gradually disappeared. Simple regression model between skin autofluorescence and age confirmed the lower values and steeper rise by increasing age in the breast-fed (y=0.013x+0.53, r=0.624, p<0.001) versus formula-fed children (y=0.010x+0.70, r=0.533, p<0.001). General linear model revealed that 41.9% of variability in skin autofluorescence is on the account of age and feeding pattern, age-dependent changes being more dominant (corrected model: F= 28.8, intercept: F=559; age: F=33.1; dietary regimen: F=22.2; p=0.001, all). Entering gestational age and/or gender into the model did not contribute significantly.

4   DISCUSSION

Intake of diets rich in MRPs exerts negative health effects (reviewed in 9). Compared to mother milk, infant formulas contain by 1 to 2 orders of magnitude higher amounts of CML (35- to 100-fold in mean)[4,5]. Here not CML, but a mixture of essentially unknown fluorescent Maillard compounds was measured, yielding a 23-fold difference between mother milk and formulas. Considering that hydrolyzed formulas have twice higher levels of CML and Maillard products[15, 4], the results obtained with fluorescence and CML analysis fit well. This observation confirms that fluorescent Maillard products are good indicators of the reaction development in food.

MRPs from infant formula are partially absorbed, and rapidly excreted via urine, as indicated by our former data on CML[4], and confirmed here, using a fluorimetric monitoring. However, the unanswered question remains whether the absorbed MRPs from infant formulas are able to contribute to glycation in vivo. Here we show that increased load with dietary MRPs in children fed with infant formulas leads to increased AGEs-associated skin fluorescence, revealing glycation of long-lived proteins, at least skin collagen. The effects were clearly linked to profound differences in MRPs intake in early childhood (infant formulas versus mother milk). With introduction of diversified diet, presumably including infant formulas or mother milk only as a supplement to mixed diet, but also cereal products rich in MRPs, difference in skin auto fluorescence between the formulas- and breast-fed groups gradually disappears. This is associated with a steeper age-dependent rise in skin autofluorescence in the weaning breast-fed babies, probably reflecting a progressive exposure to dietary MRPs in breast-fed infants up to a similar level to the formula-fed infants. The main impact of the diet is underlined by the finding that AGE-specific fluorescence of plasma and urine differs between the mother-milk- and formula-fed healthy infants significantly only during the period of life, when milk represents the sole source of nutrition.

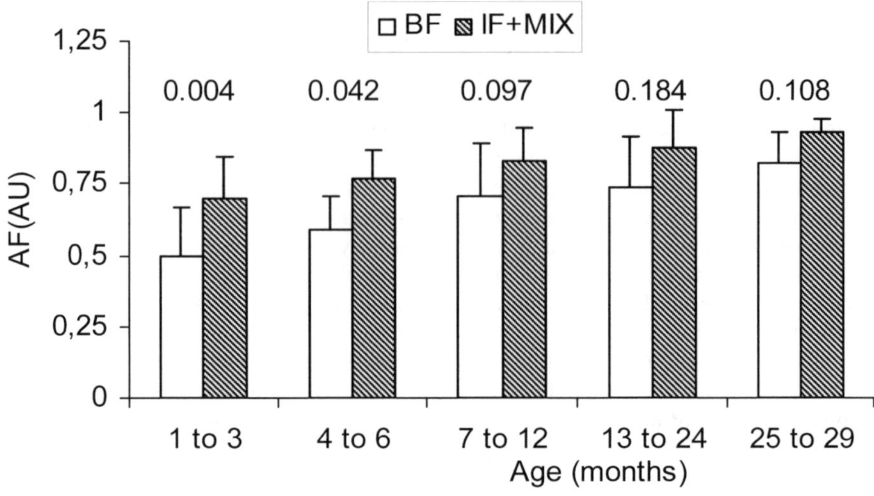

Figure 2:   *Age-dependent rise in skin autofluorescence of the breast- and formula-fed healthy infants. AF: skin autofluorescence; AU: arbitrary unit; BF: breast-fed; IF: infant formula-fed*

These data are in agreement with our previous findings of higher plasma and urinary CML levels in formula- versus breast milk-fed infants, in a smaller cohort of approximately 6 months-old infants[4]. It is to be considered that in the healthy adults, skin autofluorescence increases by 9% 2 to 3 hours post-prandially, after an intake of single meal with intermediate AGEs content[12,13]. This minor, although significant, rise is not on the account of the glycation of long-lived proteins, but reflects the acute dietary intake of MRPs, e.g. postprandial rise in circulating plasma and/or lymph MRPs levels. Mothers/legal representatives of the infants were asked not to feed the children 3 hours prior to investigation, but we did not record the exact interval between last feeding and the measurement. The mean difference in skin autofluorescence between the breast- and formula-fed infants aged 1 to 3, and 4 to 6 months was 40%, and 31%, respectively. This is much higher than could be attributed to acute postprandial effects. Although the difference between the groups decreased with aging (to 17% in 7-12-month-olds, and to 13% in the 25-29 month-olds), we suppose that it still reflects the gradual introduction of diversified diet with increasing content of MRPs to formerly exclusively breast-fed infants, not the acute postprandial rise.

As in the adults[16], skin auto-fluorescence increases age-dependently already in early childhood. In adults, body mass index (BMI) was also an independent factor contributing to skin auto fluorescence[16]. In healthy toddlers body weight or BMI can not be considered as factors independent from age, thus we considered only age as independent factor.

FAST index was confirmed as susceptible indicator of the formation of MRPs in non-hydrolyzed formulas. Interestingly, FAST index determined in the formulas did not correlate with the content of the compounds considered to have important influence on the rate of glycation, e.g. lactose, vitamin C, or iron. This is most probably due to a fact that manufacturers' data indicated on the packaging were used for calculations, e.g. the real levels of vitamin C remain unknown, because of variable initial amounts added and unexplored oxidation rates[1]. Also the form of iron could differ between formulas, accounting for variable activities. Furthermore, the precise time-temperature applied during processing is not known, and affects considerably the way the reaction develops[3]. Taken together, these data suggest that dietary MRPs are biologically available and contribute to glycation of long-lived proteins in early childhood. A long-term MRPs-rich diet is reflected by higher skin autofluorescence, and higher plasma and urinary fluorescence, already in early childhood. These differences diminish during weaning, and mainly after the introduction of diversified infants' diet. Fluorescence analysis appears a reliable and simple tool to monitor exposure and bioavailability of dietary MRPs in children. The potential biological effects that might be related to MRPs-rich diet consumption in early childhood remain to be elucidated.

Acknowledgements

This study was supported by 6th EU FP, COLL-CT-2005-516415 - ICARE (I.B-A, KŠ); and by Internal Grant of Slovak Medical University, No. 6-2007 (KK). We thank the mothers/legal representatives of the infants for participating in this study, and all nurses and pediatricians of primary care, Drs.: Bradiak F., Holanová E., Chramcová N., Kadlíčková A., Kollárová E., Križanová E., Lásková J., Lukyová M., Petríková I., Polláková V., Prokopová E., Rapošová R., Salcerová J., Sláviková E., Šimeková L., Špániková M., Uhrovič F., Vicianová K., Vitáriušová E.; Pediatric Departments (I and II) of Comenius University, and Studnička Institution in Bratislava and Trnava for recruitment of the infants and collecting the samples. We gratefully acknowledge Dr A Smitt and DiagnOptics Technologies B.V., Groningen, The Netherlands for providing the AGE-reader to perform the study.

References

1. Gliguem H, Birlouez-Aragon I. *J Dairy Sci* 2005, **88**, 891-899.
2. Leclère J., Birlouez-Aragon I. *Food Chemistry* 2002, **76**, 491-499.
3. Birlouez-Aragon I, Pischetsrieder M, Leclére J, et al., Food Chemistry 2004, **87**, 253-259.
4. Šebeková K, Saavedra G, Zumpe C, et al. Ann N Y Acad Sci 2008, **1126**, 177-180.
5. Dittrich R, Hoffmann I, Stahl P et al. *J Agric Food Chem* 2006, **54**, 6924-6928.
6. Miyata T, Maeda K, Kurokawa K, et al. *Nephrol Dial Transplant* 1997, **12**, 255-8.
7. Bierhaus A, Humpert PM, Morcos M et al, *J Mol Med* 2005, **83**, 876 – 886.
8. Somoza V, Wenzel E, Weiss C, Clawin-Radecker I, Grubel N, Erbersdobler HF. *Mol Nutr Food Res.* 2006, **50**, 833-841.
9. Šebeková K, Somoza V. *Mol Nutr Food Res.* 2007, **51**,1079-1084.
10. Mulder DJ, Water TV, Lutgers HL, et al. *Diabetes Technol Ther* 2006, **8**, 523-535.
11. Meerwaldt R, Graaff R, Oomen PH, Links TP, Jager JJ, Alderson NL, et al. Diabetologia 2004, **47**, 1324-1330.
12. Stirban A, Nandrean S, Negrean M, Koschinsky T, Tschoepe D. *Diabetes Technol Ther.* 2008, **10**, 200-205.
13. Noordzij M, Lefrandt D, de Graaff , Smit AJ. *Diabetes Technol Ther* 2009, *in press.*
14. Birlouez-Aragon I, Sabat P, Gouti N. *Int Diary J* 2002, **12**, 59-67.
15. Birlouez-Aragon I, Locquet N, De St. Louvent E et al. *Ann N Y Acad Sci* 2005, **1043**, 308-318.
16. Corstjens H, Dicanio D, Muizzuddin N, et al. *Exp Gerontol* 2008, **43**, 663-667.

NOVEL MAILLARD PIGMENTS FORMED FROM FURFURAL AND XYLOSE WITH
LYSINE UNDER WEAKLY ACIDIC CONDITIONS

M. Murata[1]

[1] Department of Nutrition and Food Science, Ochanomizu University. 2-1-1 Otsuka,
Bunkyo-ku, Tokyo 112-8610 Japan

1.   INTRODUCTION

Foods derived from plant materials contain pentose as well as hexose sugars, which
are usually formed under weakly acidic conditions. It is well known that pentose
contributes more to browning by the Maillard reaction than hexose does, because the oxo-
or reducible form of sugars is higher in pentose than in hexose. Hydroxymethylfurfural
(HMF) is one of the major decomposed products of such hexoses as glucose and fructose
under acidic conditions, while furfural is the corresponding one of pentose and is also
formed by the decomposition of ascorbic acid.
The decomposition or reaction products of hexoses, especially glucose, by the Maillard
reaction have been intensively examined. However, there is less data on the reaction
products of pentose or furfural. So far some low molecular weight yellow, red or blue
pigments formed by the Maillard reaction of furfural or pentose have been reported
(Figure 1).[1-10] For example, a yellow compound (**1**)[1] was formed by the reaction between
furfural and proline, while red compounds (**2, 3**)[1-3] were formed by the reaction between
furfural and alanine or lysine (Figure 1). A blue compound was formed by the reaction
between xylose and glycine[9, 10]. Most studies have been performed at neutral pH, although
foods derived of plants are usually weakly acidic. Recently several novel yellow pigments
were isolated from the heated solution containing furfural or xylose with lysine under
weakly acidic conditions.[11, 12, 14] . Here, we describe the isolation, identification, and color
contribution of some of these Maillard pigments.

2.   MAILLARD PIGMENTS FORMED FROM FURFURAL AND
L-LYSINE AND ITS FORMATION MECHANISM

When furfural and L-lysine are dissolved in 0.5 M acetate buffer (pH 5.0) and
heated, the solution turned brown more intensively than the solution containing glucose
and 5-hydroxymethyl furfural (HMF) with L-lysine. HMF is the major decomposition
product of hexose under acidic conditions. Low molecular weight pigments in the heated
solution containing furfural and L-lysine were examined by ODS-HPLC equipped with a
photodiode-array detector (DAD).

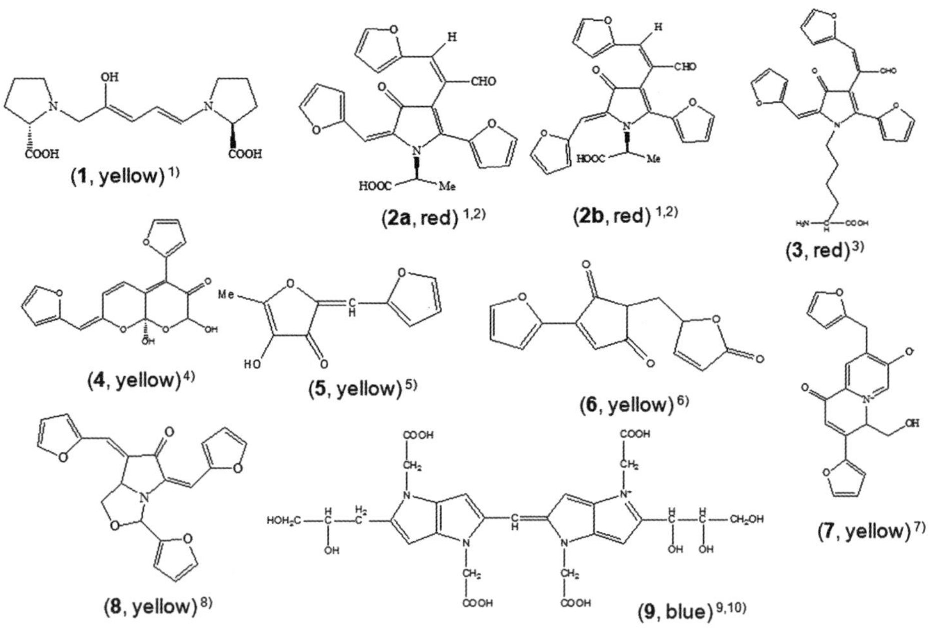

Figure 1 *Chemical structures of the coloured compounds derived from furfural and pentose with amino acids.*

Two yellow pigments (Figure 2) were isolated from the solution and identified as furpipate (10)[11] decarboxylated-furpipate (11).[12] The former compound was novel, while the latter 1 compound was reported to be synthesized from furfural and 2 2,3,4,5-tetrahydropyridine.[13] Furpipate was water-soluble and showed the absorption maxima at 365-375 nm under acidic and neutral conditions. Its neutral and acidic solution was yellow, while its alkaline solution was colorless. The detection limits of 10 and 11 in water, which was visually evaluated, were 3.0 µg/ml and 40 µg/ml, respectively. The color contributions of 10 and 11 to the brown solution prepared from 30 mM L-lysine and 90 mM furfural by being autoclaved for 30 min were respectively 25% and 3% of total color of the brown solution, which shows that 10 was the major pigment. When HMF instead of furfural was mixed with L-lysine and heated, 5-hydorxymethylfurpipate (12) and decarboxylated-5-hydorxymethylfurpipate[13] were isolated.[12] The detection limits of 12 and 13 were 4.6 µg/ml and 16 µg/ml, respectively. The color contributions of 12 and 13 to the brown solution prepared from 30 mM L-lysine and 90 mM HMF by being autoclaved for 30 min were respectively 43% and 18% of total color of the solution. 12 was the major pigment in the solution.

The formation scheme of furfural is speculated as shown in Figure 3. An aldehyde group of furfural reacted with an amino group of lysine to form an imine. The double bond was migrated to form Amadori type compound, which was hydrolyzed to form alpha-keto carboxylic acid. The formed keto carboxylic acid reacted with an intra-molecular amino group to form tetrahydro-pyridine carboxylic acid or reacted with another furfural by aldol condensation, from which furpipate was formed by dehydration of 2 molecules of water. Another possible scheme is that lysine was converted to tetrahydropyridine carboxylic acid, which reacted with furfural to form furpipate.

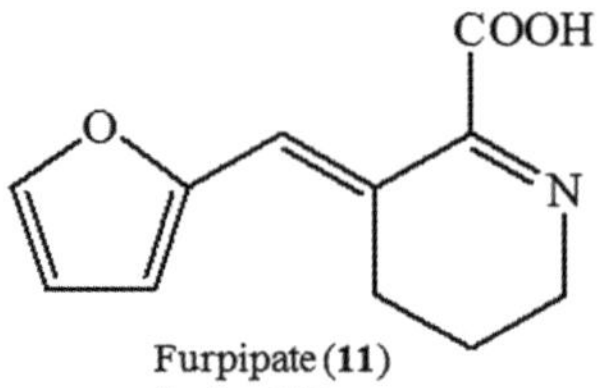

Furpipate (**11**)
furfural+Lys

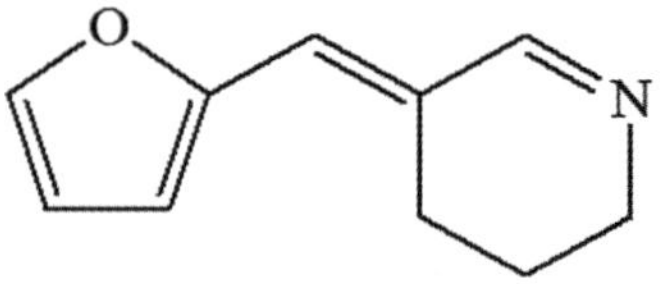

Decarboxylated-furpipate (**12**)
furfural+Lys

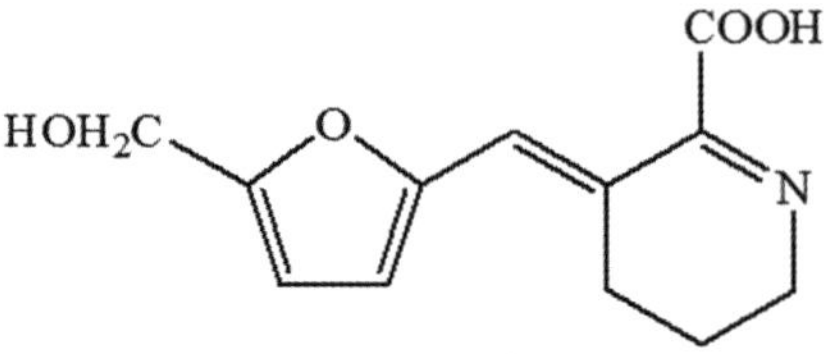

5-Hydoxymethylfurpipate (**13**)
HMF+Lys

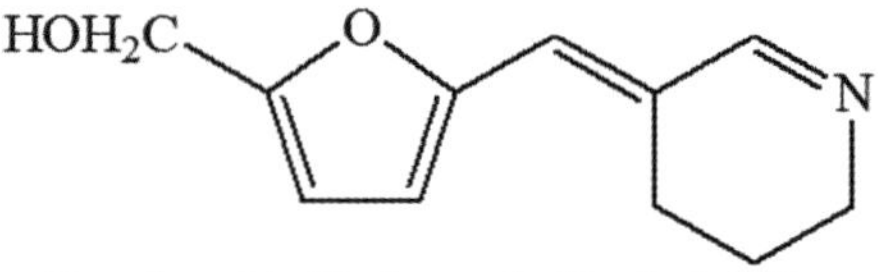

Decarboxylated-hydoxymethylfurpipate (**14**)
furfural+Lys

Figure 2 *Furpipate Derivatives.*

Miller previously synthesized decarboxylated-furpipate from furfural and 2,3,4,5-tetrahydropyridine.[13] But this scheme seemed to little contribute to the formation of furpipate, because the formation of decarboxylated-furpipate was not promoted by the addition of cadaverine (1,5-diamino-pentane) and piperidine to the reaction mixture containing furfural and that the formation of furpipate was not promoted by pipecolic acid.

3. MAILLARD PIGMENTS FORMED FROM XYLOSE AND LYSINE

Although furfural contributed more to the browning than glucose and HMF in the presence of lysine under weakly acidic conditions, pentoses such as xylose and arabinose contributed more the browning than furfural. Although several pigments from xylose and lysine were reported, there was little data on the formation of Maillard pigments under weakly acidic conditions. Then the browning of xylose with L-lysine under weakly acidic conditions were examined. An acetate buffer (pH 5.0) containing L-lysine and xylose was autoclaved at 121oC for 30 min. Figure 4 shows a typical HPLC pattern of the brown solution formed from xylose and L-lysine. Several peaks appeared. Three peaks XLW1, XLW2, and XLO1 were isolated and identified.

The brown solution was first divided into an ethyl acetate layer and a water layer. XLO1 (a pale yellow compound) was isolated from the ethyl acetate layer, which was identified as 4-hydroxy-5-methyl-2-furfurylidne 3(2H)-furanone.[5] Compound 5 showing the absorption maximum at 360 nm has been identified in the brown solution of pentose.[6] Although XLW1 and XLW2 were not extracted with ethyl acetate, they were absorbed to ODS resin. The water layer was applied to a column of ODS, which was washed with water and developed with a mixture of water and MeOH. Fractions containing pigments were applied to a column of ODS again. Eluted yellow solutions were further purified by a preparative HPLC (column, ODS; eluent, water/MeOH (8:2). Two yellow pigments soluble in water were obtained. Their UV-visible spectra were almost identical, and their absorption maxima were ca. 440 nm. Their chemical structures (Fig. 5) were elucidated by instrumental analyses as 6-[[1-[5-amino-1-carboxypentyl]-3-hydroxy-pyrrol-2-yl]- (E)-2-methylidene-5-methyl-1,2H-pyrrol-3-one-1-ly]-2-amino-hexanoic acid (dilysyldipyrrolone

A, 15) and 6-[[1-[5-amino-5-arboxypentyl]-3- hydroxy-pyrrol-2-yl]-(E)-2-methylidene-5-methyl-1,2H-pyrrol-3-one-1-ly]-2-amino-hexanoic acid (dilysyldipyrrolone B, 16).[14] They were novel pyrrolyl-methylidene-pyrrolone derivatives containing two lysine residues, which were named dilysyldipyrrolones A and B.14). Contributions of these compounds to a heated solution were estimated by the color dilution method. Dilysyldipyrrolones A and B contributed to about 5% and 10% of total color, respectively, while 4-hydroxy-5-methyl-2-furfurylidne 3(2H)-furanone contributed only to 0.05%.

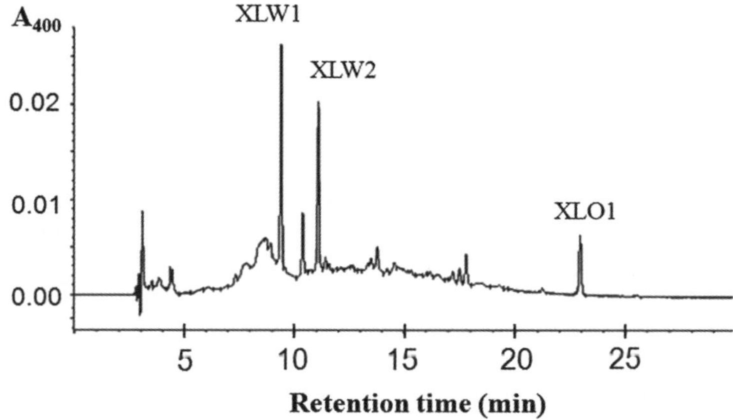

Figure 3   *Typical HPLC Patter of the Heated Solution Containing Xylose and Lysine. An acetate buffer (pH 5.0) containing 30 mM xylose and 30 mM L-lysine was autoclaved at 121oC for 30 min and applied to HPLC (column, Mightysil RP-18 (i.d. 4.6 mm x 250 mm, Kanto Chemical); eluent, a linear gradient of 2% MeOH in water and MeOH (from 100:0 to 30:70 for 30 min); flow rate, 1.0 ml/min)*

Dilysyldipyrrolones A and B were the major low-molecular weight pigments in the heated solution. A similar chromophore was reported from a reaction mixture of D-glucose and L-lysine in slightly acidic, aqueous solution.[15] Further, a new derivative from a reaction mixture containing arginine in addition to lysine was identified (data not shown). When L-arginine was added to a reaction mixture containing L-lysine and xylose, a novel peak showing a similar UV-visible spectrum to those of dilysyldipyrrolones A and B was appeared on HPLC. This peak was isolated from the reaction mixture, and a novel derivative containing the same chromophore as dilysyldipyrrolones A and B was identified.

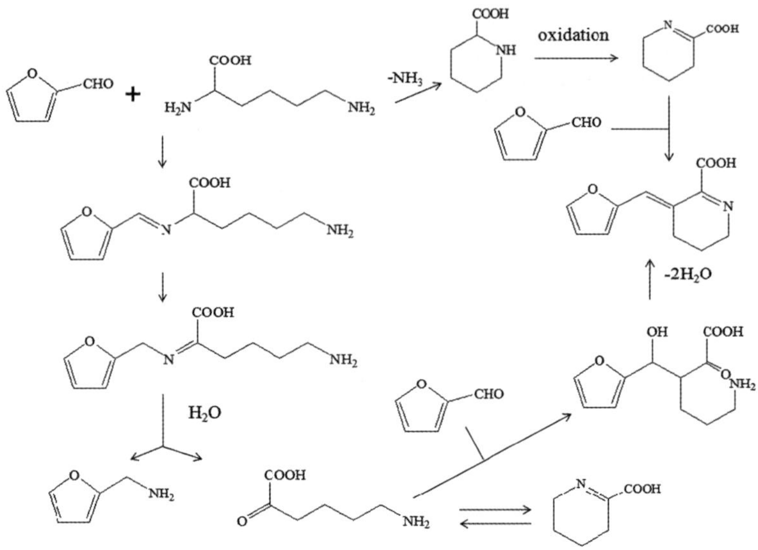

Figure 4   *Plausible Formation Scheme of Furpipate.[1]*

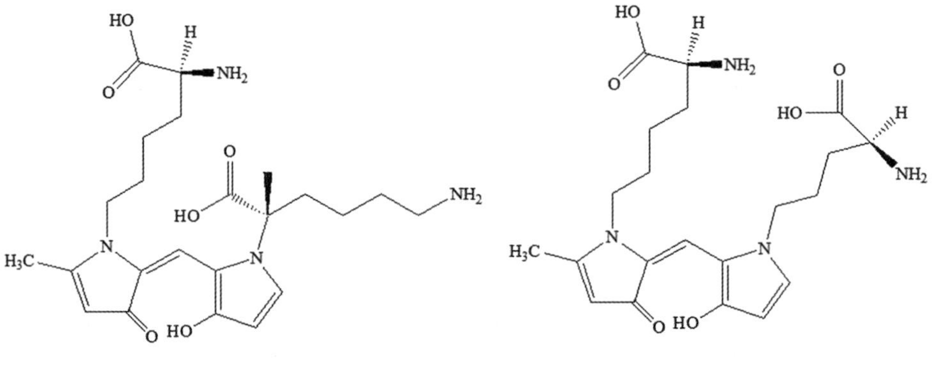

Dilysyldipyrrolone A (XLW1, **15**)   Dilysyldipyrrolone B (XLW2, **16**)

Figure 5   *Dilysyldipyrrolones A and B.*

4.   CONCLUSIONS

In conclusion, several novel Maillard pigments were identified from the reaction mixture of furfural or xylose and L-lysine under weakly acidic solution. They were low molecular weight compounds and major yellow pigments in the reaction solution.

References

1.	Hofmann, T. *J Agric Food Chem*, 1998, 46, 932-940.
2.	Hofmann, T. *Helv Chim Acta*, 1997, 80, 1843-1856.
3.	Hofmann, T. *Z Lebensm Unters Forsch*. 1998, 206, 251-258.
4.	Hofmann, T. *Carbohydr Res*, 1998, 313, 203-213.
5.	Frank, O., Jezussek, M., and Hofmann,T. *Eur Food Res Technol*, 2001, 213, 1-7.
6.	Ames, J. M., Apriyantono, A., and Arnoldi, A. *Food Chem*, 1993, 46, 121-127.
7.	Arnoldi, A., Corain, E. A., Scaglioni, L., and Ames, J. M. *J Agric Food Chem*, 1997, 45, 650-655 (1997).
8.	Hofmann, T. *J Agric Food Chem*, 1998, **46**, 3902-3911.
9.	Hayase, F., Takahashi, Y., Tominaga, S., Miura, M., Gomyo, T., and Kato, H. Biosci Biotechnol Biochem, 1999, **63**, 1512-1514.
10.	Hayase, F. Usui, T., and Watanabe, *H Mol Nutr Food Res*, 2006, **50**, 1171-1179.
11.	Murata, M., Totsuka, H., and Ono, H. *Biosci Biotechnol Biochem*, 2007, **71**, 1717-1723.
12.	Totsuka, H., Tokuzen, K., Ono, H.and Murata. M. *Food Sci Technol Res*, 2009, **15**, 45-50 (2009).
13.	Miller, R. *Acta Chem Scand*, 1987, **B41**, 208-209.
14.	Sakamoto,J., Takenaka, M., Ono, H., and Murata, M. *Biosci Biotechnol Biochem*, 2009, **73**, 2065-2069
15.	Miller, R., Olsson, K., Pernemalm, P.-A. *Acta Chem Scand*, 1984, **B38**, 689-694.

URINARY EXCRETION OF NON-TOXIC AND TOXIC CARBONYLS AS UREA DERIVATIVES

K.Suyama[1], M.Endo[1] , T. Mori[2] and T. Miyata[3]

[1] Department of Sports and Exercise Nutrition, Sendai University2-2-18 Minami Funaoka, Shibata-gun, Miyagi, Japan.
[2] Health Administration Center, Tohoku University Tohoku University Graduate School of Medicine
[3] Center for Translational and Advanced Research, Tohoku University Graduate School of Medicine

1 INTRODUCTION

Reactive carbonyls are formed as a normal part of oxidative metabolism[1-4.] The production and accumulation are enhanced in diabetes, due to a combination of chronic hyperglycemia and oxidative stress[5-9,] where they play a significant role in the development and progression of diabetic complications[10-12.] Increased accumulation of toxic carbonyls is also believed to play a role in a number of other conditions[13-17]. Most research has focused on the enzymatic detoxification pathways of toxic carbonyls [18]. However, less attention has been paid to the excretion of carbonyls in the urine.

Recently, we have isolated the urea derivative of glucose, *N*-carbamoyl-β-D-glucopyranosylamine (glucose ureide), from human urine. We hypothesise that carbonyls react with urea *in vivo* to form urea derivatives, including ureides, which are then excreted as urinary constituents. In this experiment, we have developed a systematic approach for detection of urinary carbonyls in order to verify the *in vivo* modification of carbonyls with urea.

2 MATERIALS AND METHODS

2.1 Chemicals

2,4-Dinitrophenyl hydrazine (DNPH) was purchased from Wako Pure Chemical Inc. (Osaka Japan). Silica gel TLC plate used was Silica gel 60 TLC aluminum No. 5553 sheet Merck, Germany. Acetonitrile for HPLC grade was purchased from Nacalai Tesque Inc. (Tokyo Japan). All other chemicals were of the best grade available from commercial source. DNPH reagent prepared was 0.2% DNPH in 2% HCl-methanol.

2.2   Chromatography

Urinary carbonyl-DNPH was analyzed by both Thin layer chromatography (TLC) and HPLC. TLC was conducted using an ascending system; developing solvent used was n-hexane/ethyl acetate (1:1, v/v). The HPLC analysis was performed on a Shimazu LC-10 instrument equipped with an ODS HPLC packed column (Cosmosil 5C18-MS-II, 4.6 x 250 mm, Waters type; Nacalai Tesque, Japan) with detection at 420nm. Solvent used for HPLC was acetonitrile/water/phosphoric acid (90/10/0.02, v/v). In order to exclude of excess DNPH and precipitate in the urinary reactant, silica gel packed mini column (Silycycle, Canada) was used using n-hexane/ethyl acetate (1:1, v/v) as a solvent system.

2.3   DNPH derivatization of Urinary carbonyls

Five ml of DNPH reagent was added to 5ml of human urine and then reacted at room temperature for 48 hrs. After the reaction, 5ml of ethyl acetate and 10 ml of water saturated with NaCl were added and then vigorously shaken. Upper orange colored ethyl acetate fraction separated was corrected and then charged on both TLC and HPLC chromatography. Before HPLC analysis, ethyl acetate fractions were charged on silica gel mini column to exclude of high polar DNPH, polymer DNPH and excess non-reacted DNPH reagent as described in experimental section. Standard DNPH was prepared by the reaction of DNPH reagent and standard carbonyls as the same manner as urine but using 10 ml of DNPH reagent and 0.5ml of standard carbonyls (acetone, crotonal, acetaldehyde, glyoxal and methylglyoxal), and DNPHs synthesized were crystallized from methanol.

2.4   Scavenging capability

The scavenging capability of urea to toxic carbonyl was estimated by the measurement of reactivity of toxic carbonyls, glyoxal and methylglyoxal, with urea under *quasi*-physiological conditions. Reaction mixtures (2ml) in a Pyrex test tube with Teflon-lined screw cap containing urea and each carbonyl in 0.1M PBS buffer (pH 7.4) was incubated at 37°C for 2hrs. Carbonyl and urea concentrations are given in the respective legends to figures. After incubation, the mixture was treated with 1 ml of DNPH reagent and the reaction was carried out by the same manner of human urine described above but reaction time was setting for 2min at room temperature. Amount of remaining free carbonyls after the incubation were measured by HPLC of DNPH derivatives.

3   RESULTS AND DISCUSSION

Urinary carbonyls were derivatized to corresponding hydrazones by the DNPH reagent. Interestingly, to complete the reaction to form DNPH derivatives, the reaction required a long time of over 48 hrs at room temperature. To determine the potential reasons for this long reaction time, we compared the reactivity between free carbonyls and synthetic ureides with DNPH reagent. Free methylglyoxal and authentic methylglyoxal ureide were used as model compounds for the reaction. Free methylglyoxal (1% aqueous solution) was reacts with DNPH reagent to form methylglyoxal-DNP hydrazones,

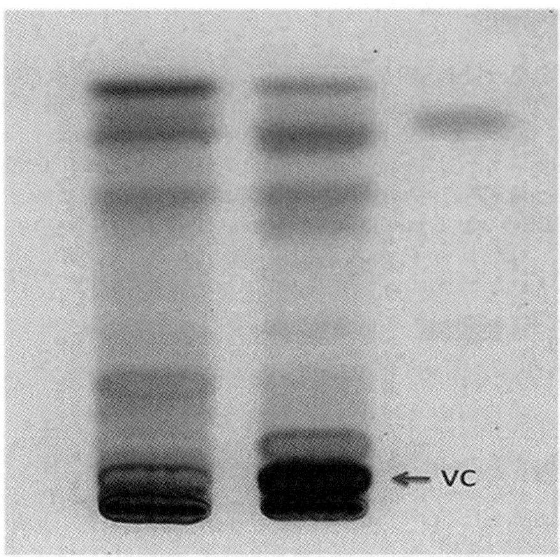

Figure 1.   *TLC of DNPH of urinary carbonyls. Experimental details were given in the text.*

theoretically, within 10 sec at room temperature, but it took over 1hrs to produce the same product from methyl glyoxal-ureide. On the other hands, the reactivity of ureides with DNPH reagent to form hydrazones was found to depend on the structure of carbonyl. To complete the reaction of DNPH reagent and authentic, acetone ureide and glyoxysal ureide polymer, reaction times required 1min and 10hr at room temperature, respectively. In case of formaldehyde/urea resin (urea resin), only part of resin reacted to form formaldehyde DNPH after 20 hr at 100°C. It was therefore concluded from the results that carbonyls in urine were inactivated as stable ureides.

When human urine was treated with ethyl ether, the majority of carbonyls were able to be detected in the aqueous fraction, with only a small amount in soluble in the organic (ethyl ether) phase. Further, when human urine was distilled by steam, a large amount of carbonyls fractionated into the non-volatile fraction. However a small but detectable amount remained in the volatile fraction. Importantly, it is generally recognized that most simple carbonyls are soluble in ethyl ether and volatiles. By contrast, ureides are insoluble in ethyl ether and not volatile. Base on these facts, we hypothesized that many carbonyls in urine might be derivatized in vivo with urea. Furthermore, these considerations imply that DNPH analysis of total urinary carbonyls may partly reflect hydrolysis of ureides as well as free dicabonyls.

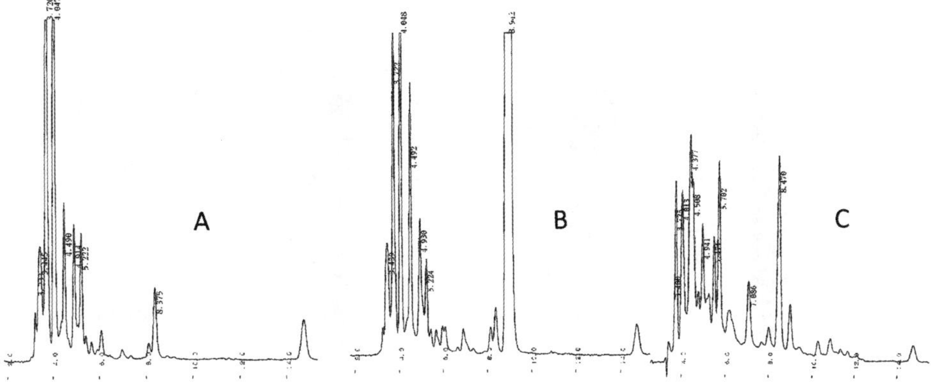

Figure 2.　　　*Typical HPLC of DNPH of urinary carbonyls obtained from 3 different persons. Experimental details were given in the text.*

Our results on the human urinary carbonyl analysis show that a large number of DNPH spots and peaks can be visualized in the TLC and HPLC, respectively. Figure 1 shows the typical result of TLC chromatogram of DNPH obtained from healthy 20 years old and 65years old males volunteers morning urine. Figure shows that there were qualitative and quantitative differences between individuals and age. Considerable differences were also found in the TLC profile of urinary DNPH between before and after drinking and after vigorous exercised (data not shown).

Figure 2A,B, C show the typical HPLC of DNPH of urinary carbonyls. The Maillard reactants such as acetone, methylglyoxal, glyoxal, crotonal, were tentatively identified among these peaks, because identification was based only on retention times of HPLC (data not shown). Figure 2 shows that there were marked qualitative and quantitative differences of urinary carbonyls by each individual. Since urine is an ultra filtrate of plasma, many factors in urine are present. Therefore, our present findings of urinary carbonyl ureides would serve as the useful guide for the discovery of novel diagnostic markers for abnormal metabolisms.

The scavenging capacity of urea to toxic carbonyls, methylglyoxal and glyoxal, was assessed by incubation of carbonyl with urea. After incubation, remaining free carbonyl was treated with DNPH reagent and then DNPH was measured by HPLC. Figure 3 shows the results of scavenging capability of urea to glyoxal and methylglyoxal as estimated the amount of residual carbonyls after incubation. As shown in Figure 3, significant amount of these toxic carbonyls were scavenged by the incubation with urea. These carbonyls may react with urea to form ureides during incubation. These results suggested that urea, which has high concentrations in normal plasma, may function as an important scavenger for toxic carbonyls *in vivo*.

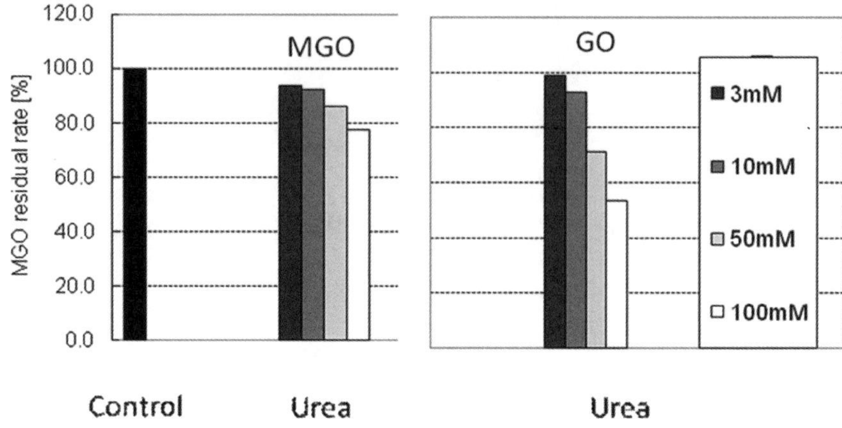

Figure 3.   *Trapping capability of toxic carbonyls, glyoxal (GO) and methylglyoxal (MGO). after the incubation of 100 μM of carbonyls with different concentration (3mM, 10mM, 50mM and 100mM) of urea at 37°C for 2hr, remaining carbonyls were determined by HPLC*

References

1.   B.Danshver, H. Frandsen, H. Autrup, L.O. Dradfstad, *Biomarkers*, 1997,**2**,117-123.
2.   M. Akagawa, K. Suyama, *Free rad. Res.*, 2002,**36**,13-21.
3.   B. Halliwell, J.M.C. Gutterridge, *Arch. Biochem. Biophys.*, 1986,**246**,501-514.
4.   M. Akagawa, K. Suyama, *Eur. J. Biochem.*, 2001,**268**,1953-1963.
5.   E.R. Stadman, *Ann. Rev. Biochem.*, 1993,**62**,797-821.
6.   P. Mococci, G. Fano, S. Fulle, U. MacGarrey, L. Shinobu, M.C. Polidori, A. Cherubini, J. Vecchiet, U. Senin, M.E. Beal, *Free Rad. Biol. Med.*, 1999,**26**,303-308.
7.   K. Uchida, S. Kawakishi, *Biochm. Biophys. Res. Commun.*, 1986,**138**,659-665.
8.   B.S. Berlett, E.R. Stadman, *J. Biol. Chem.*, 1991,**266**,2005-2008.
9.   E.R. Stadman, *Science,* 1992,**257**,1220-1224.
10. M. Akagawa, T. Sasaki, K. Suyama, *Eur. J. Biochem.*, 2002,**269**,5451-5458.
11. P.J. Thornally, A. Longborg, H.S. Minkas, *Biochem. J.*, 1999,**344**,109-116.
12. F. Hayase, R.H. Nagaraj, S. Miyata, F.G. Njoroge, V.M. Monnier, *J. Biol. Chem.*, 1989,**263**,3758-3764.
13. R. Singh, A. Barden, T. Mori, *Diabetologia,* 2001,**44**,129-146.
14. B.S. Berlett, E.R. Stadman, *J. Biol. Chem.*, 1997,**272**,20313-20316.
15. J. Choi, C.A. Malakowsky, J.M. Talent, C.C. Conrad, R.W. Gracy, *Biochem. Biophys. Res. Commun.*, 2002,**293**,1566-1570.
16. N. Ahmed, D. Dobler, M. Dean, P.J. Thonalley, *J Biol Chem.*, 2005,**280**,5724-5732.
17. N. Ahned, R. Babaei-Jadidi, S.K.Howell, P. Beisswenger, P.J. Thonalley, Diabetologia, 2005,48,1590-1603.
18. V. Faist, M. Lindernmeier, C. Geisler, H.F. Erbersdobler, T. Hofmann, J. Agrc. Food Chem., 2002,50,602-606.

ISOLATION OF GLUCOSE UREIDE, UREA DERIVATIVE OF GLUCOSE, IN HUMAN URINE

K. Suyama[1] and A.Sasaki[1]

[1]Department of Sports and Exercise Nutrition, Sendai University, 2-2-18 Fukuoka-Minami, Shibata-matchi, Miyagi, 989-1693 JAPAN

1   INTRODUCTION

N-carbamoyl-β-D-glucopyranosylamine (also known as glucose ureide ; GU) is an N-glucoside formed by the condensation of carbamoyl group of urea and hydroxyl group of the anomeric carbon of D-glucose. GU can be chemically synthesized by heating glucose and urea under acidic condition.[2] Several other glycosyl ureides, such as lactose ureide maltose ureide have been chemically synthesized for use as a dietary non-protein nitrogen source for ruminants.[4-7] In this paper, we report the isolation an identification of GU from the non-reducible sugar fraction obtained by aniline treatment of urine.[1] In addition, we explore the distribution of urea and serum GU levels in rats in the presence and absence of streptozotocin-induced diabetes.

2   MATERIALS AND METHODS

2.1 Chemicals

GU was chemically synthesized by the method of Benn and Jones2). Tetra-O-benzoyl GU was synthesized from authentic GU with benzoic anhydride in the presence of 25 mg/ml 4-dimethylaminopyridine in pyridine at 37°C for 24 hrs (mp 137-139°C, m/z 1113 [M + H+], λmax 229nm ε = 8.1037 x 107). Streptozotocin used was Sigma, and heparin Na used was Novo Nordisk. All other chemicals used were analytical grade and were purchased (Nacalai-tesque, Japan). Preparative silica-gel TLC used was Merck 5003, and TLC aluminum sheet was Merck 5553 silica gel 60F pre-coated. For analysis of creatinine and glucose in urine, commercially available Creatinine-test Wako and Glucose CII-test Wako (Wako Chemical, Japan) were used, respectively.

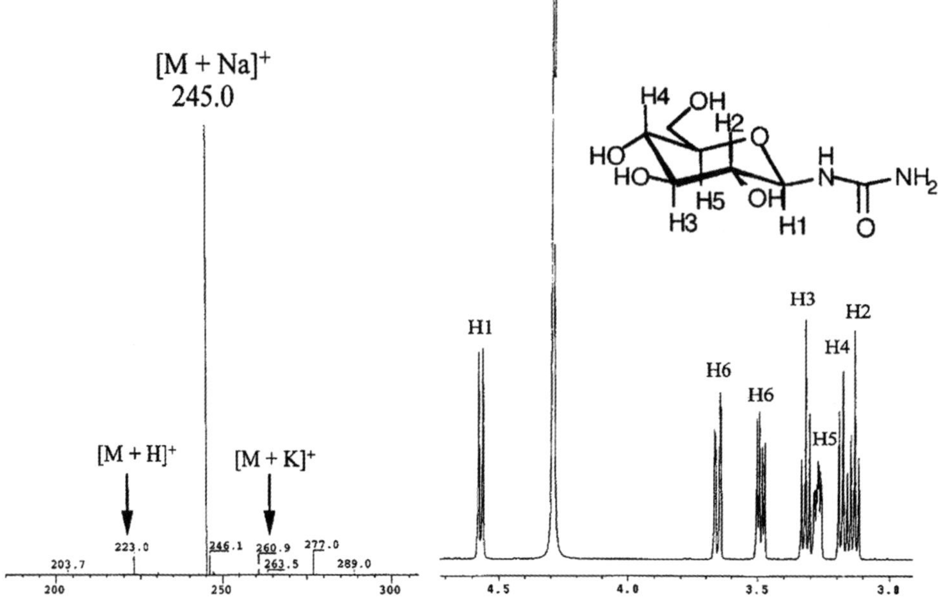

Figure 1   *Positive ESI mass spectrum (1a) ($C_7H_{14}O_6N_2$=222) and 1H-NMR spectrum (1b) of glucose ureide isolated from human urine.*

2.2 Animals

Seven-week-old male Splague-Dawley rats weighing between 220 and 230 g were divided into two groups (group1, control; group 2, diabetic). Animals were either injected with streptozotocin (40-60 mg/kg) as described in detail previously[8]. All experiments were done following the ethical procedures accepted by the Graduate School of Agricultural Science, Tohoku University.

2.3 Isolation of GU from Human Urine

Two hundred millilitres of pooled human urine sample was mixed with 500 ml of methanol and then centrifuged (1500 g, for 10 minutes). After removed the precipitate, the supernatant volume was reduced to 200 ml by rotary evaporation at 30°C. The solution was then treated with 10 ml of aniline reagent (aniline/acetic acid, 5/2 by volume) to remove reducible carbohydrate at 80oC for 2hrs as described in our previous report[2]. The reactant of urine with aniline was then washed three times of 500ml of diethyl ether to remove aniline derivatives of reducible carbohydrates and excess aniline reagent. After reduction of the water fraction sample volume to about 10ml by rotary evaporation at 40°C, each of 1ml was charged to preparative TLC (20 x 20 cm, 0.1 mm thick) and chromatography was developed with ethyl acetate/acetic acid/water (2/1/1, by volume). Fractions bands (Rf: 0.25) on preparative TLC containing GU like compounds were removed by scratch. This TLC separation was carried out on different 10 TLC plates. The residues obtained from preparative TLC were monitored by analytical TLC (10 x 10 cm) developed with the same solvent system. Visualization was done by heating at 100°C for 5 min. after spraying 1%

sulfuric acid/methanol on TLC. The fractions containing GU like compounds, which were extracted with distilled water were pooled and stored as the pre-purified sample. The pre-purified sample thus obtained was dried by rotary evaporation, and the residue was dissolved in 2ml of distilled water, and then re-applied to the preparative TLC. The carbohydrate fraction of Rf: 0.25 were collected. The fractions containing GU like compounds was isolated and extracted with distilled water, then evaporated to dryness. Crystallization of the compound was failed. The chemical structure of GU like compounds was determined by 1H-NMR (UNITY INOVA-600, Varian, 600MHz in D2O at 50°C), 13C-NMR, and FAB-MSs and ESI-MS(Mastation JSM-700). The chemical shift correlation was analyzed by 1H-1H COSY and NEOSY. The hetero-nuclear coherence was analyzed by HSQC and HMBC.

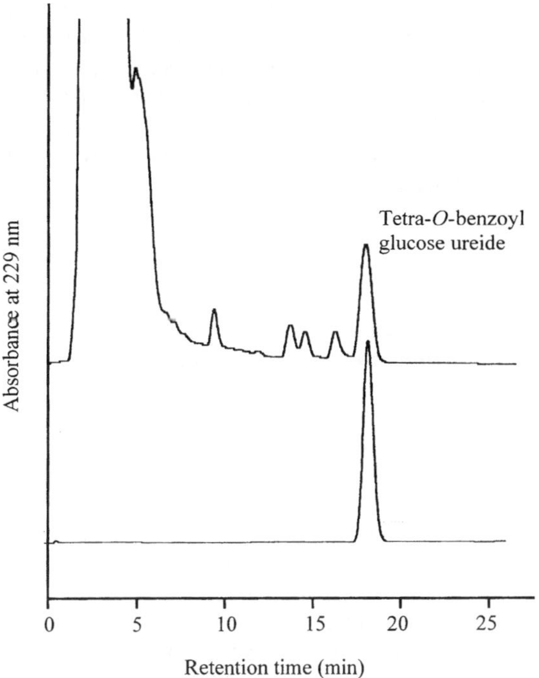

Figure 2   *HPLC chromatogram of human urine sample (top) and authentic tetra-O-benzoyl glucose ureide (bottom).*

2.4 Measurement of Urine GU level

Human urines (2,0ml) were collected from 5 (22 years old female, 22 yeas old female, 23 years old male, 25 years old male and 63 years old male) volunteers. Rat urines (1.0ml) were collected from both normal and streptozotocin induced diabetic rats. For HPLC analysis of GU, Tetra-O-benzoyl derivative of glucose ureide (mp 137-139°C, m/z 639 [M + H+]: Mastation JMS-700 in o-nitrobenzylalcohol matrix, λmax 229nm ε = 8.1037 x 107: Shimazu UV-2001) was employed. Authentic crystal tetra-O-benzyl GU was used as standard.

Samples (1 ml) were mixed with 2ml of methanol then centrifuged (1500 g, 20min). In order to remove reducible carbohydrates, supernatant thus obtained (2 ml) was treated with 1ml of aniline reagent at 80°C for 2hrs as shown in our previous report.[3] After the extraction of aniline reactive compounds composed with reducible carbohydrates with 4ml of ethyl ether 2 times, the aqueous layer was collected and evaporated to dryness. It was then dried under reduced pressure in desiccators using a di-phosphorus pentaoxide as drying reagent for overnight at room temperature. The resultants were treated with 1ml of 50mg/ml benzoic anhydride and 1ml of 25mg/ml 4-dimethylaminopyridine in dried pyridine at room temperature for 24 hrs. The reactant was evaporated to dryness by rotary evaporation, and then dissolved in 2ml of n-hexane/ethyl acetate (1/1, by volume). The sample solution was applied to a SEP-PAC Silica column (Waters, 1ml volume) equilibrated with the same solvent. After 10ml of n-hexane through out, the fraction containing tetera-O-benzoyl GU was eluted with 3 ml of ethyl acetate. Ethyl acetate fraction was then evaporated to dryness then dissolved with 0.5 ml of acetonitrile for HPLC analysis. For analysis of GU, benzoylated compounds were charged on HPLC (Hitach L-6020 system) on ODS column (Maighty-sil RP-18 GP 150 x 4.6 mm, 5μm: Kanto Kagaku, Japan) at 35oC. Solvent was acetonitrile/water (1/1, v/v) and eluted at 1ml/min flow rate at 35oC. Detection and quantification were carried out using UV detector at 229nm and Shimadzu C-R5A data station.

3   RESULTS

3.1 Isolation and Identification of GU from Human Urine

Several non-reducible carbohydrate spots were found by the first TLC separation of the human urine sample. Compared to the Rf of the authentic sample, the fractions faint amount of sugar was detected. The fraction that was developed at Rf of 0.30 (that of glucose: 0.35) was same as the authentic GU. This compound developed by preparative TLC (Rf was 0.25 that was different from analytical TLC of 0.30) was extracted and subjected to further purification by preparative TLC. A single spot was observed on TLC in this final purification, yielding 0.4mg of an apparently pure material.

The chemical structure of the isolated unknown sugar was determined by 1H-NMR and 13C-NMR and FAB/MS analysis, while COSY, NOESY, HSQC and HMBC analyses also shows this compound to be a glucoside. The positive ion FAB/MS spectrum showed a peak at m/z=223, corresponding to the molecular ($C_7H_{14}N_2O_6$) peak (Fig.1a). The 1H-NMR spectrum of this compound was indistinguishable from the authentic GU (Fig. 1b). 13C-NMR: δ=160.70(C=O), 81.10(C1), 77.22(C5), 76.67(C3), 72.12(C2), 69.58(C4), 60.86(C6).

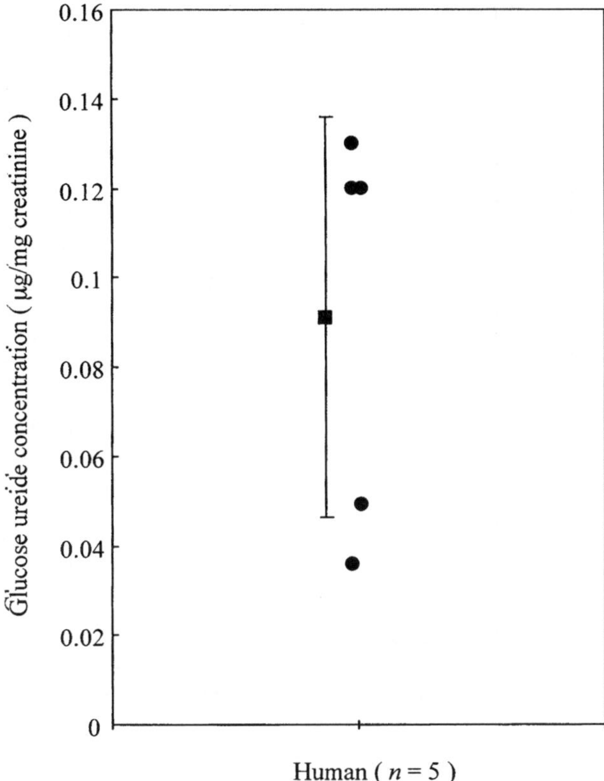

Figure 3   *Glucose ureide concentration in human urine.*

3.2 Quantitation of Urinary GU

Considering that urine contains various sugars and compounds, one milliliter of urine samples were, therefore, first applied methanol precipitation by addition of 2 ml of methanol. After centrifugation (1,500g for 15 minutes), 2 ml of supernatant was treated with aniline reagent to remove reducible carbohydrate as shown in the experimental section. Non-reducible carbohydrates were benzoylated and then GU was measured by HPLC. Figure 2 shows the typical HPLC chromatogram of sample obtained from human urine and standard tetra-O-benzoyl-GU monitored at 229 nm. Positive ion FAB/MS analysis of the compound of clear peak at 18.0 min of human urine gave m/z 639 [M+H+] of tetra-O-benzoyl-GU (data not shown). ^{1}H NMR in CDCl3, 1H-1H COSY and HSQS spectra were completely identical those of authentic tetra-O-benzoyl-GU (data not shown).

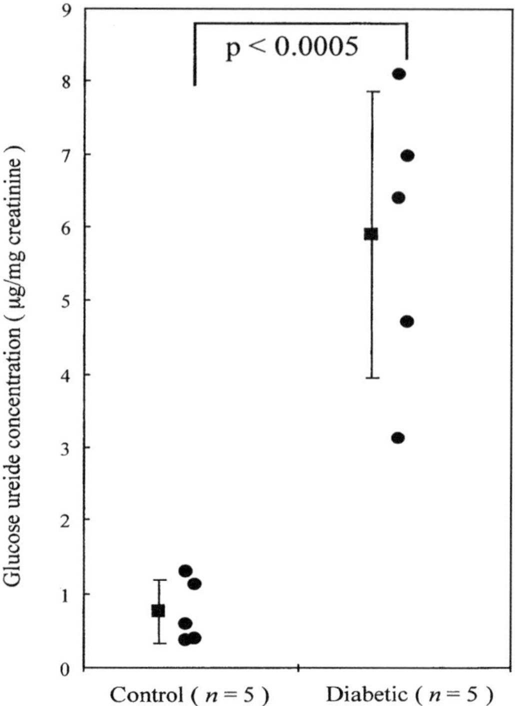

Figure 4   *Glucose ureide concentration in urines from normal and streptozotocin induced diabetic rat*

The GU levels of five healthy human urines were determined by this method. GU concentration being estimated to be 0.038 - 0.132 µg/mg creatinine with an average of 0.092 as shown in the figure 3. The result on the distributions of urea and serum GU level of normal rats and streptozotocin induced hyperglycemia rats is shown in Figure 4. The concentrations of GU in urine of streptozotocin-induced rats were significantly higher than that of normal rats.

4   CONCLUSION

These data demonstrate the presence of a urea derivative of glucose, glucose ureide (N-carbamoyl-β-D-glucopyranosylamine) in the urine of humans and rats. Since GU has not been found in foods, the presence of urinary GU cannot result from exogenous uptake. Indeed, it was shown by Heins *et al.* that 13C-labeled GU was a useful clinical marker of human intestinal transit time[3,4]. Moreover, as GU has been isolated and identified from human serum[9,10], we hypothesise that urinary GU may be synthesized in vivo by the condensation of glucose and urea *in vivo*. From these results, we hypothesise that changes in urine GU levels may be related to some pathophysiological conditions. Indeed in diabetic rats, in which the concentration of urea and glucose are increased, we demonstrated increased urinary GU levels. We speculate that urine GU levels could be utilized as a novel marker for diabetic patients.

References

1. K. Suyama, C, Hwang, M. Fujii, and S. Adachi, *Anal. Biochem.,* 1987,**162**,325-329.
2. M.H. Benn, A.S. Jones, *J. Chem. Soc.,* **1960**, 3837-3841.
3. W.E. Heine, H.K. Berthold, P.D. Klein, *Am. J. Gastroenterol,* 1995,**90**,93-98.
4. D.J. Morrison, R. Zavosy, C.A. Edwards, B. Dodson, T. Preston, L.T. Weaver, *Biochem. Soc. Trans.,* 1998,**26,** S184.
5. R.J. Merry, R.H. Smith, A.B. McAllan, *Br. J. Nutr.* 1982,**48**,287-304.
6. R.J. Merry, R.H. Smith, A.B. McAllan, *Br. J. Nutr.* 1982,**48**,305-318.
7. R.N. Coleman, L.P.Milligan, *J. Gene. Microbiol.,* 1980,**116**,445-450.
8. M. Akagawa, T. Sasaki, and K. Suyama, *Eur. J. Biochem.,* 2002,**269**,5451-8.
9. T. Hamafuji, W. Tsugawa, and K. Soda, *J. Biochem.* 2002,**6**,315-318.
10. T. Hamafuji, W. Tsugawa and K. Soda, *Renal Failure,* 2003,**25**,115-121.

SOME NATURAL PRODUCTS EXTRACTS INHIBIT THE FORMATION OF N(CARBOXYMETYL)ARGININE

Y. Fujiwara[1], M. Yoshitomi[1], K. Mera[1], M. Nagai[2], M.Takeya[1], T. Ikeda[1], R. Nagai[2]¶

[1]Graduate School of Medical and Pharmaceutical Sciences, Kumamoto University, Honjo 1-1-1, Kumamoto 860-8556, Japan.
[2]Department of Food and Nutrition, Japan Women's University, Mejirodai 2-8-1, Bunkyo-ku, Tokyo 112-8681, Japan.

1   INTRODUCTION

Long-term incubation of proteins with glucose leads, through the formation of early products such as Schiff base and Amadri products, to the formation of advanced glycation end products (AGEs) [1-4]. AGE-modified protein increase during the normal process of aging, but this is markedly accelerated in people with diabetes who have sustained hyperglycemia. Immunohistochemical studies have detected AGE modification in several pathological tissues. A recent study demonstrated that N^ε-(carboxymethyl)lysine (CML), one of the major AGE structures, accumulates in several tissue proteins including kidneys of patients with diabetic nephropathy and chronic renal failure [5], atherosclerotic lesions of arterial walls [6], amyloid fibrils in hemodialysis-related amyloidosis [7], and actinic elastosis of the photo-aged skin [8]. On the other hand, Nε-(carboxymethyl)arginine (CMA) is an acid-labile AGE and was discovered in enzymatic hydrolysate of glycated collagen [9]. Subsequently, CMA was also detected in human serum by electrospray ionization/liquid chromatography/mass spectrometry, and its level was found to be higher in diabetic patients than people without diabetes [10]. CMA is known as a major AGE structures in collagen and AGE-collagen induces apoptosis in fibroblasts through activation of reactive oxygen species and MAP kinases[11]. Therefore, CMA probably accumulates in collagen-rich tissues in vivo, and it may contribute to the development of diseases such as atherosclerosis and diabetic complications in connective tissues. The AGE inhibitors such as aminoguanidine and pyridoxamine retard the development of early renal disease in the streptozotocin-diabetic rat. Therefore, treatment with AGEs inhibitors may be a potential strategy for the prevention of clinical diabetic complications. In the present study, we investigated the inhibitory effect of natural products on CMA formation to discover the candidate agents for the new AGE inhibitors.

2   MATERIALS AND METHODS

2.1   Chemicals.

Gelatin (form porcine skin, Type A) and aminoguanidine was purchased from Sigma-Aldrich Japan (Osaka, Japan). Ribose was purchased from Wako (Osaka, Japan). Horseradish peroxidase (HRP)-conjugated goat anti-mouse IgG antibody was purchased from Kirkegaard Perry Laboratories (MD, USA). Microtitration plates (96-well, Nunc Immunoplate II) were purchased from Nippon Iner Med (Tokyo, Japan).

2.2   Preparation of crude extracts.

Zyzyphi Semen, Rehmanniae Radix, Tribuli Fructus, Epimedii Herba, Cnidii Monnieris Fructus, Zingiberis Rhizoma, Cimicifugae Rhizoma, Cnidii Rhizoma, Magnoliae Flos, Zizyphi Fructus, Aurantii Nobilis Pericarpium, Atractylodis Lanceae, Coicis Semen, Achyranthis Radix, Ophiopogonis Tuber, Hoelen, Saposhnikoviae Radix, Atractylodis Rhizoma, Armeniacae Semen, Sophorae Radix, Bupleuri Radix, Magnoliae Cortex, Aurantii Fructus Immaturus, Paeoniae Radix, Schisandrae Fructus, Schizonepetae Spica, Corydalis Tuber, Notopterygii Rhizoma, Phellodendori Cortex, Coptidis Rhizoma were extracted with MeOH (×3) by refluxing for 2h, and the extracts were concentrated in vaccuo to afford residues. The residues were loaded onto a Diaion HP-20 column and eluted by H$_2$O and MeOH. Those MeOH eluates were used as crude extracts in the AGE inhibition assay.

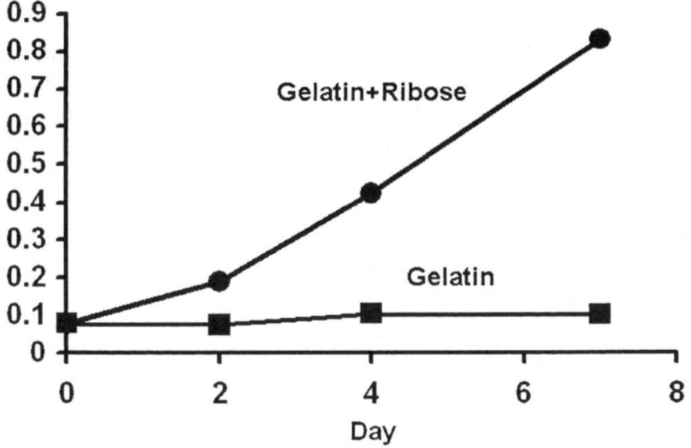

Figure 1.   *Formation of CMA during the incubation of gelatin with ribose. The samples (10 μg/mL) were coated on the immunoplate and incubated for 2 h. The wells were washed and blocked with gelatin, followed by reaction with anti-CMA monoclonal antibody, 3F5. The antibodies bound to the wells were detected by horseradish peroxidase-conjugated anti-mouse immunoglobulin G antibody.*

2.3   Inhibitory effect of natural products extracts on CMA formation.

Gelatin (2 mg/ml) and ribose (30 mM) were incubated in the presence of plant extracts (10 µg/ml) in 100 mM Sodium phosphate buffer at 37°C for 7 days, followed by the determination of CMA by a noncompetitive enzyme-linked immunosorbent assay (ELISA).

2.4   ELISA.

ELISA was performed as described previously [12]. Briefly, each well of a 96-well microtiter plate was coated with 100 µl of the sample to be tested in PBS, blocked with 0.5% gelatin, and washed three times with PBS containing 0.05% Tween 20 (washing buffer). The wells were incubated with 0.1 ml of anti-CMA antibody, 3F5, (1 µg/mL) dissolved in washing buffer for 1 hr. The wells were then washed with washing buffer three times and reacted with HRP-conjugated anti-mouse IgG antibody, followed by reaction with 1,2-phenylenediamine dihydrochloride. The reaction was terminated by addition of 0.1 ml of 1 M sulfuric acid, and the absorbance at 492 nm was read by a micro-ELISA plate reader.

2.5   Statistical analysis.

All data are expressed as the mean ±SD. Differences between groups were examined for statistical significance using the Mann-Whitney U-test. A p value less than 0.05 denoted the presence of a statistically significant difference.

3   RESULTS AND DISCUSSION

AGEs are a heterogeneous group of reaction products that form between a protein's primary amino group and a carbohydrate-derived aldehyde group. A substantial number of previous reports have indicated that AGEs exacerbate and accelerate the changes associated with aging, while also contributing to the early phases of age-related disease, including atherosclerosis, cataracts, neurodegenerative disease, renal failure, arthritis, and age-related macular degeneration [13, 14]. For these reasons, it is believed that the use of AGE inhibitors may therefore be a potentially effective strategy to prevent the pathogenesis of age-related diseases. Hammes *et al.* reported that thiamine and benfotiamine prevent intracellular AGE formation by reducing the concentration of methylglyoxal, a strong AGE-precursor, and hence inhibit diabetic retinopathy [15]. Furthermore, Babaei-Jadidi *et al.* also demonstrated that administration of thiamine and benfotiamine resulted in reduction of intracellular methylglyoxal concentration by increasing transketolase expression and prevented the development of diabetic nephropathy in diabetic rats [16, 17]. Taken together, these findings suggest that compounds that inhibit intracellular AGE formation could be potentially useful agents for the treatment of diabetic complications and atherosclerosis.

CMA is a major AGE structure in collagen and serum CMA concentration is higher in diabetic patients than people without diabetes [10]. Therefore, CMA probably accumulates in collagen-rich tissues in vivo, and it may contribute to the development of diseases such as atherosclerosis and diabetic complications in connective tissues. In the present study, we developed the assay system for CMA formation by ELISA and measured the inhibitory effect of natural products extracts on CMA formation. As shown in Figure 1, incubation of gelatin, a soluble collagen, with Ribose for 7 days resulted in the generation of CMA. Under the same conditions, natural products extracts such as *Magnoliae Cortex*, *Epimedii Herba* and *Schizonetae Spica* showed inhibitory effect of CMA formation (Fig. 2A and B). These data indicates that certain compounds contained in *Magnoliae Cortex*, *Epimedii Herba* and *Schizonetae Spica* may become candidates for the new AGE inhibitors.

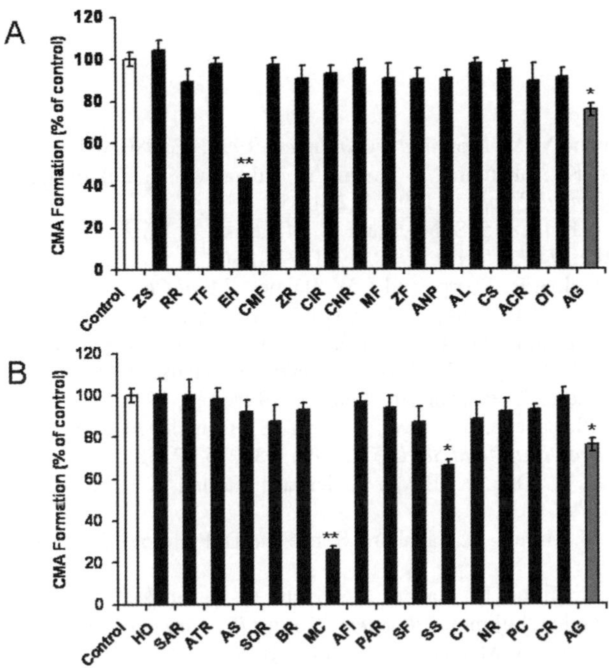

Figure 2.   *Inhibitory effect of natural products extracts on CMA formation. Gelatin (2 mg/mL) was incubated with 30 mmol/L ribose at 37°C for 7 days in the presence (10 µg/mL) or absence of natural products (ZS:Zyzyphi Semen, RR:Rehmanniae Radix, TF:Tribuli Fructus, EH:Epimedii Herba, CMF:Cnidii Monnieris Fructus, ZR:Zingiberis Rhizoma, CIR:Cimicifugae Rhizoma, CNR:Cnidii Rhizoma, MF:Magnoliae Flos, ZF:Zizyphi Fructus, ANP:Aurantii Nobilis Pericarpium, AL:Atractylodis Lanceae, CS:Coicis Semen, ACR:Achyranthis Radix, OT:Ophiopogonis Tuber, HO:Hoelen, SAR:Saposhnikoviae Radix, ATR:Atractylodis Rhizoma, AS:Armeniacae Semen, SOR:Sophorae Radix, BR:Bupleuri Radix, MC:Magnoliae Cortex, AFI:Aurantii Fructus Immaturus, PAR:Paeoniae Radix, SF:Schisandrae Fructus, SS: Schizonepetae Spica, CT:Corydalis Tuber, NR:Notopterygii Rhizoma, PC:Phellodendori Cortex, CR:Coptidis Rhizoma, AG:Aminoganidine). The yield of CMA was measured by ELISA. Bar indicates mean ± SD; *P < 0.05 , **P < 0.005 versus control.*

Acknowledgments

This work was supported in part by Grants-in-Aid for Scientific Research from the Cosmetology Research Foundation (to Yukio Fujiwara). This work was also supported in part by a Grant-in-Aid for Scientific Research (No. 21590340 to Ryoji Nagai) from the Ministry of Education, Science, Sports and Cultures of Japan.

References

1.   D. R. Sell and V. M. Monnier, *J Clin Invest*, 1990, **85**, 380-384.
2.   K. Ienaga, K. Nakamura, T. Hochi, Y. Nakazawa, Y. Fukunaga, H. Kakita and K. Nakano, *Contrib Nephrol*, 1995, **112**, 42-51.
3.   T. Niwa, T. Katsuzaki, S. Miyazaki, T. Miyazaki, Y. Ishizaki, F. Hayase, N. Tatemichi and Y. Takei, *J Clin Invest*, 1997, **99**, 1272-1280.
4.   M. U. Ahmed, S. R. Thorpe and J. W. Baynes, *J Biol Chem*, 1986, **261**, 4889-4894.
5.   K. Yamada, Y. Miyahara, K. Hamaguchi, M. Nakayama, H. Nakano, O. Nozaki, Y. Miura, S. Suzuki, H. Tuchida, N. Mimura and et al., *Clin Nephrol*, 1994, **42**, 354-361.
6.   S. Kume, M. Takeya, T. Mori, N. Araki, H. Suzuki, S. Horiuchi, T. Kodama, Y. Miyauchi and K. Takahashi, *Am J Pathol*, 1995, **147**, 654-667.
7.   T. Miyata, S. Taneda, R. Kawai, Y. Ueda, S. Horiuchi, M. Hara, K. Maeda and V. M. Monnier, *Proc Natl Acad Sci U S A*, 1996, **93**, 2353-2358.
8.   K. Mizutari, T. Ono, K. Ikeda, K. Kayashima and S. Horiuchi, *J Invest Dermatol*, 1997, **108**, 797-802.
9.   K. Iijima, M. Murata, H. Takahara, S. Irie and D. Fujimoto, *Biochem J*, 2000, **347 Pt 1**, 23-27.
10.   H. Odani, K. Iijima, M. Nakata, S. Miyata, H. Kusunoki, Y. Yasuda, Y. Hiki, S. Irie, K. Maeda and D. Fujimoto, *Biochem Biophys Res Commun*, 2001, **285**, 1232-1236.
11.   M. Alikhani, C. M. Maclellan, M. Raptis, S. Vora, P. C. Trackman and D. T. Graves, *Am J Physiol Cell Physiol*, 2007, **292**, C850-856.
12.   R. Nagai, C. M. Hayashi, L. Xia, M. Takeya and S. Horiuchi, *J Biol Chem*, 2002, **277**, 48905-48912.
13.   N. Araki, N. Ueno, B. Chakrabarti, Y. Morino and S. Horiuchi, *J Biol Chem*, 1992, **267**, 10211-10214.
14.   H. Vlassara, R. Bucala and L. Striker, *Lab Invest*, 1994, **70**, 138-151.
15.   H. P. Hammes, X. Du, D. Edelstein, T. Taguchi, T. Matsumura, Q. Ju, J. Lin, A. Bierhaus, P. Nawroth, D. Hannak, M. Neumaier, R. Bergfeld, I. Giardino and M. Brownlee, *Nat Med*, 2003, **9**, 294-299.
16.   R. Babaei-Jadidi, N. Karachalias, N. Ahmed, S. Battah and P. J. Thornalley, *Diabetes*, 2003, **52**, 2110-2120.
17.   R. Babaei-Jadidi, N. Karachalias, C. Kupich, N. Ahmed and P. J. Thornalley, *Diabetologia*, 2004, **47**, 2235-2246.

SCREENING OF AGE INHIBITORS BY ANTIBODY LIBRARY

R. Nagai[1], S. Shimasaki[1], A. Horikoshi[1], M. Nakano[1], M. Nagai[1], K. Mera[2], Y. Fujiwara[2]

[1]Department of Food and Nutrition, Laboratory of Biochemistry & Nutritional Science, Japan Women's University
[2] Faculty of Medical and Pharmaceutical Sciences, Kumamoto University

1   INTRODUCTION

Since advanced glycation end-products (AGEs) inhibitors such as pyridoxamine[1] and aminoguanidine[2] significantly inhibit the development of diabetic complications such as retinopathy and kidney failure in the streptozotocin-induced diabetic rat, treatment with AGEs inhibitors is believed to be a potential strategy for the prevention of life style-related diseases. We previously developed monoclonal antibodies for AGEs such as CML[3], carboxymethyl-cysteine (CMC)[3], N^ε-(carboxyethyl)lysine (CEL)[4], N^ω-carboxymethylarginine (CMA)[5], GA-pyridine[6], 3DG-imidazolone[7], pyrraline[8] and pentosidine[9] (Fig. 1). As shown in figure 2, these antibodies are used to demonstrate the localization and formation pathways of AGEs. In fact, we previously measured the formation pathways of CML, a major antigenic AGE structure, in order to develop new AGE inhibitors. Our immunological and instrumental analyses demonstrated that CML is generated by the oxidative cleavage of Amadori products by hydroxyl radical[10], peroxynitrite[11] and hypochlorous acid[12], thus suggesting CML to be an important biological markers of oxidative stress *in vivo*. We then applied monoclonal anti-CML antibody to screen the inhibitors for CML formation.

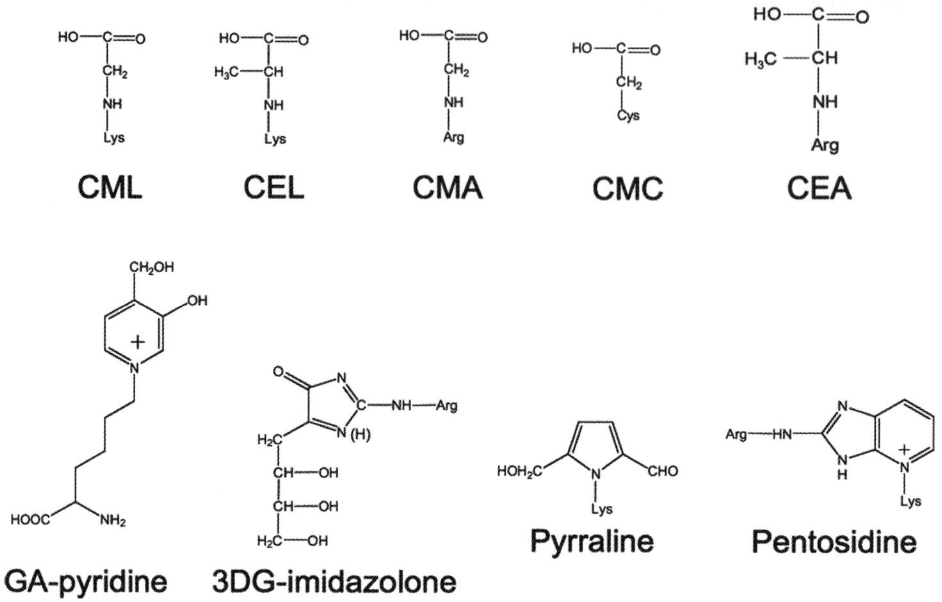

CML   **CEL**   **CMA**   **CMC**   **CEA**

GA-pyridine   **3DG-imidazolone**   **Pyrraline**   **Pentosidine**

Figure 1.   *Structure of some common AGEs.*

2   MATERIALS AND METHODS

2.1   Chemicals

Fatty acid-free bovine serum albumin (BSA) was purchased from Wako Pure Chemical Inc. (Osaka, Japan). Horseradish peroxidase (HRP)–conjugated goat anti–mouse IgG antibody was purchased from Kirkegaard Perry Laboratories (Gaitherburg, MD, USA). All other chemicals were of the best grade available from commercial sources.

2.2   Preparation of natural compounds

Acteoside was isolated from *Verbena brasiliensis*. Sophoradiol was isolated form *Puerariae Radix*. Lupeol was isolated from *Radix Isatidis*. Quercetin was isolated from *Sophorae Flos*. Quercetin 3-sambubioside was isolated from *Helwingia japonica*.

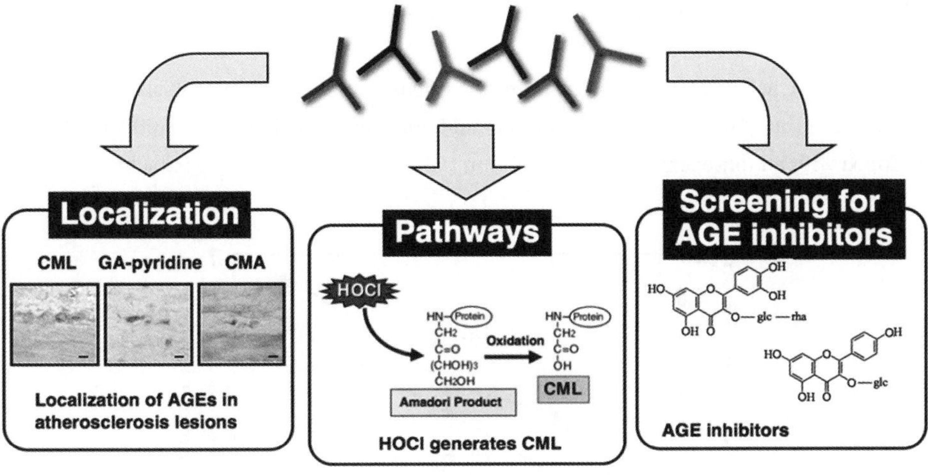

Figure 2.　　　*Usefulness of antibody library for detection of AGE structures.*

2.3　Determination of the inhibitory effect of plant extracts and natural compounds on CML formation.

BSA (2 mg/ml) and ribose (33 mM) were incubated with natural compounds (1 mM) in PBS at 37°C for 7 days, followed by the determination of CML by a noncompetitive enzyme-linked immunosorbent assay (ELISA).

2.4　ELISA.

ELISA was performed as described previously[6]. Briefly, each well of a 96-well microtiter plate was coated with 100 µl of the sample to be tested in PBS, blocked with 0.5% gelatin, and washed three times with PBS containing 0.05% Tween 20 (washing buffer). The wells were incubated with 0.1 ml of anti-CML antibody, 6D12, (0.5 µg/mL) in washing buffer for 1 hr. The wells were then washed with washing buffer three times and reacted with HRP-conjugated anti-mouse IgG antibody, followed by reaction with 1,2-phenylenediamine dihydrochloride. The reaction was terminated by addition of 0.1 ml of 1 M sulfuric acid, and the absorbance at 492 nm was read by a micro-ELISA plate reader.

2.5   Detection of CML by HPLC.

The CML content of the samples was quantified by acid hydrolysis with 6 N of HCl for 24 hr at 110°C, followed by amino acid analysis on a Hitachi L-8500A instrument equipped with an ion-exchange HPLC column (#2622 SC, 4.6 x 80 mm; Hitachi)[13].

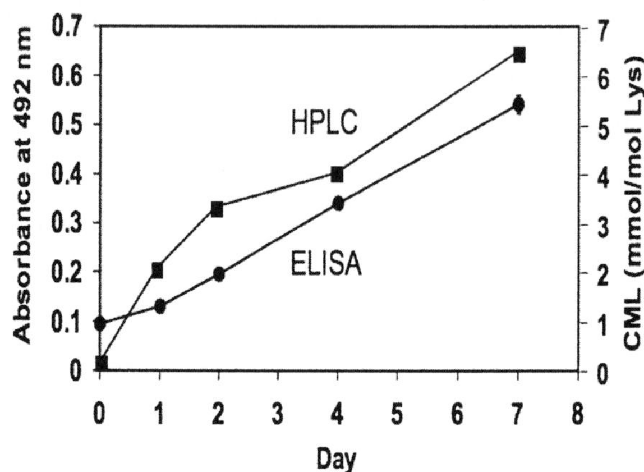

Figure 3.   *Time dependency of CML formation. BSA and ribose were incubated in phosphate buffer for 7 days followed by determination of CML by noncompetitive ELISA and HPLC analysis. For ELISA, the samples (10 µg/mL) were coated on the ELISA plate and then were incubated for 1 hr. The wells were washed and blocked with gelatin, followed by reaction for 1 hr with anti-CML antibody. The antibodies bound to wells were detected by HRP-conjugated anti-mouse IgG antibody.*

3   RESULTS AND DISCUSSION

As shown in figure 3, the reactivity of anti-CML antibody increased in a time dependent manner during incubation of BSA (2 mg/mL) with 33 mM ribose in PBS at 37°C. A similar tendency was observed when CML content was measured by HPLC (Fig. 3). We then measured the inhibitory effect of many natural products on CML formation. Incubation of BSA with ribose increased CML formation. Under the assay conditions employed, lupeol and sophoradiol inhibited CML formation, whereas some compounds such as quercetin 3-sambubioside, quercetin, acteoside enhance CML formation (Fig. 4). Since Akagawa *et al.*[14] demonstrated that some polyphenols such as epicatechin gallate and epigallocatechin gallate generate hydrogen peroxide, quercetin 3-sambubioside, quercetin

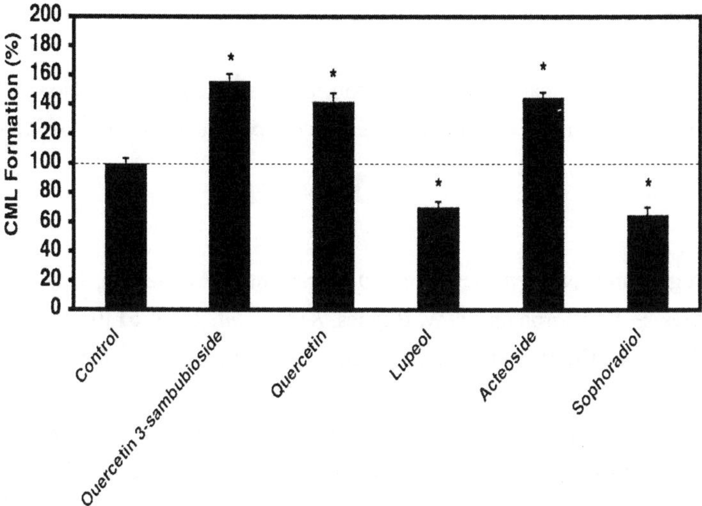

Figure 4.   *Effect of natural compounds on CML formation. BSA and Ribose were incubated with 1 mM natural products in phosphate buffer for 7 days followed by determination of CML by noncompetitive ELISA as described in the Material Methods. Data are presented as the mean ± SD. *, P < 0.05 vs. control.*

and acteoside may have enhanced CML formation by producing hydrogen peroxide. A substantial number of previous reports have indicated that AGEs exacerbate and accelerate the changes associated with aging, while also contributing to the early phases of age-related disease, including atherosclerosis, cataracts, neurodegenerative disease, renal failure, arthritis, and age-related macular degeneration. For these reasons, it is believed that the use of AGE inhibitors may therefore be a potentially effective strategy to prevent the pathogenesis of age-related diseases. The interaction of AGEs/AGE receptors are believed to enhance the progression of the diseases. Therefore, treatment with AGE inhibitors is also believed to be a potential strategy for the prevention of clinical diabetic complications. However, the screening of new AGE inhibitors by instrumental analyses such as high performance liquid chromatography (HPLC) is required for several preparation steps, and practically difficult to estimate the inhibitory effects of a great number of candidate compounds. Although measurement of CMA content by HPLC after standard hydrolysis is difficult because of its acid instability, CMA content is estimated rapidly by ELISA without any sample preparations[5]. Under the assay conditions employed, we found that several compounds inhibit CML formation, whereas some compounds enhance CML formation. Taken together, antibody libraries for AGEs are thus considered to be a powerful tool for measuring AGE content and screening of new AGE inhibitors.

Acknowledgments

This work was supported in part by Grants-in-Aid for scientific Research (No. 18790619 to Ryoji Nagai) from the Ministry of Education, Science, Sports and Cultures of Japan.

References

1.   T. P. Degenhardt, N. L. Alderson, D. D. Arrington, R. J. Beattie, J. M. Basgen, M. W. Steffes, S. R. Thorpe and J. W. Baynes, *Kidney Int*, 2002, **61**, 939-950.

2.   T. Soulis, M. E. Cooper, D. Vranes, R. Bucala and G. Jerums, *Kidney Int*, 1996, **50**, 627-634.

3.   K. Mera, M. Nagai, J. W. Brock, Y. Fujiwara, T. Murata, T. Maruyama, J. W. Baynes, M. Otagiri and R. Nagai, *J Immunol Methods*, 2008, **334**, 82-90.

4.   R. Nagai, Y. Fujiwara, K. Mera, K. Yamagata, N. Sakashita and M. Takeya, *J Immunol Methods*, 2008, **332**, 112-120.

5.   K. Mera, Y. Fujiwara, M. Otagiri, N. Sakata and R. Nagai, *Ann N Y Acad Sci*, 2008, **1126**, 155-157.

6.   R. Nagai, C. M. Hayashi, L. Xia, M. Takeya and S. Horiuchi, *J Biol Chem*, 2002, **277**, 48905-48912.

7.   T. Jono, T. Kimura, J. Takamatsu, R. Nagai, K. Miyazaki, T. Yuzuriha, T. Kitamura and S. Horiuchi, *Pathol Int*, 2002, **52**, 563-571.

8.   T. Jono, R. Nagai, X. Lin, N. Ahmed, P. J. Thornalley, M. Takeya and S. Horiuchi, *J Biochem (Tokyo)*, 2004, **136**, 351-358.

9.   K. Miyazaki, R. Nagai and S. Horiuchi, *J Biochem (Tokyo)*, 2002, **132**, 543-550.

10.   R. Nagai, K. Ikeda, T. Higashi, H. Sano, Y. Jinnouchi, T. Araki and S. Horiuchi, *Biochem Biophys Res Commun*, 1997, **234**, 167-172.

11.   R. Nagai, Y. Unno, M. C. Hayashi, S. Masuda, F. Hayase, N. Kinae and S. Horiuchi, *Diabetes*, 2002, **51**, 2833-2839.

12.   K. Mera, R. Nagai, N. Haraguchi, Y. Fujiwara, T. Araki, N. Sakata and M. Otagiri, *Free Radic Res*, 2007, **41**, 713-718.

13.   R. Nagai, T. Araki, C. M. Hayashi, F. Hayase and S. Horiuchi, *J Chromatogr B Analyt Technol Biomed Life Sci*, 2003, **788**, 75-84.

14.   M. Akagawa, T. Shigemitsu and K. Suyama, *Biosci Biotechnol Biochem*, 2003, **67**, 2632-2640.

TASTE MODULATING MAILLARD REACTION PRODUCTS OF CREATININE

C. Kunert[1], T. Sonntag[1], A. Walker[1] and T. Hofmann[1]

[1] Food Chemistry and Molecular Sensory Science, Technische Universität München, Lise-Meitner-Strasse 34, D-85354 Freising-Weihenstephan, Germany

1 INTRODUCTION

Nonenzymatic reactions of reducing sugars and the amino group of amino acids, as well as amines, peptides and proteins have been investigated by multiple studies in the past decades. The so-called Maillard reaction is long known to play a key role in the formation of key odorants during thermal food processing such as roasting of beef (*1*) and malt (*2*). Moreover, several Maillard-derived taste compounds have been identified in recent years such as, e.g. the bitter tasting 3-(2-furyl)-8-[(2-furyl)methyl]-4-hydroxymethyl-1-oxo-1*H*,4*H*-quinolizinium-7-olate (quinizolate) (*3*), the umami- and sweetness enhancing *N*-(1-carboxyethyl)-6-(hydroxymethyl)pyridinium-3-ol inner salt ((*S*)-alapyridaine) (*4-5*), the umami tasting glycoconjugates, *N*-(D-glucos-1-yl)-L-glutamate and *N*-(1-deoxyfructos-1-yl)-L-glutamic acid (*6*).

About 10 years ago *N*-(1-methyl-4-hydroxy-3-imidazolin-2,2-ylidene)alanine was isolated from beef broth and reported to impart a brothy, thick sour taste as characteristic taste impression of beef bouillon (*7*). Although this compounds has been suggested to be formed from the reaction between creatinine and lactic acid, neither the formation pathways, nor the human threshold concentration for the sensory activity of that molecule was yet confirmed. In order to answer the question as to whether this taste modulator might be formed by Maillard-type reactions of creatinine, the objective of the present investigation was to screen for taste modulating creatinine glycation products in heated creatinine/ribose solutions and to investigate the sensory activity of these derivatives. Furthermore, LC-MS/MS studies should be performed to verify the natural occurrence of the identified taste modulators in beef stock.

2   MATERIALS AND METHODS

2.1   Chemicals.

Chemicals and solvents were purchased from Merck KGaA (Darmstadt, Germany), Fluka (Neu-Ulm, Germany), and Sigma-Aldrich (Steinheim, Germany). For sensory analysis, bottled water (Evian) was adjusted to pH 5.9 with trace amounts of formic acid. Model Broth was prepared with Gistex X-II LS Yeast obtained from FID (Werne, Germany). Solvents were of HPLC grade (Fisher Scientific, Schwerte, Germany), and deuterated solvents were supplied by Euriso-Top (Saarbruecken, Germany). Beef stock were obtained from the local supermarket (Rinderfond, Lacroix, Lübeck, Germany)

2.2   Identification of Taste-Modulating Creatine Glycation Products in Thermally Treated Creatinine/Ribose-Solutions.

Creatinine (4 mmol) and D(-)-Ribose (4 mmol) were dissolved in 40 mL phosphate buffer solution (pH 7.0; conc.) and heated in at 100°C for 4h. After cooling down to room temperature, the solution was lyophilized, the lyophilisate was dissolved in acetonitrile/water (60/40 v/v, 10 mL) and the soluble compounds were separated by means of semi-preparative hydrophilic interaction liquid chromatography (HILIC) on a 300 x 21.5 mm i.d., 10 µm, TSKgel Amide-80 column (Tosoh Bioscience, Stuttgart, Germany). Using a flow rate of 6 mL/min, chromatography was performed by isocratic elution with a mixture (40/60, v/v) of aqueous formic acid (0.1% in water) and acetonitrile. The major peaks were collected separately (**Figure 1**), freed from solvent under vacuum, and lyophilized. LC-MS/MS and 1D/2D-NMR experiments led to the identification of 2-*N*-(1methyl-4-hydroxy-3-imidazolin-2,2-ylidine) butanoic acid (**1**) in fraction 1, 2-*N*-(1-methyl-4-oxoimidazolidin-2-ylideneamino) propionic acid (**2**) in fraction 2, 4-hydroxy-2-*N*-(1methyl-4-hydroxy-3-imidazolin-2,2-ylidine) butanoic acid (**3**) in fraction 3, 4-hydroxy-2-*N*-(1-methyl-4-oxoimidazolidin-2-ylideneamino) pentanoic lactone (**4**) in fraction 4, and 4,5-hydroxy-2-*N*-(1-methyl-4-oxoimidazolidin-2-ylideneamino) pentanoic acid (**5**) in fraction 5 (**Figure 2**). The MS and NMR data of these compounds will be reported separately. Comparison of retention times and co-chromatography led to the identification of unreacted creatinine in the late eluting fraction collected after about 45 min.

2.3   Analytical Sensory Experiments.

Prior to sensory analysis, the isolated compounds were suspended in water, and after removal of the volatiles under high vacuum (< 5 mPa), each compound was freeze-dried twice. The pH value of all samples was adjusted to 5.9 with trace amounts of either aqueous formic acid (1 g/100 g) or sodium hydroxide (0.1 mo 1 /L). Taste detection thresholds concentrations for the taste modulating activity were determined by nine trained panelists (four males, five females, ages 26-40 years) in several sessions using the triangle test (*8*) in water as well as model broth solution.

2.4   LC-MS/MS Identification of Creatinine Glycation Products in Beef Stock.

An aliquot (1 mL) of a commercially available beef stock was membrane-filtered (0.45 µm) and then aliquots (10 µL) were injected into the HPLC-MS/MS system equipped with a 300 x 7.8 mm i.d., 5 µm TSKgel Amide-80 column (Tosoh Bioscience, Stuttgart,

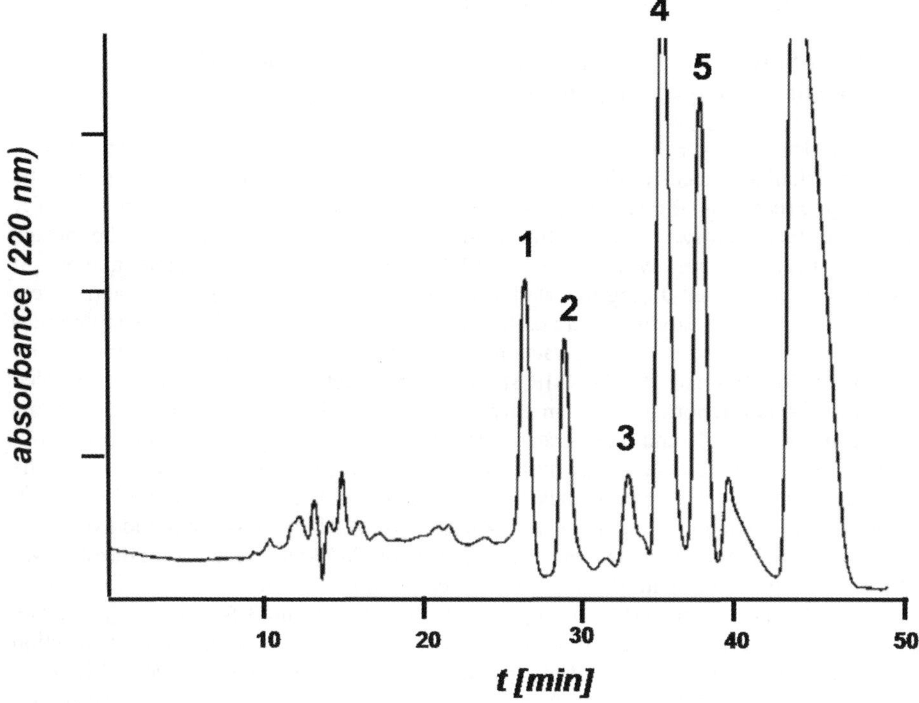

Figure 1. *HILIC chromatogram of a creatinine/ribose solution (pH 7.0) thermally treated for 4h at 100°C.*

Germany). Chromatography was performed at a flow rate of 1 mL/min using acetonitrile containing 1% mixture of 80% A and 20% B for 5 min, the amount of solvent B was increased to 100% within 30 min and then hold for additional 10 minutes. After chromatographic separation, the effluent was split in a ratio of 1:5 to reduce the effluent volume entering the mass spectrometer. After tuning the mass spectrometer with the purified creatinine derivatives, compounds **1-5** were analyzed using the following mass transitions: 2-*N*-(1-methyl-4-oxoimidazolidin-2-ylideneamino) butanoic acid (**1**, *m/z* 200→58), 2-*N*-(1-methyl-4-oxoimidazolidin-2-ylideneamino) propionic acid (**2**, *m/z* 186→87), 4-hydroxy-2-*N*-(1-methyl-4-oxoimidazolidin-2-ylideneamino) butanoic acid (**3**, *m/z* 216→74), 4-hydroxy-2-*N*-(1-methyl-4-oxoimidazolidin-2-ylideneamino) pentanoic lactone (**4**, *m/z* 228→87) and 4,5-dihydroxy-2-*N*-(1-methyl-4-oxoimidazolidin-2-ylideneamino) pentanoic acid (**5**, *m/z* 246→87), respectively. The comparison of both, the retention time and the individual mass transition undoubtedly confirmed the identification of the creatinyl-compounds in the beef stock.

3 RESULTS AND DISCUSSION

3.1 Identification of Creatinine Glycation Products from a Thermally Treated Creatinine/D(-)-Ribose Solution.

In order to answer the question as to whether *N*-(1-methyl-4-hydroxy-3-imidazolin-2,2-ylidene)alanine, recently identified as a taste modulator in beef (*7*), is formed by Maillard-type reactions of creatinine, a solution of creatinine and the ribose was thermally treated and then analyzed for creatinine glycation products. Analysis of the heated creatinine/ribose mixture by means of HILIC demonstrated a remarkable number of reaction products formed during the Maillard-type reaction (**Figure 1**). The quantitatively predominating reaction products were collected separately in several chromatographic runs and, after removing the solvent in vacuum, their chemical structure was determined by means of LC/MS/MS and 1D/2D-NMR studies. The spectroscopic experiments revealed the previously not reported 2-*N*-(1-methyl-4-oxoimidazolidin-2-ylideneamino) butanoic acid (**1**) in fraction 1 and the recently reported *2-N*-(1-methyl-4-oxoimidazolidin-2-ylideneamino)propionic acid (**2**) as the major compound in fraction 2. Furthermore, the previously unknown compounds 4-hydroxy-2-*N*-(1-methyl-4-oxoimidazolidin-2-ylideneamino) butanoic acid (**3**), 4-hydroxy-2-*N*-(1-methyl-4-oxoimidazolidin-2-ylideneamino)pentanoic lactone, and 4,5-dihydroxy-2-*N*-(1-methyl-4-oxoimidazolidin-2-ylideneamino)pentanoic acid (**5**) were identified in fractions 3, 4, and 5, respectively (**Figure 2**). To the best of our knowledge, the compounds **1** and **3-5** were not previously reported in literature. On the basis of the structures of these compounds, a reaction pathway leading to the formation of the glycation products **1-5** was proposed in **Figure 3**. Upon thermal treatment, the dicarbonyls 2-oxobutanal, 2-oxopropanal, 4-hydroxy-2-oxobutanal, and 3-deoxypentosulose, generated by pentose breakdown, react via their hydrate form with the exocyclic imino function of creatinine to give a Schiffs base, which upon enolization gives rise to the imino acids **1-5**, respectively.

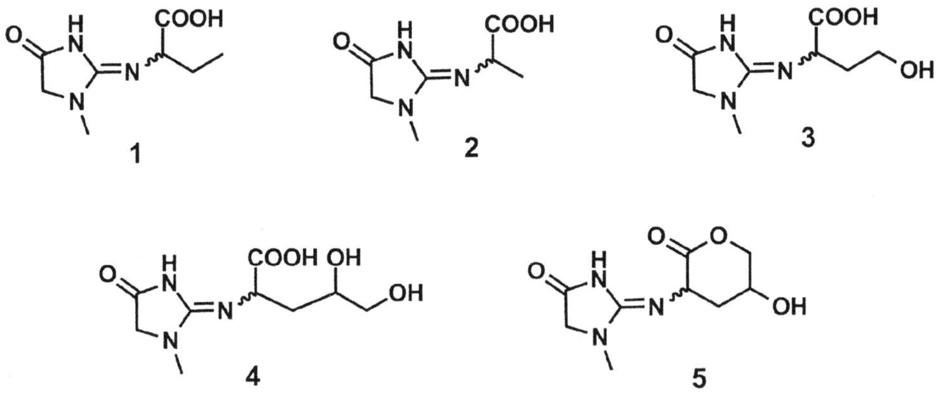

Figure 2.   *Chemical structures of creatinine glycation products 1-5 identified in the thermally treated creatinine/ribose solution.*

3.2 Taste Modulating Activity of Creatinine Glycation Products.

In order to determine the taste modulating activity of the identified compounds, triangle tests were performed with the purified compounds dissolved in water (for intrinsic taste) as well as in a model broth solution (for taste modulatory activity), respectively. None of the compounds **1-5** showed any intrinsic taste up to a maximum concentration of 1000 µmol/kg, but all these creatinine derivatives imparted taste modulating activity in model broth by enhancing its thick-sour, brothy taste. The lowest threshold concentration of 76 µmol/kg was found for 4-hydroxy-2-*N*-(1-methyl-4-oxoimidazolidin-2-ylideneamino) butanoic acid **(3)**, whereas the highest threshold level of 489 µmol/kg was determined for 4-hydroxy-2-*N*-(1-methyl-4-oxoimidazolidin-2-ylideneamino)pentanoic lactone **(4)** (**Table 1**).

Table 1. *Taste threshold concentration of the identified creatinine glycation products*

Compound[a]	Taste threshold concentration in water (µmol /L)	Taste modulating threshold concentration (µmol / L)[b]
1	> 1000	129 (± 55)
2	> 1000	209 (± 83)
3	> 1000	76 (± 29)
4	> 1000	489 (± 105)
5	> 1000	159 (± 83)

a. Compound numbering refers to **Figure 2**.
b. Taste modulating threshold concentrations were determined in model broth solution

3.3 Identification of Creatinine Glycation Products in Commercial Beef Stock.

To verify the presence of the taste modulatory creatinine glycation compounds in beef products, a commercial beef stock was screened for these glycated compounds by means of LC/MS/MS operating in the MRM mode. Each purified target compound was tuned for its characteristic mass transition and, then, a chromatography method resulting with a baseline separation of each compound was determined. Thereafter, an aliquot of the beef stock was injected into the HILIC-MS/MS system equipped with a TSKgel Amide-80 column. Comparison of the retention time and the characteristic mass transition of each reference compound, followed by co-chromatography allowed the unequivocal identification of the creatinine glycation products **1-5** in the beef stock, thus demonstrating the natural occurrence of these novel taste modulators in meat products. Further studies are currently on-going to quantify these compounds and to investigate the contribution to the typical taste of meat products on the basis of dose/activity considerations.

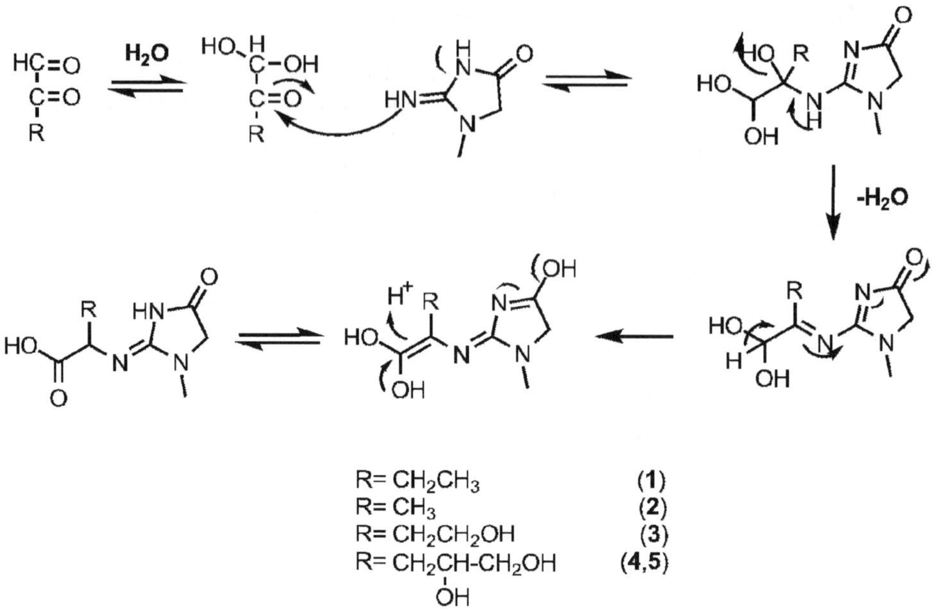

R= CH$_2$CH$_3$ (1)
R= CH$_3$ (2)
R= CH$_2$CH$_2$OH   (3)
R= CH$_2$CH-CH$_2$OH   (4,5)
         OH

Figure 3.   *Proposed reaction pathway showing the formation of the creatinine glycation products 1-5.*

Due to its desirable taste profile, beef juice is highly appreciated as a sapid base for savory dishes all over the world. Previously, *N*-(1-methyl-4-hydroxy-3-imidazolin-2,2-ylidene)alanine was reported to impart a brothy, thick sour taste as characteristic taste impression of beef bouillon. Although this compounds has been suggested to be formed from the reaction between creatinine and lactic acid, neither the formation pathways, nor the human threshold concentration for the sensory activity of that molecule was yet confirmed. In order to answer the question as to whether the Maillard reaction of creatinine is involved in the generation of such taste modulators and to identify additional thermally generated key molecules responsible for the thick-sour meaty orosensation in beef meat, thermally treated solutions of the putative Maillard reaction precursors creatinine and D-(-)-ribose were separated by means of hydrophilic interaction liquid chromatography (HILIC) and the reaction products formed were investigates by means of LC-MS and 1D/2D-NMR studies as well as analytical sensory studies. A series of hydrophilic creatinine glycation products were determined in their chemical structures and, although being tasteless in water, found to enhance the thick-sour meaty orosensation in a model broth matrix. For the first time, human detection thresholds for the taste modulatory activity of these Maillard-modified creatinine derivatives were determined and their natural occurrence in beef stock was verified by means of HILIC-MS/MS-MRM experiments.

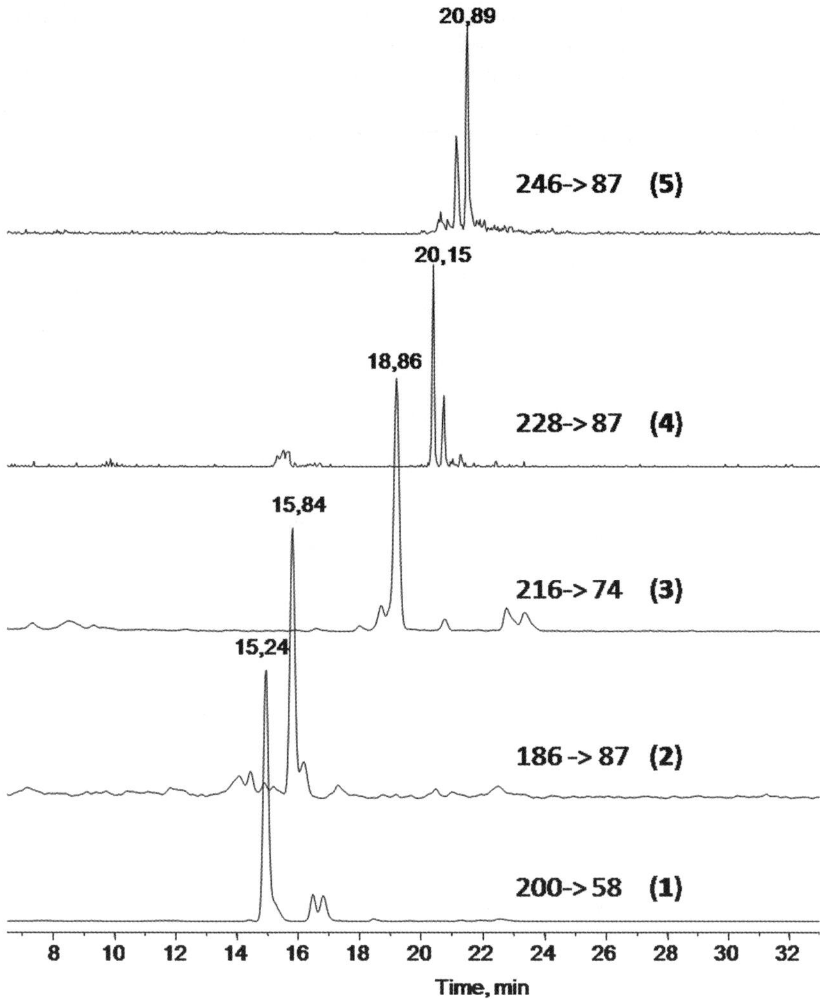

Figure 4. *HILIC-MS/MS-MRM screening for the creatinine glycation products 1-5 in a commercially available beef stock.*

References

1   Cerny, C.; Grosch, W. Z. Lebensm. Unters. Forsch. 1992, **194**, 322-325.
2   Fickert, B.; Schieberle, P. Food 1998, **42**, 371-375.
3   Frank, O.; Ottinger, H. Hofmann, T. J. Agric. Food Chem. 2001, **49**, 231-238.
4   Ottinger, H.; Soldo, T.; Hofmann, T. J. Agric. Food Chem. 2003, **51**, 1035-1041.
5   Ottinger, H.; Hofmann, T. J. Agric. Food Chem. 2003, **51**, 6791-6796.

6   Beksan, E.; Schieberle, P.; Robert F.; Blank, I.; Fay, L.B.; Schlichtherle-Cerny, H.; Hofmann, T. *J. Agric. Food Chem.* 2003, **51**, 5428-5436.
7   Shima, K.; Yamada, N.; Suzuki, E.; Harada, T. *J. Agric. Food Chem.* 1998, **46**, 1465-1468.
8   Wieser, H.; Belitz, H.-D. *Z. Lebensm. Unters. Forsch.* 1975, **159**, 65-72.

IMPROVED EXTRACTION OF ACRYLAMIDE AND ITS QUANTIFICATION IN PORK SAUSAGES

M. Plotkowiak[1], S. Floyd[1] and L. Farmer[2]

[1]Queens University Belfast, University Road, Belfast BT7 1NN
[2]Agri-Food and Biosciences Institute, Newforge Lane, Belfast BT9 5PX, UK

1   INTRODUCTION

Acrylamide (2-propenamide) has many commercial uses. As a monomer it is used in adhesives and grouts [1] whereas as a polymer is it used in polyacrylamide gels or as flocculants for wastewater treatment, in the mining industry and various other processing industries. It is thus understandable that the discovery of acrylamide in foods in 2002 was startling. Tareke et al first suspected acrylamide formation in cooked food in 2000 [2] and confirmed his findings in 2002 [3]. Acrylamide has been classified as a potential carcinogen in rodents [1, 4] and is also considered a neurotoxin, genotoxin, and a reproductive toxin [5]. Therefore, this led to a number of studies measuring the acrylamide concentration in foods [3, 6-12] to assess the danger. However, in most of these studies, very little data has been reported on the concentrations of acrylamide in meat and meat products such as sausages.
In this study, we present the results of quantification of acrylamide in a set of commercial and non-commercial sausage samples as well as an improved method to extract acrylamide from protein and fat rich samples such as sausages.

2   MATERIALS AND METHODS

2.1   Reagents

HPLC grade methanol (Chromasolv) was purchased from from Sigma-Aldrich (Poole, UK) while double distilled water was obtained using a Bibby Aquatron (A4000D; Stone, UK) water purification system. Formic acid (reagent grade, $\geq 95\%$) was also purchased from Sigma-Aldrich. Acrylamide (puriss, $\geq 99.8\%$) and D_3-labelled Acrylamide ($\geq 98.5\%$) were purchased from Fluka (Poole, UK) and CDN-Isotopes (Quebec, Canada).

2.2   Analytical methods

Labelled and unlabelled acrylamide were separately dissolved in double distilled water forming stock solutions of 100 μg/mL, which were frozen until further use. From the stock, fresh solutions were prepared each day at concentrations of 1 μg/mL and 10 μg/mL for both labelled and unlabelled acrylamide, which were then used for standard curves or as internal standard. The standard solutions were prepared at five concentrations between 2.5 ng/mL and 50 ng/mL of acrylamide combined with D_3-labelled acrylamide at the concentration of 20 ng/mL. To minimize matrix effects, standard solutions were made up in raw sausage extract rather than water. This sausage extract was prepared the same way as the sample material except for the use of internal standard.

2.3   Samples

Six types of sausages were analysed for acrylamide; these were two types prepared experimentally and four commercial brands. The experimental sausages (E5 and E6) were prepared from pork shoulder, pork fat (purchased from Dunbia Ltd, NI), rusk (purchased from Spices Nice, Lisburn, NI), salt (Tesco, Dundee, UK) and spices (purchased from Spices Nice, Lisburn, NI) in a 28mm diameter collagen casing (Devro Handlink, Devro, Moodiesburn, UK). The commercial sausages included independent brands of medium (BM) and premium (BP) quality and supermarket own brands from their value (SV) and premium (SP) ranges. In addition to the analysis for acrylamide, the sausages were analyzed for free fat concentration using the British Standard soxhlet method[13]. Furthermore the protein content was determined using the Dumas combustion using a LECO FP-2000 Analyser (Stockport, UK) according to the British Standard method[14] .The final composition of these and the purchased sausages is given in Table 2. The sausages were analysed for acrylamide either raw, oven cooked to a "golden colour" or "overcooked". In a preliminary trial, the cooking times for "golden colour" and "overcooked" for each type of sausage were determined by cooking the sausages at 180°C to a previously agreed colour using a reference sample. The cooking times used as well as the approximate size of the sausages are given in Table 1.

For the acrylamide analyses, fresh, defrosted raw sausages or cooled cooked sausages were cut into smaller pieces and minced into a homogeneous paste using a food processor (Kenwood Chef with Multi Mill attachment). A sample of 2 g raw or cooked minced sausage was placed into a 50 mL centrifuge tube (Apex®, tube material: PP, cap material: PE, sterile). Internal standard spikes consisted of 40 μL of D_3-labelled acrylamide solution (10 μL/mL). Double distilled water (20 mL) was added to the test portions and the samples were homogenized for ca. 30s with an Ultra-Turrax homogeniser (T18 basic, IKA, Staufen, Germany). The samples were then centrifuged for 20 min at 2000 rpm (3-16K Sigma Centrifuge (Osterode am Harz, (Germany)) and 2 x 1.5 mL of the supernatant (avoiding the top fat layer) were transferred into reaction tubes and centrifuged again for 10 min at 4000 rpm. The supernatant was then filtered through 0.45 μm syringe filters (E0 non pyrogenic hydrophilic 0.45μm, Sartorius Minisart).

The cleanup method was adapted from Andrzejewski et al. [15]. OASIS-HLB columns (Waters, 6 cc (300 mg), International Sorbent Technology Ltd) were conditioned with 3.5 mL methanol followed by 3.5 mL water. The solvents used for the column conditioning were discarded. The cartridge was loaded with 1.5 mL sample extract which was allowed to pass completely through the sorbent bed without drying the column out and was followed by 0.5 mL water. The column eluent was discarded. To elute the acrylamide 1.5 mL eluent (90% water, 0.1% formic acid, 10% methanol) was added onto the sorbent and the eluate was collected for LC-MS/MS analysis.

Standards and sample portion extracts (10 µL) were analyzed on an Agilent 1100 HPLC (Agilent Technologies, Palo Alto, CA) with a vacuum degasser, binary pump, autosampler coupled to a Micromass Quattro Ultima Pt. quadrupole mass spectrometer (Waters/Micromass, Milford, MA) with an electrospray source and Masslynx version 4.1 software with a Hypercarb column (50 x 2.1 mm, 5 µm; ThermoHypersil, Waltham, USA) together with a Hypercarb Guard precolumn (10 x 2 mm; ThermoHypersil) at ambient temperature. The mobile phase was 100% water with a flow rate of 0.2 mL/min for 5.0 min (analytes recorded), washing with 80% aqueous acetonitrile (0.4 mL/min) from 5.1 – 9min reconditioning with water (0.2 mL/min) from 9.1min to 20min between sample injections. The electrospray source had the following settings: capillary voltage 1 kV, cone voltage 35 V, source temperature 130°C, desolvation temperature 350°C, cone gas flow 70 L/h and desolvation gas flow 650 L/h. Argon at 2.5mBar was used as collision gas. Acrylamide was identified by multiple reaction monitoring (MRM). For all MRM transitions the dwell time was 0.48 s and the instrument settings were: acrylamide transition 72.1 > 55.1, collision energy 6 eV, D3 Acrylamide transition 75.1 > 58.1, collision energy 6 eV.

3   RESULTS AND DISCUSSION

3.1   Method Development

A great number of varying analytical methods for acrylamide have been published since the discovery of acrylamide in food. These may be grouped into methods using liquid chromatography (LC) [15-20] and gas chromatography (GC) [7, 10, 17, 21, 22] usually coupled with mass spectrometry (MS). Liquid chromatography with tandem mass spectrometry (LC-MS/MS) has been used in this study following an adaptation of the method suggested by Tareke et al in 2002 [3] and the US Food and Drug Administration[23]. The goal of this study was to develop a method suitable for a fat and protein rich product such as pork sausages. British pork sausages contain rusk, a filler material similar to breadcrumbs. This was expected to be the main source of

Table 1   *Cooking times and sizes of sampled sausages*

	Types of sausages[a]					
	E5	E6	BM	BP	SV	SP
Cooking time: "golden" (mins)	20	20	14	16	15	25
Cooking time "overcooked" (mins)	40	40	28	36	28	40
Length x diameter (mm x mm)	20 x 60	20 x 60	20 x 80	17 x 130	20 x 80	23 x 100

a   E5 and E6 are experimental sausages, BM and BP are independent brands of medium and premium quality while SV and SP are supermarket own brands from their value and premium ranges, respectively.

acrylamide for pork sausages as rusk is high in starch and during its production has been exposed to high temperatures which would enhance acrylamide formation [24]. Furthermore, during cooking, the sausages were exposed to high temperatures, which could enhance further acrylamide formation.

The absolute recoveries using the method of Tareke[3] and Eriksson [25] were very low and not satisfactory as only ca. 10% of the internal standard has been recovered. In this method, the homogenate was cleaned up by passing it through two frequently used columns, OASIS HLB [16, 18] and ISOLUTE Multimode [3, 26]. Tests conducted with both columns indicated that the majority of the acrylamide was retained in the Isolute Column, up to 75% whereas ca. 50% was retained in the Oasis Column. It was decided to only continue using the OASIS column for further extractions. The use of Carrez I and II solutions [27, 28] as well as defatting steps were suggested in a number of publications but proved unsuccessful in our trials. Likewise, homogenizing the sample in a 2:1 chloroform/methanol solvent did not notably improve the absolute recoveries as they remained at about 50%.

It has been observed however, that the matrix effects of sausage extract could be avoided by making up the standard solutions in sausage extract. In this extraction process, eluting the column with 0.1% formic acid in 90% water and 10% methanol [23, 29] also improved recoveries as did changing the mobile phase from 1% acetonitrile to pure water. These changes improved the absolute recoveries to an average of 76%. The limit of detection (LOD) was 2.2 µg/kg and the limit of quantification (LOQ) was 7.25 µg/kg.

3.2   Acrylamide in sausages

The results of the compositional analysis of the sausages are presented in Table 2 and the results of the analysis for acrylamide are shown in Figure 1. The rusk used for making the experimental sausages (E5 and E6) was found to contain acrylamide at a concentration of 167 µg/kg. The results indicate that, in raw sausages, the amount of acrylamide is related, in part, to the carbohydrate content. Previous research has suggested that the formation of acrylamide occurs mainly in carbohydrate rich foods through the reaction of reducing sugars and asparagine[30]. The highest levels were found in SV sausages, which were high in carbohydrate and relatively low in protein. The high acrylamide levels in these sausages, even when raw, suggest that the rusk used itself contained unusually high levels of acrylamide.

Table 2   *Sample make-up of commercial and non-commercial sausages (* through calculation)*

Sausage Type[a]	Water (%)	Protein (%)	Fat (%)	Carbohydrate* (%)
E5	62	14	13	11
E6	68	16	6	10
BM	50	11	24	15
BP	52	15	25	8
SV	59	9	16	16
SP	62	13	20	5

a   See footnote to Table 1

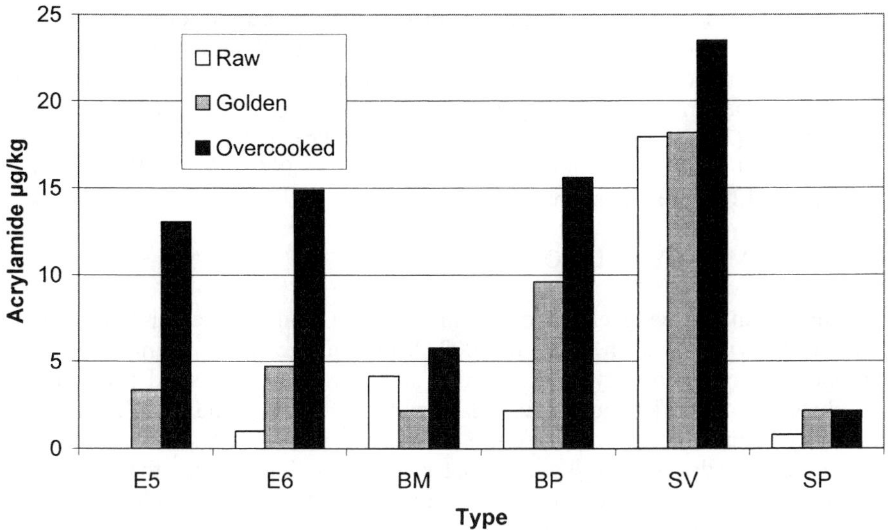

Figure 1   *Acrylamide content in commercial and non-commercial sausages*

Upon cooking, acrylamide is formed in all sausages, but at different rates. The greatest increase in acrylamide levels are observed in E5, E6 and BP. These were the sausages with the higher protein levels, but this difference does not appear to be sufficient to explain the elevated acrylamide levels. The formation of acrylamide could also be influenced by other ingredients present in commercial sausages as has been shown in other foods[31]. However, preservatives and antioxidants were present in almost all commercial sausages but not in E5 and E6, which may partly explain the high formation rates in these two sausages but not in the BP type.

One possible contributory reason for the different formation rates may be the size and shape of the sausages (Table 1). For example, BP sausages were thinner and longer than SP sausages having thus a higher surface area, which in turn means a greater exposure to high temperatures, facilitating acrylamide formation. However, this does not apply to E5 and E6. Furthermore, trial experiments in this study showed that the internal temperature of the samples did not exceed 105°C, a temperature that has been described as too low for extensive acrylamide formation [6, 30, 32]. Studies investigating the relationship between core and surface temperature have shown similar results for potato chips and French fries, indicating that acrylamide synthesis is mainly limited to the surface of the heated sample[33] and increases with prolonged heating time and temperature[34].

The sausages were cooked to achieve a comparable degree of browning rather than cooking them all the same time, which would result in different cooking degrees, and this could affect acrylamide formation. However, this would suggest the highest acrylamide formation in SP sausages, which in fact show the lowest acrylamide levels. Some research also suggested a decrease in acrylamide after a cooking time of ca. 20min [35] while there is also evidence suggesting that acrylamide is continuously formed through the cooking process [36]. All of these factors may contribute to the formation of acrylamide in these sausages but which of these has the most effect will have to await further investigation.

The results of this investigation suggest that British sausages, even at a very overcooked state, contain acrylamide in only small quantities, up to 25 µg/kg. With the exception of one of the sausages analysed, most contain less than 10µg/kg when cooked to a normal endpoint. These quantities are similar to or less than the amount found in bread (31 µg/kg), breaded chicken (24 µg/kg), pizza (20 µg/kg) or pie (22 µg/kg) [31] and compare favourably with 200 µg/kg -12000 µg/kg in French fries and 800 µg/kg -1200 µg/kg in crispbread[5].

4   SUMMARY AND CONCLUSIONS

The toxicity of the levels of acrylamide found in food has been highly debated since its discovery in food in 2002. In 2008, Wolfram Parzefall [37] summarised the recent findings on the toxicity of acrylamide and quoted "no observed level effect levels" between 80 µg/kg bodyweight/day and 500 µg/kg bodyweight/day for neurotoxicity. These levels would require an average person (80kg) to eat at least 640kg sausages per day at 10 µg/kg in sausages. These findings suggest that sausages, cooked to a normal cooking endpoint, are likely to contribute very little to human exposure to acrylamide and subsequently to the risk of cancer.

References

1.   IARC, in *Monographs on the evaluation of carcinogenic risks to humans*, Lyon, 1994, vol. 60, p. 389.
2.   E. Tareke, P. Rydberg, P. Karlsson, S. Eriksson and M. Tornqvist, *Chemical Research in Toxicology*, 2000, **13**, 517-522.
3.   E. Tareke, P. Rydberg, P. Karlsson, S. Eriksson and M. Tornqvist, *Journal of Agricultural and Food Chemistry*, 2002, **50**, 4998-5006.
4.   K. A. Johnson, S. J. Gorzinski, K. M. Bodner, R. A. Campbell, C. H. Wolf, M. A. Friedman and R. W. Mast, *Toxicology and Applied Pharmacology*, 1986, **85**, 154-168.
5.   M. Friedman, *Journal of Agricultural and Food Chemistry*, 2003, **51**, 4504-4526.
6.   A. Becalski, B. P. Y. Lau, D. Lewis and S. W. Seaman, *Journal of Agricultural and Food Chemistry*, 2002, **51**, 802-808.
7.   J. S. Ahn, L. Castle, D. B. Clarke, A. S. Lloyd, M. R. Philo and D. R. Speck, *Food Additives and Contaminants*, 2002, **19**, 1116-1124.
8.   J. Rosén and K.-E. Hellenäs, *Analyst* 2002, **127**, 880-882.
9.   A. P. Arisseto, M. C. Toledo, Y. Govaert, J. Van Loco, S. Fraselle, E. Weverbergh and J. M. Degroodt, *Food Additives and Contaminants*, 2007, **24**, 236-241.
10.   H. Ölmez, F. Tuncay, N. Özcan and S. Demirel, *Journal of Food Composition and Analysis*, 2008, **21**, 564-568.
11.   Y. Zhang, J. Jiao, Y. Ren, X. Wu and Y. Zhang, *Analytica Chimica Acta*, 2005, **551**, 150-158.
12.   Y. Zhang, Y. Ren, H. Zhao and Y. Zhang, *Analytica Chimica Acta*, 2007, **584**, 322-332.
13.   British Standards Institute, British Standard BS 4401-5, 1996.
14.   British Standards Institute, British Standard BS EN ISO 16634-1, 2008.
15.   D. Andrzejewski, J. A. G. Roach, M. L. Gay and S. M. Musser, *Journal of Agricultural and Food Chemistry*, 2004, **52**, 1996-2002.

16.	J. A. G. Roach, D. Andrzejewski, M. L. Gay, D. Nortrup and S. M. Musser, *Journal of Agricultural and Food Chemistry*, 2003, **51**, 7547-7554.
17.	K. Hoenicke, R. Gatermann, W. Harder and L. Hartig, *Analytica Chimica Acta*, 2004, **520**, 207-215.
18.	J. A. Rufián-Henares and F. J. Morales, *Food Chemistry*, 2006, **97**, 555-562.
19.	C. T. Kim, E.-S. Hwang and H. J. Lee, *Food Chemistry*, 2007, **101**, 401-409.
20.	Y. Zhang, J. Jiao, Z. Cai, Y. Zhang and Y. Ren, *Journal of Chromatography A*, 2007, **1142**, 194-198.
21.	A. Pittet, A. Périsset and J.-M. Oberson, *Journal of Chromatography A*, 2004, **1035**, 123-130.
22.	Y. Zhang, Y. Dong, Y. Ren and Y. Zhang, *Journal of Chromatography A*, 2006, **1116**, 209-216.
23.	S. Musser, *US FDA draft analytical method*, www.ifst.org/documents/misc/FDAdraftanalyticalmethod.pdf 2008.
24.	E. Bråthen and S. H. Knutsen, *Food Chemistry*, 2005, **92**, 693-700.
25.	S. Eriksson and P. Karlsson, *LWT - Food Science and Technology*, 2006, **39**, 393-399.
26.	A. Becalski, B. P. Y. Lau, D. Lewis and S. W. Seaman, *Journal of Agricultural and Food Chemistry*, 2003, **51**, 802-808.
27.	T. Delatour, A. Perisset, T. Goldmann, S. Riediker and R. H. Stadler, *Journal of Agricultural and Food Chemistry*, 2004, **52**, 4625-4631.
28.	V. Gökmen and H. Z. Senyuva, *Journal of Chromatography A*, 2006, **1120**, 194-198.
29.	Y. Zhang, G. Zhang and Y. Zhang, *Journal of Chromatography A*, 2005, **1075**, 1-21.
30.	D. S. Mottram, B. L. Wedzicha and A. T. Dodson, *Nature*, 2002, **419**, 448-449.
31.	M. Friedman and C. E. Levin, *Journal of Agricultural and Food Chemistry*, 2008, **56**, 6113-6140.
32.	R. H. Stadler, L. Verzegnassi, N. Varga, M. Grigorov, A. Studer, S. Riediker and B. Schilter, *Chemical Research in Toxicology*, 2003, **16**, 1242-1250.
33.	D. Taubert, S. Harlfinger, L. Henkes, R. Berkels and E. Schomig, *Journal of Agricultural and Food Chemistry*, 2004, **52**, 2735-2739.
34.	V. Gökmen, T. K. Palazoglu and H. Z. Senyuva, *Journal of Food Engineering*, 2006, **77**, 972-976.
35.	P. Rydberg, S. Eriksson, E. Tareke, P. Karlsson, L. Ehrenberg and M. Tornqvist, *Journal of Agricultural and Food Chemistry*, 2003, **51**, 7012-7018.
36.	T. M. Amrein, B. Schonbachler, F. Escher and R. Amado, *Journal of Agricultural and Food Chemistry*, 2004, **52**, 4282-4288.
37.	W. Parzefall, *Food and Chemical Toxicology*, 2008, **46**, 1360-1364.

CHEMISTRY OF PIGMENTS AS INTERMEDIATE OF MELANOIDINS

F. Hayase[1*], Y. Sshirahashi[1], T. Machida[1], T. Ito[1], T. Usui[1], and H. Watanabe[2]

[1]Department of Agricultural Chemistry Meiji University, and 1-1-1 Higashi-mita, Tama-ku, Kawasaki, Kanagawa 214-8571, Japan
[2]Department of Life Science, Meiji University, 1-1-1 Higashi-mita, Tama-ku, Kawasaki, Kanagawa 214-8571, Japan

1   INTRODUCTION

Melanoidins, the final products of the Maillard reaction, are brown nitrogen-containing polymers that are difficult to decompose. They have physiologically positive effects such as antioxidative activity, including strong scavenging activity against reactive oxygen species.[1] Many researchers have studied the partial structure of melanoidins in various ways.[2] Tressel's group reported that N-substituted pyrrole, 2-furaldehyde, and N-substituted 2-formylpyrroles, formed in pentose and hexose Maillard reaction systems, were identified as components of extraordinary polycondensation activity.[3] Hofmann identified novel red-brown and red compounds formed in the Maillard reaction system of arginine with glyoxal and furan-2-carboxaldehyde,[4] pentoses and primary amino acids,[5] and hexoses and primary or secondary amino acids.[6] However, above-mentioned compounds have fewer hydrophilic properties, indicating significant differences in the hydrophilic properties of melanoidins. Miura and Gomyo[7] and Gomyo et al.[8] found the formation of colored compounds formed in the Maillard reaction between D-xylose and glycine. Our group[9-

[11] has reported identification of the novel blue pigments designated Blue-M1 (Blue-M1, blue Maillard intermediate-1) and Blue-M2. These blue pigments have a methine proton between two pyrrolopyrrole rings. Blue-M1 and Blue-M2, involving the pyrrolopyrrole structure, were readily changed to a brown polymer. Accordingly, blue pigments are assumed to be generated independently and to be polymerized to form melanoidins. In this paper, we reviewed on the kinetic studies of Blue-M1, the generation of the other blue compounds in D-xylose-glycine reaction system, similarity between blue compounds and melanoidins, precursors of blue pigments, blue and red pigments from the other Maillard reaction, and the formation of melanoidins from blue pigments.

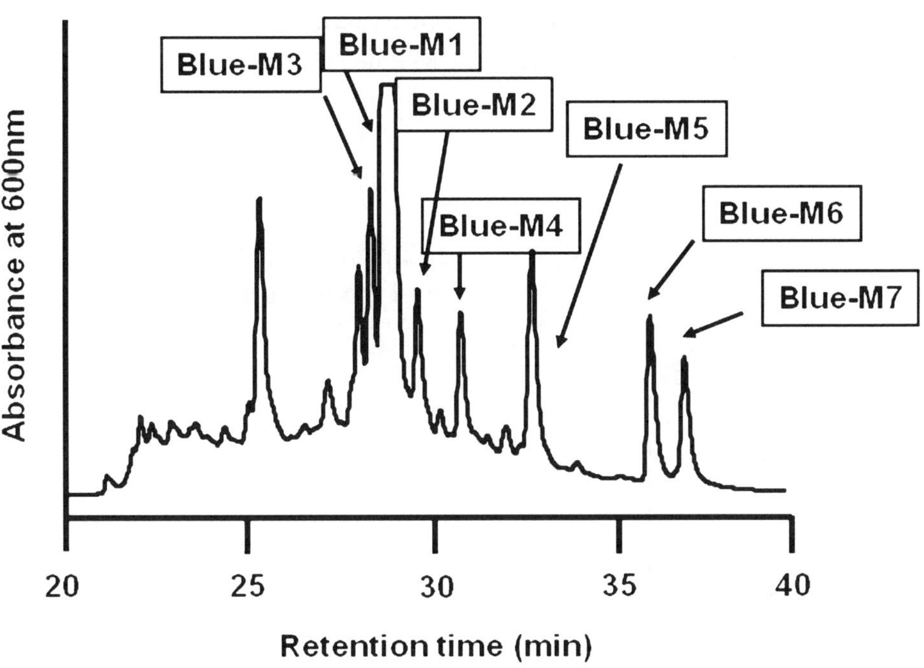

Figure 1. *Reversed Phased HPLC pattern of Blue Pigments Generated by the D-Xylose-glycine Reaction System.*

2   FORMATION AND KINETICS OF BLUE PIGMENTS

Blue pigments were firstly separated by using anion exchange column from the reaction mixture of D-xylose and glycine dissolved in 60% ethanol (starting at pH 8.1) under nitrogen at 26.5°C for 48h or 2°C for 96h.[9] Secondly, blue pigments were isolated by reversed phased HPLC column. Figure 1 shows the reversed phased HPLC pattern of blue pigments. Blue-M1, a major blue pigment, revealed its novel chemical structure that has two different side chains being dihydroxypropyl and trihydroxypropyl chains (Fig. 3).[9] Blue-M1 had two pyrrolopyrrole rings coupled with methine bridge. Blue-M1 independently has a polymerizing activity and changes to brown color. The chemical character and antioxidative activities of Blue-M1 is similar to those of melanoidins.[12] Accordingly, Blue-M1 would be a key intermediate in melanoidin formation. Blue-M2 which we already identified is assumed to be formed from Blue-M1 and di-xylulose glycine as an Amadori rearrangement product.[10,11] We tried to study kinetic aspects of the Blue-M1 and identified the Blue-M3, M4, M5, M6, and M7.

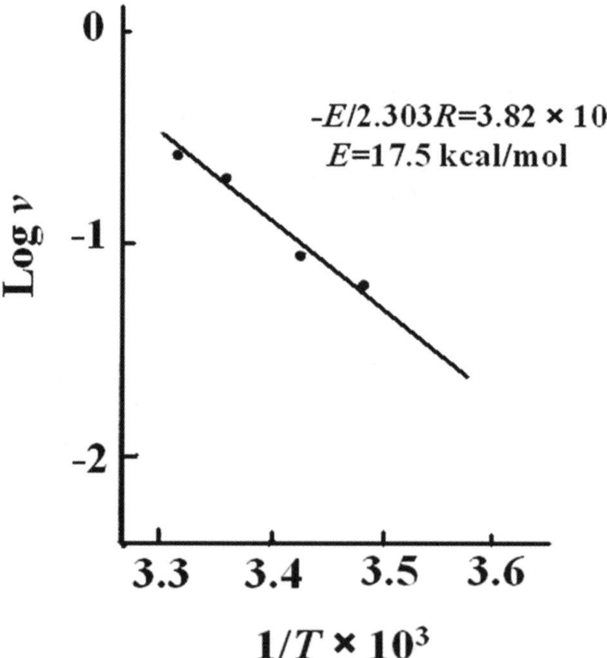

Figure 2.   *Arrhenius Plots for the Rate of Coloration.*

Blue-M1 dominated the coloration in the early stage when the steady state of the reaction was realized. Two linear curves were obtained from the studies on the dependence of the reaction rate on the concentration of D-xylose in excess of glycine, and of glycine in excess of D-xylose.[8] The reaction rate was determined on the basis of the slope of the linear portion in the reaction curve. The reaction was found to obey good second-order kinetics with respect to both D-xylose and glycine. Accordingly, the rate equation for the production of Blue-M1 could be expressed as follows:

$$v=k[\text{D-xylose}]^2 \cdot [\text{glycine}]^2 \quad (1)$$

In the equation, v and k are the reaction rate and the rate constant, respectively. Figure 2 shows Arrhenius plots relating to the coloration rate in terms of the absorption at 630nm, where $-E/2.303R$ was found to be 3.82 x 10^3. Accordingly, the activation energy (E) for the production of Blue-M1 (Mw=619) could be calculated as 17.5 kcal/mol. This value is low when compared with other organic reactions in an aqueous medium, indicating that room temperature was enough to produce Blue-M1 from the reaction system consisting of D-xylose and glycine. In the linear portion of the reaction curve as shown in Fig. 2, the reaction rate (v) at 20 °C in terms of molar concentration could be calculated as follows:

$$v=d(\text{Abs}/\varepsilon)/(dt)=1.593 \times 10^{-8}\text{mol} \cdot l^{-1} \cdot \text{min}^{-1} \quad (2)$$
$$k=3.983 \times 10^{-7}\text{mol}^{-3} \cdot l^3 \cdot \text{min}^{-1} \quad (3)$$

In conclusion, free energy ΔF, activation enthalpy ΔH, and activation entropy ΔS were calculated as follows.[8]

$$\Delta F^* = 22.0 \text{ kcal/mol}; \Delta H^* = 16.9 \text{ kcal/mol}; \Delta S^* = -17.4\text{cal/deg/mol}$$

The activation entropy ΔS provided a negative value, which indicates a decrease in the degree of freedom corresponding to the intermediate structure at the transient state. This suggests that a specific molecular orientation took place in the activation process of the reaction.Figure 3 shows the chemical structure of isolated blue compounds. Blue-M3 was isomer of Blue-M1 due to the asymmetric carbon as indicated with asterisk (unpublished data). Blue-M5 was composed of symmetrical structure, having two side chains of the 2,3-dihydroxypropyl group and two pyrrolopyrrole structures.[13] Blue-M6 and M7 were isomers of ethanol adducts of Blue-M1 and M3, respectively (unpublished data). Blue-M4 is assumed to be formed from Blue-M1 or M3 and C-3 compounds by NMR and high resolution FAB-MS data (data not shown).

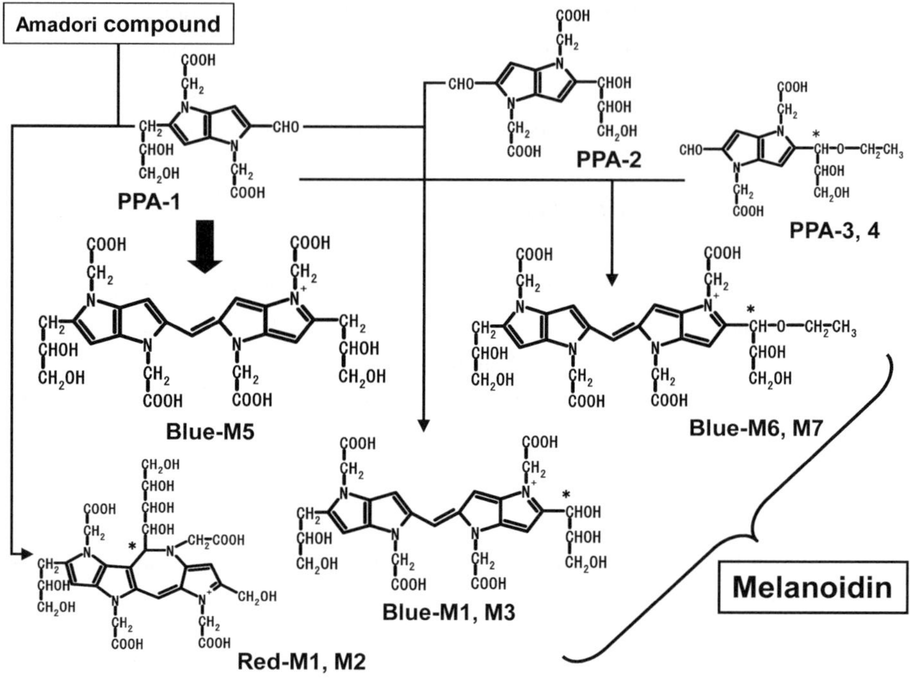

Figure 3.   *Proposed Formation Pathway of Blue and Red Pigments.*

3   PRECURSORS OF BLUE PIGMENTS

Colored compounds from D-xylose and glycine independently changed to brown color during incubation, respectively. These pigments were composed of common pyrrolopyrrole structure as mentioned above. Then we tried to detect the pyrrolopyrrole compounds as precursor of pigments. We isolated pyrrolopyrrole compound designated PPA-1 in the reaction mixture. UV-VIS and fluorescent spectra of PPA-1 showed maximum peak at 348nm. FAB-MS for PPA-1 showed the presence of the $[M+H]^+$ ion at m/z 325 and $[M+Na]^+$ ion at m/z 347. PPA-1 was identified as pyrrolopyrrole aldehyde composed of side chain of dihydroxypropyl group by NMR data (Fig. 3).[13] We found the other pyrrolopyrrole aldehyde in the same reaction mixture. Figure 3 shows the identified structure of the pyrrolopyrrole aldehyde designated PPA-2. The compound is composed of the side chain of a trihydroxypropyl group (unpublished data). Figure 4 indicates the proposed formation pathway of PPA-1 and PPA-2. 1,2-Enaminol would be found from D-xylose and glycine via Schiff base and Amadori compound, followed by 3-dexyxylosone (3-DX). 3-DX might react with glycine to give pyrrole-2-carbaldehyde. The pyrrole-2-carbaldehyde is assumed to react with imine of 3-DX to generate PPA-1.Figure 3 shows the proposed formation pathway of Blue-M1 and melanoidins. Blue-M1 and Blue-M3 are probably formed from PPA-1 and PPA-2 with decarboxylation. The proposition was confirmed by the experimental results that Blue-M5 was generated when PPA-1 alone was incubated at 26.5°C for 1 day.[13] Moreover, we identified also two isomers of ethanol adducts (PPA-3 and PPA-4) as shown in Figs. 3 and 4. PPA-3 and PPA-4 are assumed to be formed from PPA-2 and ethanol in the reaction solution. Blue-M6 and Blue-M7 are similarly generated from PPA-3 or 4 (Fig. 4).These blue pigments which involved pyrrolopyrrole structure were readily changed to brown polymer as shown in Fig. 3.[2,11-13] Accordingly, blue pigments are assumed to be generated independently and polymerized to form melanoidins.

4   SIMILARITY BETWEEN BLUE PIGMENTS AND MELANOIDINS

The chemical character of Blue-M1 was compared to that of a nondialyzable melanoidin preparation (Melanoidin-1) obtained from the reaction mixture of D-xylose and butylamine neutralized with acetic acid in methanol incubated at 50°C for 7 days. The results indicated strong resemblance between both substances.[12] Chemical composition of both Blue-M1 and Melanoidin-1 were calculated from experimental data and/or theoretical ones. Higher sugar moiety and lower hydroxyl content in Melanoidin-1 would be responsible to the involvement of non-nitrogeneous products such as furfural. Dehydration ratio of furfural, however, was similar to that of pyrrole-2-carbaldehyde. Dehydrogenation occurred also in Melanoidin-1 as well as Blue-M1. The results seem to indicate some resemblances between Blue-M1 and Melanoidin-1. As mentioned above, N-substituted pyrrole-2-carbaldehyde condenses with Schiff base of 2-ketoaldehyde to form a pyrrolopyrrole ring.[12,13] We have already reported that D-glucose-butylamine reaction system produced smaller amounts of N-butylpyrrole-2-carbaldehyde in addition to its 5-hydroxymethyl derivative (pyrraline).[14] If D-glucose and proteins could produce N-substituted pyrrole-2-carbaldehyde, there would be a possibility of the formation of pyrrolopyrrole in colored AGEs.

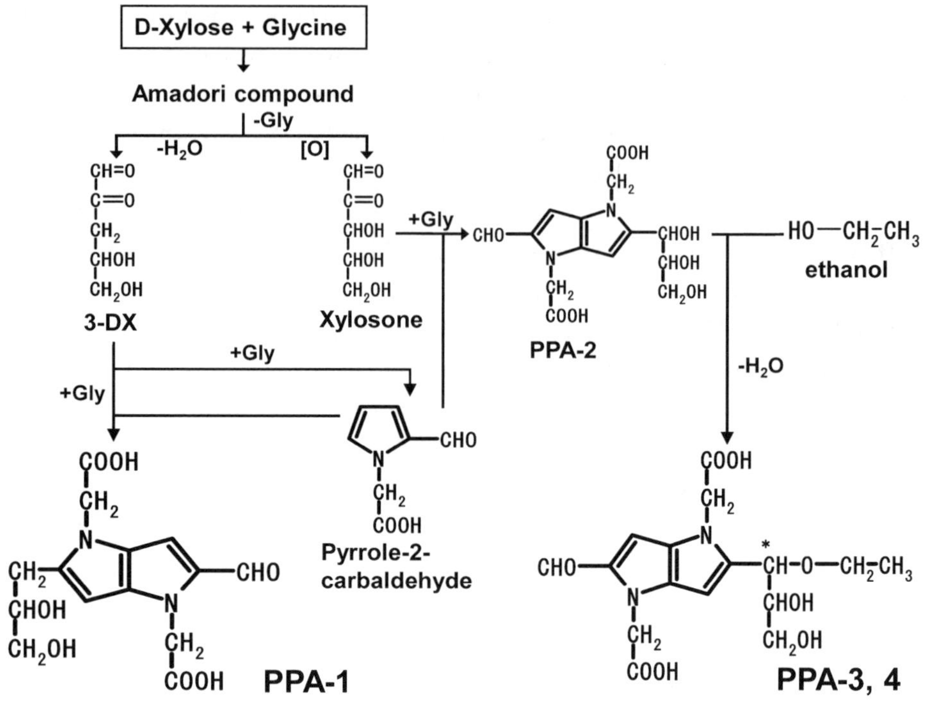

Figure 4   *Proposed Formation Pathway of Precursors of Melanoidins.*

Blue-M1 as well as melanoidins effectively suppressed the peroxidation of linoleic acid. The scavenging activity toward of Blue-M1 on hydroxyl and DPPH radicals was also as strong as that of melanoidins.[2,15] Furthermore, Blue-M1 prevented the oxidative cell injury. Therefore, Blue-M1 will be an antioxidant, which protect against the oxidative stress in biological systems.

5   BLUE AND RED PIGMENTS FROM AMINO COMPOUNDS AND REDUCING SUGARS

We tried to detect blue pigments from D-glucose-glycine reaction system. Some blue products were formed by the D-glucose reaction system as well as D-xylose system. The major blue compound was Blue-G1. Blue-G1 was composed of two pyrrolopyrrole structures similar to Blue-M5, and had two trihydroxybutyl groups as side chains.[13] Moreover, blue compounds were detected in the D-xylose-β-alanine reaction system. The major blue pigment named Blue-B1 was identified as having a similar structure to Blue-M1. Blue-B1 had four carboxyethyl groups instead of carboxymethyl groups of Blue-M1.[13] In the case of D-xylose-methylamine reaction system, the major blue pigment was identified as having a similar structure to Blue-M1 except N-side chain having four methyl groups (unpublished data). We recently isolated two major red pigments (Red-M1 and Red-M2) generated from D-xylose and glycine reaction system by reversed phased HPLC.[16] Figure 3 shows that Red-M1 is composed of pyrropyrrole, pyrrole, and xylulose-glycine. Various NMR spectra of Red-M1 support the structure of Red-M1 having azepine ring. We confirmed that Red-M1 and Red-M2 were identified as isomers due to the asymmetric carbon as shown in the asterisk mark. Figure 3 indicates the proposed formation pathway of Red-M1. Red-M1 might be formed from PPA-1, hydroxymethylpyrrole, and Amadori compounds.

6   FORMATION OF MELANOIDINS FROM BLUE PIGMENTS

When the Blue-M1 (200 µM) was incubated in 50% Ethanol at pH8.1 with NaHCO$_3$ (0.1M), color of Blue-M1 changed to brown after 5 days. Absorption at 625 nm was decreased and absorption at 350 nm was proportionally increased with incubation. Blue-M1 was decreased for 2 days, and absorption at 420 nm indicating browning was increased. The reversed phased HPLC pattern of incubated Blue-M1 for 2 days indicated that PPA-1 and PPA-2 were generated as the products of Blue-M1 (unpublished data). The other products were named BM1P-3, 4, and 5. PPA-1 and PPA-2 were increased for 2 days. These findings suggest that the methine bridge of Blue-M1 is cleaved to form PPA-1 and PPA-2. BM1P-3 and P-5 were not identified but BM1P-4 was tentatively identified as a novel compound having dimmer structure of pyrrolopyrrole including formyl group. These products were increased and then decreased with incubation. In conclusion, Blue-M1 is degraded to PPA-1, 2, BM1P-3, 4, and 5. Moreover, these products are polymerized each other to form melanoidins.

7   CONCLUSIONS

Some blue and red pigments generated in D-xylose-glycine reaction system are postulated to be intermediate oligomers in the generation of melanoidins. Secondly, blue pigments were also generated in D-glucose-glycine reaction system as well as D-xylose-methylamine and D-xylose-β-alanine reaction systems. Thirdly, major blue-pigment named Blue-M1 was assumed to be generated from PPA-1 and PPA-2 which are pyrrolopyrrole aldehydes. Finally, these pigments which have pyrrolopyrrole structures were readily changed to brown polymers with decomposition.

References

1. F. Hayase and H. Kato, *Comments on Agric. Food Chem.*, 1994, **3**, 111-128.

2. F. Hayase, T. Usui, and H. Watanabe, *Mol. Nutr. Food Res.* 2006, **50**: 1171-1179.

3. R. Tressl, GT Wondrak, L-A Garbe, R-P Krüger, and D. Rewicki, *J. Agric. Food Chem.*, 1998, **46**, 1765-1776.

4. T. Hofmann, *J. Agric. Food Chem.*, 1998, **46**, 3896-3901.

5. T. Hofmann, *J. Agric. Food Chem.*, 1998, **46**, 3902-3911.

6. T. Hofmann, *J. Agric. Food Chem.*, 1998, **46**, 3918-3928.

7. M. Miura and T. Gomyo *Nippon Nôgeikagaku Kaishi (in Japanese)*, 1982, **56**, 417-425.

8. T. Gomyo, L. Haiyan, M. Miura, F. Hayase, and H. Kato, *Agric. Food Chem.*, 1989, **53**, 949-957.

9. F. Hayase, Y. Takahashi, S.Tominaga, M. Miura, T. Gomyo, and H. Kato, *Biosci. Biotechnol. Biochem.*, 1999, **63**: 1512-1514.

10. S. Sasaki, Y. Shirahashi, K. Nishiyama, H. Watanabe, and F. Hayase, *Biosci. Biotechnol. Biochem.*, 2006. **70**: 2529-2531.

11. F. Hayase, T. Usui, K. Nishiyama, S. Sasaki, Y. Shirahashi, N. Tsuchiya, N. Numata, H. Watanabe, *Ann. N. Y. Acad. Sci.*, 2005, **1043**, 104-110.

12. H. Kato and F. Hayase, *The Maillard Reactions in Food Chemistry and Medical Science: Update for the postgenpmic Era,* Vol. 1245, S. Horichi, N. Taniguchi, F. Hayase, T. Kurata, and T Osawa (eds), Elsevier, Amsterdam, 2002, 3-7.

13. F. Hayase, T. Usui, Y. Ono, Y. Shirahashi, T. Machida, T. Ito, N. Nishitani, K. Shimohira, and H. Watanabe, *Ann. N.Y. Acad. Sci.*, 2008, **1126**, 53-58.

14. F. Hayase, M. Sato, H. Tsichida, and H. Kato, *Agric. Biol. Chem.,* 1982, **46**, 2987-2996.

15. S. Shizuuchi and F. Hayase, *Biosci. Biotechnol. Biochem.*, 2003, **67**, 54-59.

16. Y. Shirahashi, H. Watanabe, and F. Hayase, *Biosci. Biotechnol. Biochem.*, 2009. **73**, 2287-2292.

Subject Index